The Potential of U.S. Grazing Lands to Sequester Carbon and Mitigate the Greenhouse Effect

The Potential
of U.S. Grazing Lands
to Sequester Carbon
and Mitigate the
Greenhouse Effect

R.F. Follett • J.M. Kimble • R. Lal

LEWIS PUBLISHERS

A CRC Press Company
Boca Raton   London   New York   Washington, D.C.

Library of Congress Cataloging-in-Publication Data

The potential of U.S. grazing lands to sequester carbon and mitigate the greenhouse effect / edited by R.F. Follett, J.M. Kimble, and R. Lal.
    p. cm.
Includes bibliographical references.
ISBN 1-56670-554-1 (alk. paper)
1. Soils--Carbon content--United States. 2. Carbon sequestration. 3. Greenhouse effect, Atmospheric--United States. 4. Greenhouse gases--Environmental aspects--United States. 5. Rangelands--United States. I. Follett, R. F. (Ronald F.), 1939-II. Kimble, J. M. (John M.) III. Lal, R.

S592.6.C35 P68 2000 2001
631.4'1 —dc21

00-042832

Visit the CRC Press Web site at www.crcpress.com

© 2001 by CRC Press LLC
Lewis Publishers is an imprint of CRC Press LLC

No claim to original U.S. Government works
International Standard Book Number 1-56670-554-1
Library of Congress Card Number 00-042832
Printed in the United States of America   3 4 5 6 7 8 9 0
Printed on acid-free paper

Executive Summary

The Potential of U.S. Grazing Lands to Sequester Carbon and Mitigate the Greenhouse Effect, edited by R.F. Follett, J.M. Kimble, and R. Lal, describes grazing lands, the areas they occupy, and their important role in sequestering C to help mitigate the greenhouse effect. The editors and 36 other authors prepared the 17 chapters, which each includes extensive references. Chapter 16 provides a summary and overview of C sequestration in grazing land soils and estimates the overall potential of U.S. grazing land to sequester C, while Chapter 17 considers research needs and priorities.

Grazing lands represent the largest and most diverse single land resource in the U.S. and in the world. In the U.S., rangelands and pastures together make up about 55% of the total land surface, and more than half of the earth's land surface is grazed. Grazing lands occupy an even larger area than cropland in the U.S. — 212 Mha (524 million acres) of privately owned and over 124 Mha (300 million acres) of publicly owned land, or more than twice the area of cropland. The large area grazing land occupies, its diversity of climates and soils, and the potential to improve its use and productivity all contribute to its great importance for sequestering C and mitigating the greenhouse effect and other aspects of climate change.

Section 1: The Extent, General Characteristics, and Carbon Dynamics of U.S. Grazing Land

Chapter 1 provides an introduction to the book. It describes the authors' objectives, the background which prompted them to prepare the book, and the wide range of questions involved. It also describes (and provides photographs of) the wide range of ecosystems grazing land includes. It summarizes the negative effects of human intervention on grazing lands and the best management practices needed to reverse those effects and effectively sequester C and prevent its further loss.

Chapter 2 provides the broad-scale perspective on the extent, distribution, and characteristics of grazing lands. It also provides the analyses of the National Resources Inventory (NRI) for the areas and descriptions of the privately owned U.S. grazing lands (rangeland and pasture) used in this book.

Chapter 3 describes the organic C pools and cycles and soil organic C (SOC) sequestration, while Chapter 4 describes the soil inorganic C (SIC) pools and cycles in grazing land soils.

Section 2: Soil Processes, Plant Process, and Carbon Dynamics on U.S. Grazing Land

Chapter 5 considers heterogeneity, long-term sustainability, and feedback loops, including structural features recognizable at the plant, plant community, and landscape scales as they relate to rangeland areas with different C sequestration potentials. Chapter 6 explains in greater detail the important biological roles provided by plant roots and soil microbial biomass for organic C sequestration.

Chapter 7 gives experimentally measured CO_2-flux data across a transect of grazing land sites extending from north to south in the Great Plains. These data can be extrapolated to a large part of the Great Plains and show a net sequestration of C into the soil and vegetation in this important region. The estimates are similar in magnitude to those made independently by the authors of Chapter 11.

Chapter 8 considers the very complex plant and soil processes related to C sequestration in tundra, alpine, and mountain meadow systems. Understanding the feedback relationships between biogeochemical processes and human activities in these cold environments is increasingly important because human concerns and opinions and public policy decisions can severely affect these fragile ecosystems.

Section 3: Managerial and Environmental Impacts on U.S. Grazing Land

Chapter 9 addresses soil erosion and C dynamics, provides estimates of land cover and the factors affecting soil erosion, and estimates areas of highly erodible grazing land and C losses and emissions resulting from soil erosion. It shows how restoring eroded grazing land soil can sequester a large amount of SOC.

Chapter 10 covers the management of soil's physical characteristics such as bulk density, crusting, aggregation, and infiltration characteristics and capacity, which decrease soil erosion and encourage soil C sequestration.

Follett, Kimble, and Lal, editors

Chapter 11 discusses important considerations for the soil C dynamics of rangeland. It also discusses the effects of grazing management on changes in rangeland plant communities and the potential of rangelands to sequester C.

Chapter 12 discusses how pasture management options provide an excellent opportunity to sequester SOC. Although occupying a smaller total area than rangelands, pastures generally have more favorable climatic conditions and respond well to the planting of improved species, grazing, and fertility management.

Chapter 13 evaluates the effects of fire on soil C sequestration, especially for tall grass prairie. Many grassland areas have coevolved with fire and grazing. When properly managed, fire can promote growth of many of the grass species and help to control the invasion of unwanted species. The increase in plant growth that results from fire can compensate for the loss of aboveground C with little or no change in SOC.

Section 4: Using Computer Simulation Modeling to Predict Carbon Sequestration in Grazing Land

Because of the importance of plant productivity to the sequestration of C, Chapter 14 focuses on the complex interactions between system components and how computer simulation modeling can help evaluate the effects of environmental change and of management strategies to improve rangeland production.

Chapter 15 describes soil C responses to environmental change in grassland systems and the factors that control soil C dynamics. In addition, Chapter 15 also demonstrates the predictive capability of computer modeling by simulating changes in aboveground plant production and SOC during a 50- to 75-year period after initiation of altered climate conditions and a doubling of CO_2 levels in the Great Plains.

Section 5: Summary and Overview and Research and Development Priorities

Chapter 16, the synthesis chapter, summarizes information from all the other chapters and presents an overview of U.S. grazing lands' potential to sequester C. Chapter 17 addresses issues that require research and identifies some of the priorities scientists, policy makers, and producers need to consider. Chapter 16 concludes that C sequestration under grazing land (pasture and rangeland) results from both managed and unmanaged activities. The total soil C sequestration potential for U.S. grazing land is 29.5 to 110.0 (mean = 69.9) million metric tons

(MMT) C/yr and represents a sink 1.6 times the size of the CO_2-C emission from all U.S. agriculture (42.9 MMTC/yr; Lal et al. 1998) and 4.4 times the CO_2-C emission from all grazing land agriculture.

This total includes 45% from land conversion and restoration, 7% from unmanaged rates of C sequestration, and 48% due to adoption of improved practices. Much can be accomplished to decrease the CO_2 emitted from grazing land through land conversion and restoration practices (17.6 to 45.7 MMTC/yr), specifically by decreasing soil erosion (6.2 to 24.6 MMTC/yr) and with the Conservation Reserve Program (CRP), which has the potential to sequester SOC (7.6 to 11.5 MMTC/yr). Other conversion and restoration practices include mined land restoration and land conversion to pasture, which cumulatively result in the sequestration of an additional 2.6 to 7.8 MMTC/yr.

We also estimate that nonintensively managed grazing land currently sequesters between –4.1 and 13.9 MMTC/yr and includes sequestration of inorganic carbonates on extensive rangeland areas in the western U.S. Intensification on U.S. rangeland through improved management has the potential to sequester 5.4 to 16.0 MMTC/yr, while more intensive grazing management, improved species, and fertility and manure management of U.S. pastureland has the potential to sequester a total of 10.5 to 34.3 MMTC/yr.

Little information is available concerning the response of high-altitude, cold region grazing land in the U.S. to management inputs, except for mountain meadows. Mountain meadows are estimated to potentially sequester 0.05 to 0.10 MMTC/yr with improved fertility. Because mining provides economic resources whereby inputs of fertility, reseeding, and other land restoration practices can be made available, it also has the highest potential rates of SOC sequestration (> 1000 kg C/ha/yr).

The productivity and sustainability of grazing lands requires that high soil quality (e.g., biomass productivity potential and environmental quality) be maintained. Both nutrients and water-holding capacity of soil directly relate to SOC content. High-quality soils lead to C sequestration and result in low levels of CO_2 and other greenhouse gas emissions. A wide range of options exist for sequestering C in soil. Appropriate options differ for different soils and ecoregions, and no single option universally applies. Site-specific adaptation and total ecosystem management is needed, and the most appropriate combination of options must be selected. Thus, depending upon the ecosystems, some promising options include:

- Growing species with deeper root systems and perhaps those that contain high lignin content, especially in their roots and residues.
- Managing rangelands and pastures through controlled grazing, vegetation management, and other recommended management practices. Pasture management options also include fertility inputs and improved species.
- Using strategies that increase biomass yields.

Follett, Kimble, and Lal, editors

- Using strategies to conserve water and soil resources.
- Converting marginal and degraded lands to restorative land uses (e.g., establish perennial cover, such as with CRP or afforestation).
- Restoring degraded soils.
- Enhancing biological N fixation.

Numerous gaps and uncertainties exist related to obtaining exact estimates for both the emissions of greenhouse gases from grazing lands and for narrowing the range for the estimated potential of grazing land's agricultural activities to sequester C. In spite of these gaps, the authors of this book clearly have established that significant amounts of C are sequestered in grazing land soils and, with careful and appropriate management, even more C can be sequestered. The diversity of climates and soils of grazing lands and the potential to improve their use and productivity all contribute to their great importance for sequestering C.

There are 212 Mha of privately owned and over 124 Mha of publicly owned grazing land in the U.S. — more than twice the area (155 Mha) of cropland soils. Because of the benefits to producers, the public, and society in general, sequestering C in soil is a win-win scenario with the potential to enhance soil quality and productivity, restore degraded grazing land resources, and mitigate the greenhouse effect. Improved management practices, restoration of degraded soils and ecosystems, and enhanced biomass productivity on grazing lands can provide a significant contribution to offsetting U.S. emissions of CO_2 while helping to mitigate the greenhouse effect.

As Chapter 17 states, however, grazing land systems can contribute to both the environmental and economic well-being of U.S. agriculture while also sequestering C — provided that our national strategic goals include enhancing (1) grazing land condition, (2) environmental quality, and (3) the economic viability of producers. Realizing this vast potential depends on (1) obtaining research data on soil carbon dynamics, (2) assessing the cost of soil C by monitoring additional costs associated with adopting recommended management practices, (3) identifying policy issues that facilitate/encourage adopting the desired practices, and (4) putting in place the mechanisms needed.

Soil C sequestration research has been ad hoc and piecemeal. We need to develop a systematic research program based on identified knowledge gaps and priorities of issues. The program has to be interdisciplinary and involve soil scientists, agronomists, economists, and policy makers. Similarly, we need to develop interdisciplinary research teams (including biophysical and social scientists) to identify appropriate policies for soil C sequestration. Various governmental bodies may need to identify and implement appropriate policies to facilitate widespread adoption of recommended practices.

An important component of the interdisciplinary program is the land manager. Farmers and land managers need to be involved in program planning and policy development, from the beginning. A research program developed without involving farmers and land managers would be a counterproductive and futile effort.

Close cooperation between agricultural scientists (crop and range specialists) and foresters in their work also is crucial. Choosing appropriate land uses (whether agriculture or forestry) depends on identifying criteria that cover the entire spectrum from seasonal crops to perennials, and this broad spectrum is a continuum without disciplinary boundaries.

Research must strengthen the databases for land use and soil characteristics and processes. Databases for land use and SOC and SIC pools and fluxes need to include assessments of the impacts of land use and management on the SOC and SIC pools and on the physical, chemical, and biological properties of grazing land soils. Soil erosion, especially as it affects C dynamics and rates of soil C loss from grazing land systems, also has not been addressed adequately. The effects of soil restoration and restorative measures on rates of SOC and SIC sequestration are most important. Such measures improve soil quality and biomass productivity; we need to understand the characteristics of soil resilience so that we can predict how and where restorative efforts will have the greatest chance to succeed.

We also must establish criteria for determining the value of soil C and treating it as a commodity. Those criteria must account for both the on-site benefits associated with soil quality and the off-site benefits to society, which include mitigating the greenhouse effect, enhancing water and air quality, improving wildlife habitat, and helping maintain biodiversity.

Follett, Kimble, and Lal, editors

Contents

Tables

Figures

Follett, Kimble, and Lal, editors

About the Editors

Dr. R.F. Follett is Supervisory Soil Scientist, USDA-ARS, Soil Plant Nutrient Research Unit, Fort Collins, CO. He previously served 10 years as a National Program Leader with ARS headquarters in Beltsville, MD. Dr. Follett is a Fellow of the Soil Science Society of America, American Society of Agronomy, and Soil and Water Conservation Society. He twice received the USDA Distinguished Service Award (USDA's highest award).

Dr. Follett organized and wrote the ARS Strategic Plans both for *Ground-Water Quality Protection — Nitrates* and for *Global Climate Change — Biogeochemical Dynamics*. He has served as editor or coeditor of several books and as a guest editor for the *Journal of Contaminant Hydrology*. His scientific publications include topics about nutrient management for forage production, soil-N and -C cycling, groundwater quality protection, global climate change, agroecosytems, soil and crop management systems, soil erosion and crop productivity, plant mineral nutrition, animal nutrition, irrigation, and drainage.

Dr. J.M. Kimble is a Research Soil Scientist at the USDA, Natural Resources Conservation Service, and National Soil Survey Center, Lincoln, NE, where he has been for the last 19 years. Previously, he was a Field Soil Scientist in Wyoming for three years and an Area Soil Scientist in California for three years. He has received the International Soil Science Award from the Soil Science Society of America.

While in Lincoln, Dr. Kimble worked for 15 years with the U.S. Agency for International Development on a project to help developing countries with their soil resources. For the last 10 years, the major part of his work has been related to Global Climate Change and the role soils can play in this area. His scientific publications deal with topics related to soil classification, soil management, global climate change, and sustainable development. He has worked in many different ecoregions, from the Antarctic to the Arctic and all points in between.

Dr. R. Lal is a Professor of Soil Science in the School of Natural Resources at Ohio State University. Before joining Ohio State in 1987, he served as a soil scientist for 18 years at the International Institute of Tropical Agriculture, Ibadan, Nigeria. His research interests include soil degradation and its effects on productivity in relation to erosion, compaction, and anaerobiosis; sustainable management of soil and water resources; and soil carbon sequestration through conservation tillage, residue management, and cover cropping. Prof. Lal is a fellow of the Soil Science Society of America, the American Society of Agronomy, the Third World Academy of Sciences, the American Association for Advancement of Sciences, and the Soil and Water Conservation Society.

Dr. Lal is a recipient of the International Soil Science Award, the Soil Science Applied Research Award of the Soil Science Society of America, the International Agronomy Award of the American Society of Agronomy, and the Hugh Hemmend Bennett Award of the Soil and Water Conservation Society. He is past president of the World Association of Soil and Water Conservation and of the International Soil Tillage Research Organization.

Above, left to right, the editors: R.F. Follett, USDA, ARS, Fort Collins, CO; J.M. Kimble, USDA, NRCS, Lincoln, NE; and R. Lal, The Ohio State University, Columbus, OH.

Follett, Kimble, and Lal, editors

Other Books Related to Soil and Global Climate Change

These books related to soil and global climate change were edited or written by the editors (see photograph on previous page) and, in some cases, their colleagues.

Currently Available

Soil Management and Greenhouse Effect
  R. Lal, J.M. Kimble, E. Levine, and B.A. Stewart
Management of Carbon Sequestration in Soil
  R. Lal, J.M. Kimble, R.F. Follett, and B.A. Stewart
Soil Processes and the Carbon Cycle
  R. Lal, J.M. Kimble, R.F. Follett, and B.A. Stewart
Global Climate Change and Pedogenic Carbonates
  R. Lal, J.M. Kimble, and B.A. Stewart
Global Climate Change and Tropical Ecosystems
  R. Lal, J.M. Kimble, and B.A. Stewart
The Potential of U.S. Cropland to Sequester Carbon and
  Mitigate the Greenhouse Effect
  R. Lal, J.M. Kimble, R.F. Follett, and C.V. Cole

Forthcoming

Global Climate Change: Cold Regions' Ecosystems
  R. Lal, J.M. Kimble, and B.A. Stewart
Methods of Assessment of Soil Carbon
  R. Lal, J.M. Kimble, R.F. Follett, and B.A. Stewart

Acknowledgments

The editors have drawn on material and data from numerous sources, especially on the information the authors of the chapters in this book presented.

We are thankful to scientists of the Jornada Experimental Range and faculty of the New Mexico State University for their help in organizing the workshop which began this effort.

Special thanks are due to Drs. Jeff Herrick, Curtis Monger, Joel Brown, and Kris Havstad. We especially appreciate the support and encouragement to proceed with this work that Dr. Herman Mayeux, former ARS National Program Leader for Rangelands and Global Change Research, gave us. All of the authors did a commendable job in compiling state-of-the-art information and responding promptly to editorial comments and questions.

Over a dozen colleagues reviewed Chapter 16, the synthesis chapter, and we gratefully acknowledge their helpful suggestions. Their time, effort, and comments helped us immensely to improve the final manuscript. We also appreciate the assistance Ms. Sandy Hayes of ARS Information Staff and many others for their willingness to provide pictures of grazing lands.

We thank Ms. Lynn Everett of the Ohio State University for her help in organizing the workshop and in conducting the review process. We also thank Maria Lemon, Ph.D., of the Editor Inc., for her many comments and suggestions and for her work in editing the manuscript and creating the camera readies for printing.

We appreciate the contribution of the front cover photos from (top to bottom) the USDA Agricultural Research Service and Dr. Richard Hart, Rangeland Scientist, High Plains Grasslands Research Station, Cheyenne, Wyoming.

Thanks are also due to the staff of Lewis/CRC Press for their efforts in publishing this information on time to make it available to the overall scientific community, land managers, and policy makers.

The Editorial Committee

Follett, Kimble, and Lal, editors

Preface

The U.S. is blessed with an abundance of naturally fertile and productive soil. Too often, the importance of soil to major facets of human life is poorly understood, taken for granted, or ignored. Soil is central to the abundance of food, feed, and fiber production in the U.S. Although most Americans recognize this, they sometimes must be reminded.

However, the important role of soil as it interfaces and interacts with the atmosphere (air), hydrosphere (water), and biosphere (life on the earth's surface) is less well recognized and often results in an attitude of "treating soil like it is only dirt." Something to compact under highways or houses or walk on without ever recognizing that soil is a living substance that contains untold billions of individual microorganisms — the microflora and micro- and macrofauna that help cleanse the water percolating through the soil, recycle plant nutrients, decompose organic and sometimes toxic substances, and form symbiotic relationships with plants to capture nitrogen from the atmosphere and improve the fertility and productivity of the soil for plant growth.

A few short years ago, one of the least understood facts about soil was its critical role in storing carbon (C) as soil organic matter (SOM) and soil inorganic carbonates (SIC) and in exchanging the atmospheric C held as carbon dioxide (CO_2). The CO_2 in the atmosphere is one of the most important of several greenhouse gases. If concentrations of CO_2 increase too much, then the earth's temperature can be increased. Stashing the C from atmospheric CO_2 into the C in SOM, through the intervention of photosynthesis by green plants, both helps prevent an excess buildup of the concentration of CO_2 in the atmosphere and enriches the SOM level in soil. The SOM in the pedosphere enhances the capability of soil to cleanse water, store more soil water for plant growth, recycle and improve the plant nutrients in the soil, aid in the decomposition of organic wastes, resist soil erosion, and maintain the land's fertility and productivity.

In 1998, we published *The Potential of U.S. Cropland to Sequester Carbon and Mitigate the Greenhouse Effect*. Policy makers, scientists, and the public have read and used that book, to better understand the important role of the 155 million hectares (Mha) (383 million acres) of U.S. cropland soils. This new volume, on grazing land soils, is intended both to complement the previous book and to allow those who read it to become more aware of the related and great importance of the grazing lands of the U.S.

Grazing lands occupy an even larger area than cropland — 212 Mha (524 million acres) of privately owned and over 124 Mha (300 million acres) of publicly owned land in the U.S., or more than twice the area of cropland. The large area grazing land occupies, its diversity of climates and soils, and the potential to improve its use and productivity all contribute to its great importance for sequestering C. This book is a first comprehensive evaluation of the potential of these important, diverse, and majestic land resources to sequester C. We hope that we have provided the reader with a new appreciation of grazing land soils and their potential importance in helping to maintain and improve the global environment.

Ronald F. Follett
John W. Kimble
Rattan Lal

Foreword

Climate change remains at the center of a debate over the nature and severity of potential impacts of a number of global environmental changes. The scientific evidence clearly indicates that atmospheric concentrations of greenhouse gases are rising to levels which may alter the earth's energy balance and influence temperatures, precipitation patterns, and weather variability, albeit in ways that we cannot yet predict with confidence at the regional and smaller scales relevant to agriculture or other industries.

Atmospheric concentrations of the greenhouse gas CO_2 have doubled since the end of the last Ice Age and have increased another 30% since the beginning of the Industrial Revolution, about 200 years ago, to the current level of about 370 ppm. This recent increase is apparently the result of human activities, such as the conversion of natural ecosystems to agricultural ones and the combustion of fossil fuels. An atmospheric level of 700 ppm is a probably reasonable expectation at some point later in this century, and even higher levels are likely before concentrations peak or are reduced by international efforts to limit greenhouse gas emissions.

Other trace gases, like CH_4, are increasing steadily in atmospheric concentration, help trap heat in the atmosphere, and contain C, but none plays as significant a role as CO_2 in C storage and other phenomena of global change. The C stored in biomass and soils, whether on croplands, forested lands, or grazing lands, originated as atmospheric CO_2, which plants assimilated and converted to organic matter (OM). Changes in the atmospheric concentration of CO_2 influence the growth and the water balance of plants. These direct effects may alter certain ecosystem processes which, in turn, control the structure and productivity of plant communities.

Although the scientific community's overarching goal related to global climate change is to resolve the uncertainty about its impacts, we must also provide policy makers with as many options and as much flexibility as possible to deal with those impacts. We thus must know where the C sinks are, which land uses or land resource types can store the most C, and which do that most readily in response to manipulation. We are focused mostly on what can be done to store C in northern, temperate, and tropical forests because they occupy about a third of the world's land surface, they support large amounts of biomass per unit area in relation to other biomes and land uses, and management can enhance their productivity.

In the last few years, we also have focused on intensively managed croplands, having determined that changes in management and implementation of new conservation and production practices allow relatively rapid rates of C storage, as much as 1 MTC/ha/yr on some lands. The realization that such was the case was made possible because of long-standing interest in the chemistry and other characteristics of cropland soils. Scientists concerned with sustainable agricultural production and the role of organic C (OC) in soil quality have accumulated much data on changes in OC contents of cropland soils. Studies have monitored changes in soil C contents in response to management on croplands for over a hundred years at a few places like Sanborn Field on the campus of the University of Missouri, the Morrow plots at the University of Illinois, and the Broadbalk winter wheat experiment at Rothamsted Experimental Station in England.

In comparison, interest in the "health" of grazing lands and their soils developed very recently. Data on responses of soil C to use and management of grazing lands are sparse, and only recently have scientists sought out opportunities to document soil C contents and study C dynamics on grazing lands maintained for many decades under different management regimes. Compared to data on croplands and forests, we have little information on the effects of use and management of any of the various kinds of grazing lands on soil OC (SOC) contents. Similarly, mechanisms of SOC dynamics are not studied extensively on our highly diverse grazing lands, as they have been on croplands. Perhaps as a consequence, we long have assumed, and written, that the C budgets of grazing lands are in equilibrium, are stable over time. As certain chapters in this book suggest, in all likelihood we will revise that opinion as we accumulate additional information.

What are grazing lands? The word refers more to a set of highly diverse land resources than it does a land use. Grazing lands include the relatively undisturbed rangelands of the West (grasslands, savannas, and shrublands) and the intensively managed pastures that occur in every state, including those of the humid Southeast. This huge resource includes the annual grasslands of California, tundra in Alaska, the hot deserts and shrublands of the Southwest, the temperate deserts and sagebrush steppe of the Pacific Northwest, the cold deserts of the Great Basin, the prairies of the Great Plains, and native grasslands in the South and Atlantic states. Grazed forests often are included.

Grazing lands represent the largest and most diverse single land resource in the U.S. and in the world. In the U.S., rangelands and pastures together make up about 55% of the total land surface, and more than half of the earth's land surface is grazed. According to the Natural Resources Conservation Service, privately owned rangelands and pastures in the U.S. comprise a total of 212 Mha. Much of the 124 Mha of publicly owned lands in the West properly are classified as grazing lands. Millions of additional hectares of cropland throughout the nation are planted to forages, and these forages often are grazed as well as hayed or harvested as other forms of stored forage.

Grazing lands provide a wide array of goods and services of considerable economic, environmental, and social importance. In terms of agriculture, grazing lands provide the basis for the nation's beef, dairy, and sheep industries. More than 85% of the publicly owned lands in the West are grazed. A recent analysis indicates that 106 Mha of western public lands support about 3 million beef cattle. Grazing lands usually are stocked with more than 60 million cattle and 8 million sheep, supporting a livestock industry that annually contributes $78 billion in farm sales to the U.S. economy.

Grazing lands also support 20 million deer, 500,000 pronghorn antelope, 400,000 elk, 55,000 wild horses and burros, and many other wildlife species. But the importance of grazing lands is not limited to the animals and vegetation found on them. Grazing lands also function as watersheds; much of our water supply originates as rainfall or snowmelt on grazing lands. Along with forests, our grasslands and other grazing lands are the principal repositories of biological diversity. Grazing lands also provide many opportunities for recreational activities. Sequestration of atmospheric C appears to be another way in which these lands could contribute to the environmental and economic health of our nation.

Grazing lands might be considered potentially important as sinks for atmospheric C because of the magnitude of the land area involved. Even if rates of C storage are low on an areal basis, or maximum amounts of storage are limited and soon attained, the total mass could be quite large. Some portion of our grazing lands, like arid rangelands in the West, can be expected to offer little potential for C storage because of low seasonal productivity and shallow soils. But other areas, even some in the West, are highly productive and have deep, well developed soils.

Annual production of plant biomass on many native perennial grasslands and most pastures exceeds that of croplands, and intensively managed pastures planted to annual grasses or legumes on deep soils in many regions might store C at rates equal to those of annual crops. Some of the highest rates of C storage to be found in the scientific literature were measured on perennial grass pastures, especially after establishment on former cropland, even when the stands were not managed for high productivity but were left undisturbed, as in lands enrolled in the Conservation Reserve Program.

Vegetation on grazing lands may have attributes other than high primary productivity that promote relatively high rates of C storage. The vegetation of most grazing lands is predominantly perennial, unlike that of croplands, or consists of a mixture of annuals and perennials. Grazing land vegetation often consists of a mixture of cool-season (C_4) and warm-season (C_3) species, again unlike that of cropland. These characteristics may maximize the length of time during the year in which rates of C assimilation exceed respiration, at least in mesic to humid areas. Grassland species, especially perennial bunchgrasses, partition much higher proportions of total plant biomass into belowground tissues than do trees, most shrubs, or annual crops. Also, there may be processes under way on complex rangeland landscapes, yet unrecognized, that promote organic and even inorganic soil C contents.

Little consideration has been given to the possibility that grazing lands can be managed to enhance C storage rates. Improvements in the management of grazing, sometimes defined as manipulating the timing, duration, and intensity of use of the vegetation, often increase the productivity of grazing land vegetation and may contribute to higher C storage rates. The same is true of the trend toward conservative or sustainable use, implying lower stocking rates and removal of less forage, leaving more residue. Soil C contents of pasturelands should benefit from more widespread fertilization, irrigation, and other practices which increase C content of cropland soils.

Rates of change in OC content of croplands were the subject of an earlier book, *The Potential of U.S. Cropland to Sequester Carbon and Mitigate the Greenhouse Effect*, authored by the editors of this present effort. The "black book," as it commonly is called, appeared in 1998, and it presents estimates of potential C storage for various actions or scenarios on arable lands, including a total potential sequestration which ranges from 75 to 208 MMTC/yr. That volume originated in increasingly frequent requests from policy makers, mostly in Washington, D.C., for better information concerning the value of U.S. farmland for storing C. Their need to know in large part was associated with efforts to include Article 3.4 in the Kyoto Protocol of December 1997, which expanded the definition of sinks to include agricultural lands.

This volume has similar origins and will be as relevant. It begins to establish the extent to which grazing lands should be considered, along with forests and croplands, as potential sinks. It, too, presents an estimate of potential sequestration, as well as information on C dynamics of grazing lands and the processes which control it. However, it differs from the earlier volume in that it is less of a synthesis and analysis of information and more a presentation of current research findings, reflecting the wide differences in what we know about soil C dynamics on croplands and grazing lands.

Along with what we know about the C dynamics of forests, the two books provide a basis for shaping policy and targeting research. We will have to establish priorities, given the limits to funding for C cycle and C storage research. Shall scientists and society focus on forests, croplands, grazing lands, wetlands, or other land uses as having the highest potential for C storage? If one or more is superior to the others in terms of forestalling climate change, to what extent will we have to balance that use against society's requirements for grains and other crops, for forest products, for forage and the red meat produced on grazing lands, or for the other goods and services provided by the nation's croplands and forests and rangelands?

Herman Mayeux
El Reno, OK
February 3, 2000

The Extent, General Characteristics, and Carbon Dynamics of U.S. Grazing Lands

CHAPTER 1

Introduction: The Characteristics and Extent of U.S. Grazing Lands

J.M. Kimble,[1] R.F. Follett,[2] and R. Lal[3]

Introduction

For the purposes of this book, the definition of grazing land is the Society for Range Management's, "a collective term that includes all lands having plants harvested by grazing without reference to land tenure or other land uses, management, or treatment practices." The U.S. contains 212 million hectares (Mha) of privately owned grazing lands.

Figure 1.1 shows the areas of private U.S. grazing land in 1997. Most of it is in the western states and is predominately rangeland, with a soil moisture regime of acidic to dry ustic (water limited). However, large areas of grasslands are in the wet ustic and udic moisture regimes of the Dakotas, Nebraska, Kansas, Oklahoma, Texas, and Missouri. Substantial areas of grazing lands also are improved pasture found in many of the eastern states, where water is not a limiting factor. Figure 1.2 shows the distribution of rangeland, and Figure 1.3 shows the distribution of pastureland.

Grazing lands are essential ecosystems containing 10 to 30% of the world's soil organic carbon (SOC) (Eswaran et al., 1993). Batjes (1999) estimates soils to a 1-m depth within grazing lands (grassland/steppe, extensive grasslands, hot deserts, scrubland, and savannas) may contain 306 to 330 Pg SOC and 470 to 550 Pg soil inorganic carbon (SIC).

Waltman and Bliss (1997) estimate that about 5% of the world's SOC is in soils of the U.S., with 15.3 to 16.5 Pg in grazing lands. This amount reflects sig-

[1] Research Soil Scientist, USDA-NRCS-NSSC, Fed. Bldg. Rm 152 MS 34, 100 Centennial Mall North, Lincoln, NE 68506-3866, phone (402) 437-5376, fax (402) 437-5336, e-mail john.kimble@nssc.nrcs.usda.gov.

[2] Supervisory Soil Scientist, USDA-ARS, P.O. Box E, Fort Collins, CO 80522.

[3] Prof. of Soil Science, School of Natural Resources, The Ohio State University, Columbus, OH, e-mail lal.1@osu.edu.

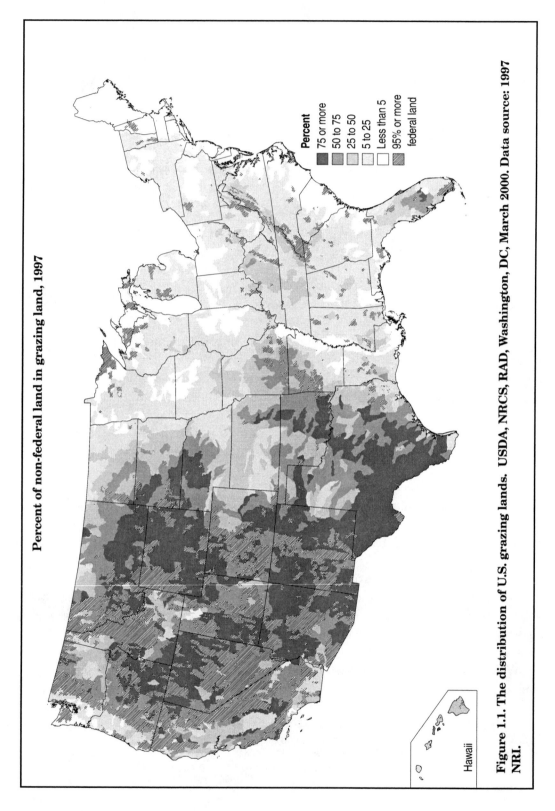

Percent of non-federal land in grazing land, 1997

Percent
75 or more
50 to 75
25 to 50
5 to 25
Less than 5
95% or more federal land

Hawaii

Figure 1.1. The distribution of U.S. grazing lands. USDA, NRCS, RAD, Washington, DC, March 2000. Data source: 1997 NRI.

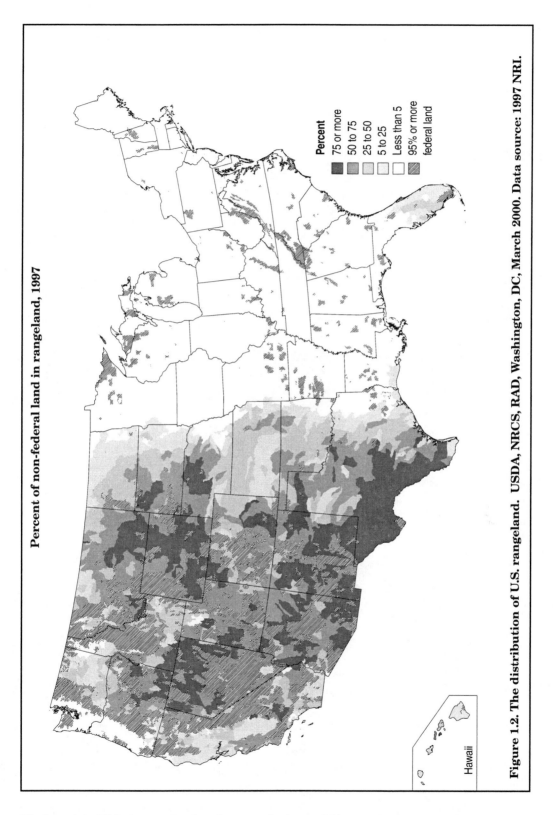

Percent of non-federal land in rangeland, 1997

Percent
- 75 or more
- 50 to 75
- 25 to 50
- 5 to 25
- Less than 5
- 95% or more federal land

Hawaii

Figure 1.2. The distribution of U.S. rangeland. USDA, NRCS, RAD, Washington, DC, March 2000. Data source: 1997 NRI.

The Potential of U.S. Grazing Lands to Sequester Carbon and Mitigate the Greenhouse Effect

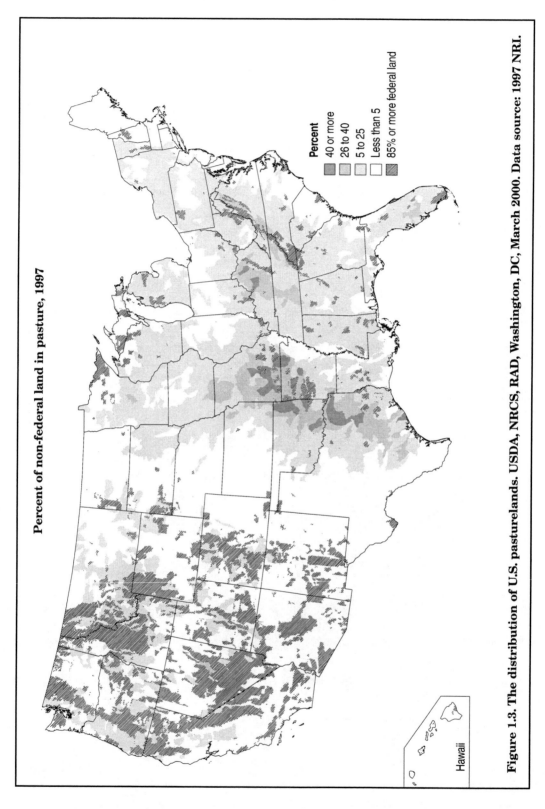

Percent of non-federal land in pasture, 1997

Percent
- 40 or more
- 26 to 40
- 5 to 25
- Less than 5
- 85% or more federal land

Hawaii

Figure 1.3. The distribution of U.S. pasturelands. USDA, NRCS, RAD, Washington, DC, March 2000. Data source: 1997 NRI.

nificant losses over time caused by such human activities as overgrazing of animals, introduction of new and different plant and animal species, and varying management practices. Oldeman (1994) reported that overgrazing, salinization, alkalization, and acidification have degraded large areas of the grasslands of the world, and overgrazing is the greatest cause of their degradation (Oldeman et al., 1991).

In cropland, 20 to 50% of the original SOC in the zone of cultivation has been lost (Lal et al., 1998a). The loss in grazing lands probably is not this great in many systems but, in severely degraded systems, it could be greater. Because many grazing land soils are degraded, still poorly managed, or not managed at all, and also receive low levels of inputs, they have a high potential to sequester C if we apply best management practices (BMPs) and increase inputs to them.

Conservation tillage and no-till have a positive effect on organic C (OC) sequestration in the semiarid regions of the U.S. (Potter et al., 1997; Lal and Kimble, 1997). Grazing lands can be considered a form of no-till. Grasses, legumes, and other forbes are left without cultivation and, with proper management, we can expect sequestration of C to increase. Enhanced management may require the addition of more fertilizer, better grazing management, water management, introduction of improved species, and the overall intensification of grazing lands. In reality, many of the same practices described in Lal et al.'s earlier book (1998a) apply to grazing lands.

Background

In 1997, Ron Follett, Rattan Lal, and John Kimble met with the Under Secretary of Agriculture to discuss the potential of U.S. cropland and grazing lands to help reduce the greenhouse effect by fixing CO_2 through photosynthesis. (After fixation into the plant biomass, C can be sequestered as SOC or SIC.) We decided to complete a book on cropland, which resulted in *The Potential of U.S. Cropland to Sequester Carbon and Mitigate the Greenhouse Effect* by R. Lal, J. M. Kimble, R. F. Follett, and C. V. Cole (1998). Over 3000 copies of this book have been sold, showing the extreme interest in C sequestration and the role that agriculture can play. This book stimulated a great deal of discussion related to the sequestration of C in cropland soils and has rekindled the discussion about the role that grazing lands could play.

In December of 1997, 165 countries adopted the Kyoto Protocols, which, under Article 3.4, deal with sequestering C in soil sinks, including those in agricultural soil and those resulting from land use, land use change, and forestry. The Intergovernmental Panel on Climate Change (IPCC) is producing a special report for the United Nations Framework Convention on Climate Change (UNFCC) to address possible sinks, including in grazing lands. This activity spurred the idea of this second book on C sequestration, with specific attention to grazing lands.

Objectives

We decided to do this second book in a slightly different format, inviting authors to address assigned topics while we introduced (Ch. 1) and synthesized (Ch. 16) the material and outlined research needs (Ch. 17). Our original objectives were to:

1. Focus on SOC, SIC, and CO_2.
2. Cover all grazing lands except grazed forestland.
3. Collate and synthesize the available information on the net contribution of U.S. grazing lands to the greenhouse effect.
4. Assess the role of soil restorative processes and BMPs on C sequestration in grazing land.
5. Identify policy and management options that enhance C sequestration and mitigate the greenhouse effect.
6. Identify (a) the current base of soil C, and (b) with present management practices, the potential to increase the levels of SOC and SIC. Chapters 2 through 15 discuss this information in detail; Chapter 16 synthesizes that information for policy makers and others interested in the potential of grazing lands to sequester C; Chapter 17 prioritizes research needs.

Questions

Some of the basic questions to be addressed include why are we concerned with grazing lands and C sequestration? What is the overall potential for C sequestration? How did we get where we are now, and where do we go from here?

Research on rangeland has been conducted on different sites for many years (Child and Frasier, 1992). These activities at times were completed cooperatively and at other times only at specific sites. Because of the growing interest in global climate change, a Rangeland Carbon Dioxide Flux Project (Svejcar et al., 1997) was initiated to measure rangeland CO_2 flux. Results of the flux measurements will be tied to long-term productivity and changes in vegetative patterns and will help us understand the effects of various management practices on these fluxes. In some cases, change may be a result of management and, in other cases, a result of global climate change.

In addition to looking at the potential to sequester C in soils, we must consider the effect that global climate change, particularly elevated CO_2, will have on soil properties and plant life. Even without global warming, elevated CO_2 levels may alter ecosystems in both their structure and function (Bazzaz and Fajer, 1992).

Follett, Kimble, and Lal, editors

What role will CO_2 play in the C sequestration process? ARS is addressing this through the Flux project.

Global climate change can affect many soil properties, not only in cropland, but also in grazing lands. SOC positively affects soil structure, soil erodibility, crusting, compaction, infiltration rates, runoff, salinity, and the cycling of plant nutrients (Kimble et al., 1998) and thus helps prevent or reverse degradative processes. Therefore, anything that can increase or maintain SOC will have a positive effect on soil quality.

Book organization

This book is divided into five main sections. Section 1 discusses what grazing lands are, their distribution and potential to sequester C, and the dynamics of SOC and SIC. Section 2 covers soil and plant processes and carbon dynamics. Section 3 deals with managerial and environmental impacts on grazing lands. Section 4 presents information about using computer simulation modeling to predict C sequestration. Section 5 synthesizes all the information and other data to give readers an overall perspective on the potential of grazing lands to sequester C, and it outlines research priorities.

The Importance of Management Practices

Grazing lands are a bridge between natural ecosystems and agroecosystems. In many areas, they still contain native vegetation but have been modified by human interventions. Because of this, few if any systems are truly native.

Human interventions have changed the mix of animals within grazing land systems. In much of the western U.S., cattle (Fig. 1.4) and sheep (Fig. 1.5) have been introduced on the rangelands, causing the mix of native grazers to change and grazing intensity to increase. Although little has been done to introduce new plant species, water development has changed grazing intensity in areas where water is limited (Fig. 1.6). Windmills pumped water to allow grazing animals in areas where animals had been limited, and the overall biodiversity of the rangeland systems thus has been changed. Thus the distinction between native and agroecosystems is hard to define.

Rangelands have not undergone the same interventions as croplands, but they have lost much of their native C because of excessive grazing and poor management. This has led to erosion by water, as in Figure 1.7, which shows the formation of gullies in Wyoming. Figure 1.8 shows a typical wind-eroded landscape in New Mexico. Figure 1.9 shows the combined effects of both wind and water, resulting in a highly degraded landscape.

Figure 1.4. Cattle are rounded up on the Fort Keogh Livestock and Range Research Station in Montana. Photo is from the ARS Photo Library.

Figure 1.5. Sheep grazing in a mountain meadow near Odell Lake on the ARS Sheep Experiment Station in the Centennial Mountains of southwestern Montana. Photo is from the ARS Photo Library.

Figure 1.6. Stock water development in northeastern Wyoming, through the use of a windmill.

Figure 1.7. Water erosion in northern Converse County in Wyoming.

Figure 1.8. Wind erosion in southern New Mexico.

Figure 1.9. The effects of the combination of wind and water erosion in northeastern Wyoming.

Overgrazing and poor management lead to a loss of system C, and the overall productivity of the land decreases as a result. Kemper et al. (1998) showed that deep-rooted grasses can increase the C deep in the profile in eastern gamagrass (*Trysicum notatum*) systems. When grasses are grown on sloping land, more C is sequestered and erosion is controlled (Lal et al., 1998b).

Since many of the grazing lands in the U.S. are in arid and semiarid environments, the sequestration of SIC may play a larger role in C sequestration there than in cropland. Much of the C in drier areas may be in the form of pedogenic carbonates. Zuberer et al. (1996) reported that much of the C in soils studied in the Desert Project in New Mexico was in relatively unavailable forms. It is imperative for management practices not to allow more C in these soils to become active and emitted as CO_2.

When we consider grazing, we are not concerned only with domesticated animals, but also with the effects that management will have on wildlife (Fig. 1.10). How private and public lands are managed can and does affect wildlife and is important in the overall approach to grazing land management. Grazing intensities need to reflect both domesticated animals and the pressures of native species.

Figure 1.10. Buffalo grazing in the Theodore Roosevelt National Park in North Dakota.

Conclusion

Swift et al. (1996) discussed biodiversity and agroecosystems, focusing on the idea that systems influenced by human activities have a lower biodiversity, with this being the case in many grazing land systems. One way of increasing SOC is through agricultural intensification (Lal et al., 1998a). Much of what is discussed in this book could be considered grazing land intensification. As many of these techniques are used, they will need to be examined holistically, including their impact on the overall biodiversity of grazing lands.

Grazing lands are complex ecosystems. They range from the permafrost-affected tundra in Alaska (Fig. 1.11), where the main grazer may be reindeer, to the highly managed pastures in the southeastern U.S. (Fig. 1.12). Large areas in many parts of the U.S. also are used for dairy production (Fig. 1.13).

Alpine meadows (Fig. 1.14) are grazed in many mountain states in the western U.S. Figure 1.15 represents the grazing lands in many areas of California, where the landscapes often are covered by a mix of trees on more northern slopes and by open savannas on southern exposures. These are areas with cool winter rains and hot dry summers.

Figure 1.11. Arctic tundra on the north slope of Alaska where the main grazers are reindeer. This is a very fragile ecological zone, where the effects of climate change could be dramatic.

Figure 1.12. Angus cattle grazing on highly managed pasture. Photo is from the ARS Photo Library.

Figure 1.13. Grazing on forested lands converted to pasture in the dairy area of Vermont.

Figure 1.14. Deer grazing in a mountain meadow in the Bighorn Mountains of Wyoming.

Figure 1.15. Open grazing lands on southern exposures in the coast range of northern California.

Figure 1.16 shows a typical range area in northern Wyoming that receives 25 to 35 cm of moisture a year. Figure 1.17 shows a typical range area in the dry Southwest. These systems are very fragile. Management practices have a major impact on areas in the lower parts of the watershed where these alpine meadows are found. For example, applied management practices on meadows in Wyoming, Colorado, and Montana may have an influence on the whole Missouri watershed.

The chapters in this book point out the vast and diverse knowledge required to properly use grazing lands and to increase the sequestration of C in soil as SOC or SIC. Their recommendations are not meant to imply that we should graze or produce large numbers of animals to produce manure to help sequester C, but they do recognize that land will be grazed and fertilizers applied, and farmers, ranchers, and land managers will continue to use grazing lands to produce fiber. This needs to be done in an environmentally friendly way. BMPs will improve water quality, overall grazing land quality, and wildlife habitat, while simultaneously sequestering C.

Grazing lands are a significant part of the land coverage in the U.S. How we manage them can and will have major impacts on many different areas. We need to manage them holistically, even on a scale as large as the Missouri River watershed, which drains almost 1/3 of the U.S. We must strive to maintain biodiversity

Figure 1.16. Typical range in northeastern Wyoming in the 25- to 35-cm precipitation zone.

Figure 1.17. Degraded rangeland in the desert southwest of New Mexico.

of both animals and plants, and we must increase the sequestration of C through many of the techniques this book discusses. By doing so, we will improve overall soil quality and have a net positive impact on our entire agroecosystem.

References

Batjes, N.H. 1999. *Management options for reducing CO_2 concentrations in the atmosphere by increasing carbon sequestration in the soil. Report no.: 410 200 031.* IS-RIC, Wageningen, Netherlands.

Bazzaz, F.A., and E.D. Fajer. 1992. Plant life in a CO_2-rich world. *Sci. Am.* Jan. 1992:68-74.

Child, R.D. and G.W. Fraiser. 1992. ARS research. *Rangelands* 14:17-32.

Eswaran, H.E., Van den Berg, and P. Reisch. 1993. Organic carbon in soils of the world. *Soil Sci. Soc. Am. J.* 57:192-194.

Kemper, W.D., E.E. Alberts, C.D. Foy, R.B. Clark, J.C. Ritchie, and R.W. Zobel. 1998. Aerenchyma, acid tolerance, and associative N fixation enhance carbon sequestration in soil. *In* R. Lal, J.M. Kimble, R.F. Follett, and B.A. Stewart (eds), *Management of Carbon Sequestration in Soils*, CRC Press, Boca Raton, FL, pp. 221-234.

Kimble, J.M, R. Lal, and R.B. Grossman. 1998. Alteration of Soil Properties Caused by Climate Change. *In* H.P. Blume, H. Eger, E. Fleischhauer, A. Hebel, C. Reij, and

K.G. Steiner (eds), *Towards Sustainable Land Use. Furthering Cooperation Between People and Institutions,* Vol 1. *Advances in Geoecology 31,* Caten Verlag, Germany, pp. 175-184.

Lal, R., and J.M. Kimble. 1997. Conservation tillage for carbon sequestration. *Nutri. and Cycling in Agroecosys.* 49:243-253.

Lal, R., J.M. Kimble, R.F. Follett, and C.V. Cole. 1998a. *The Potential of U.S. Cropland to Sequester Carbon and Mitigate the Greenhouse Effect.* Ann Arbor Press. Chelsea, MI.

Lal, R., P. Henderlng, and M. Flowers. 1998b. Forages and row cropping effects on soil organic carbon and nitrogen contents. *In* R. Lal, J.M. Kimble, R.F. Follett, and B.A. Stewart (eds), *Management of Carbon Sequestration in Soils,* CRC Press, Boca Raton, FL, pp. 365-379.

Oldeman, L.R. 1994. The global extent of soil degradation. *In* D.J. Greenland and I. Szabolcs (eds), *Soil Resilience and Sustainable Land Use,* CAB International, Wallingford, U.K., pp. 99-117.

Oldeman, L.R., V.W.P. van Engelen, and J.J.M. Pulles. 1991. The extent of human-induced soil degradation. *In World Map of the Status of Human Induced Soil Degradation: An Explanatory Note,* 2nd Rev. ed, International Soil Reference and Information Center, Wageningen, Netherlands, Tables 1-7.

Potter, K.N., O.R. Jones, H.A. Torbert, and P.W. Unger. 1997. Crop rotation and tillage effects on organic carbon sequestration in the semiarid southern great plains. *Soil Sci.* 162:140-147.

Svejcar, T., H. Mayeux, and R. Angell. 1997. ARS Research. *Rangelands* 19(5):16-18.

Swift, M.J., J. Vandermeer, P.S. Ramakrishnan, J.M. Anderson, C.K. Ong, and B.A. Hawkins. 1996. *In* H.A. Mooney, J.H. Cushmand, E. Medina, O.E. Sala, and E.D. Schulze (eds), *Functional Roles of Biodiversity: A Global Perspective,* John Wiley and Sons, London, UK, pp. 261-298.

Waltman, S.W. and N.B. Bliss. 1997. *Estimates of SOC Content.* NSSC. Lincoln, NE.

Zuberer, D.A., C.T. Hallmark, and L.P. Wilding. 1996. *Processes and Forms of Carbon Sequestration in Calcareous Soils of Arid and Semi-arid Regions.* Project Report, NRCS Global Change Program. Texas A&M University, College Station, TX.

CHAPTER **2**

A Broad-Scale Perspective on the Extent, Distribution, and Characteristics of U.S. Grazing Lands

T.M. Sobecki,[1] D.L. Moffitt,[2] J. Stone,[3] C.D. Franks,[4] and A.G. Mendenhall[5]

Introduction

Broadly scaled information about the extent and character of U.S. grazing lands provides a context for considering the more finely scaled processes driving carbon (C) cycling on grazing lands. Such information also is necessary for estimating the potential for grazing lands to sequester atmospheric C at local, regional, national, or global scales, relative to such other general land cover/use categories as cropland and forest. The terms "broad" and "fine" scale appear often throughout this chapter, and we define them as did Turner and Gardner (1991). *Broad scale* implies coarse spatial resolution of land attributes and correspondingly large areas of land; *fine scale* refers to a greater degree of spatial resolution of land attributes and smaller areas of land. The land attributes central to our discussion are existing and potential vegetative land cover and soil organic carbon (SOC) content.

Our objectives in this chapter are to present a broad-scale overview of the areal extent, regional distribution, and ecological character of U.S. grazing land and

[1] Landscape Ecologist, USDA-Natural Resources Conservation Service, 1400 Independence Ave., S. W., Room 6821, Washington, DC 20250, Phone: 202-690-3719, e-mail: terry.sobecki@usda.gov.

[2] Program Analyst, USDA-Natural Resources Conservation Service, 1400 Independence Ave., S. W., Room 6827, Washington, DC 20250, Phone: 202-720-2426, e-mail: dave.moffitt@usda.gov.

[3] Senior Resource Assessment Specialist, USDI-Bureau of Land Management, 1620 L Street N. W., Mail Stop 1075LS, Washington, DC 20240, Phone: 202-452-7786, e-mail: jstone@wo.blm.gov.

[4] Research Soil Scientist, USDA-Natural Resources Conservation Service, Federal Building, Room 152, Mail Stop 41, 100 Centennial Mall North, Lincoln, NE 68508-3866, Phone: 402-437-5316, e-mail: carol.franks@nssc.nrcs.usda.gov.

[5] Range Management Specialist, USDA-Natural Resources Conservation Service, Federal Building, Room 152, Mail Stop 36, 100 Centennial Mall North, Lincoln, NE 68508-3866, Phone: 402-437-4176, e-mail: arnold.mendenhall@nssc.nrcs.usda.gov.

to assess the implications of this information for estimating U.S. grazing lands' potential to store C. A significant part of the picture is a description of these lands' spatial heterogeneity at a scale appropriate for national assessments. Their heterogeneity includes variety in biophysical environments, vegetative land cover type, and changes resulting from patterns of land use.

Defining and Classifying Grazing Land

Numerous national and global figures are available on the extent, distribution, and characteristics of grazing land. To interpret such figures, however, one must recognize the heterogeneous nature of such a general land cover/use category as *grazing land*.

The definition of *grazing land* for a particular inventory has a significant impact on how one can interpret and use that inventory. To view grazing lands from the standpoint of C cycling and their ability to sequester atmospheric C (CO_2-C), we also must differentiate the concepts of *land cover* and *land use*.

Land use

The definition of *grazing*, "feeding on growing grass or herbage" (*The American Heritage Dictionary of the English Language*, 1979), suggests that the term *grazing land* is fundamentally a land use term. *Land use* is "the human activity associated with a specific piece of land" (Lillesand and Kiefer, 1987) and it is the "purpose of human activity on the land" (NRCS, 1992). From strictly a land use viewpoint, *grazed land* (NRCS, 1997) might be more appropriate than *grazing land*. One might also wish to include herbivory by wildlife, in addition to that of livestock grazing for direct human benefit, as a land use in a broader description of grazed land.

To characterize U.S. grazing lands adequately for their potential to sequester atmospheric C, however, one must go beyond a land use definition. Implicit in the definition of *grazing* is the presence of a certain type of vegetative land cover suitable for use as forage or browse: grasses, forbs, and shrubs. A shift from purely a land use perspective to a vegetative land cover perspective changes the picture of the extent and distribution of grazing land (Daugherty, 1989).

Land cover

Land cover (a.k.a. *earth cover*) refers to the natural or artificial material, including vegetation, present on the earth's surface (Lillesand and Kiefer, 1987; NRCS, 1992). Unlike *land use, land cover* is not defined based on human activity. Since production of vegetative biomass is the fundamental driver of C dynamics in grazing lands, the character of grazing land's vegetative land cover becomes a primary consideration. Changes in the character of the vegetative land cover, due to natural plant succession, natural catastrophic events, or human land use, alter the potential of the land to sequester atmospheric C.

Addressing a range of resource issues on grazing lands, Joyce (1989) also recognized the utility of characterizing U.S. grazing lands for both land use *and* vegetative cover. She applied the term *resource value rating* to characterize the societal benefit derived from the human *use* of grazing lands. She reserved the term *ecological status* as a use-independent characterization of grazing land's *vegetation* that spoke to its inherent productive potential. We also use the term *ecological status* that way, to refer to the character and condition of grazing land's vegetative cover in the landscape, relative to a particular function or process and in relation to a defined ecological reference state.

Vegetation

Since *land use* usually depends on a particular type of *land cover*, the two concepts commonly combine in joint land cover/use classification systems. For example, the Anderson "land use and land cover" classification system (Anderson et al., 1976) is the basis for several national and global digital data layers the U.S. Geological Survey developed (USGS) (USGS, 1998). The U.S. Department of Agriculture (USDA) uses a "land cover/use" system in its National Resources Inventory (NRI) program to characterize the U.S. landscape nationally (NRCS, 1992).

Most such systems contain fundamental distinctions between classification units, based on differences in vegetation. Vegetation is a fundamental component not only of land cover/use systems, but also of more integrated biogeographic ecological unit classifications, such as those of Bailey (1997) and Driscoll et al. (1984). Biogeographic systems are useful for describing the biophysical settings that drive and constrain many landscape processes and that lend a particular character to an area in terms of climate, geology, vegetation, soils, etc.

Many of the class names in these systems use vegetation classification terms (FGDC, 1997). The next two sections provide some information about these terms.

Physiognomy

Vegetation is classified in terms of both physiognomy and floristic characteristics. *Physiognomy* refers to the outward appearance or structure of the vegetation, its "growth form" or "life form" (Barbour et al., 1987; FGDC, 1997), which is a fundamental plant community attribute (Barbour et al., 1987; Kuchler, 1964).

The basic classification unit of a physiognomic classification system is the *formation* or a hierarchical generalization thereof (UNESCO, 1973). Table 2.1 gives

Table 2.1. Examples of taxa names for different classification levels of the UNESCO vegetation classification system (UNESCO, 1973).

Formation Class	Formation Subclass	Formation Group	Formation
Closed Forest (crowns interlocking)	Mainly Deciduous Forest	Cold deciduous (summer green) forests without evergreen trees	Temperate lowland and submontane broad-leaved cold deciduous forest
Woodland (open stand of trees, > 40% cover)	Deciduous Dwarf Scrub	Cold deciduous dwarf shrubland	Cold deciduous cushion dwarf shrubland
Scrub (shrubland or thicket)	Mainly Deciduous Scrub	Cold deciduous scrub	Temperate deciduous thicket (or shrubland)
Herbaceous Vegetation	Tall Graminoid	Tall grassland with tree synusia covering 10-40%[1]	Woody sysnusia broad-leaved deciduous

[1]*Synusia is an association of plant species of similar life form and habitat requirements (FGDC, 1997).*

an example of some of the taxa from the UNESCO vegetation classification hierarchy (Ellenberg and Mueller-Dombois, 1967; UNESCO, 1973).

The formation or higher levels are what most people would recognize as a particular type of vegetative land cover characteristic of a particular region or biophysical setting (Barbour et al., 1997; FGDC, 1997). Vegetation classification systems based on physiognomy are very amenable to generalized mapping of vegetation at broad scales and are relatively easy to relate to ecological conditions (Ellenberg and Mueller-Dombois, 1967). We rely heavily on a generalized physiognomic classification of vegetation to develop our broad-scale picture of the ecological status of U.S. grazing lands.

Floristics

Classifications based on floristics are concerned with the taxonomic or species characteristics of vegetation. Floristic composition also is considered a basic plant community attribute (Barbour et al., 1987). The fundamental unit of classification is the *plant association,* which is a particular type of plant community of defined floristic composition and uniform physiognomy characteristic of a particular habitat (Barbour et al., 1987). For example, one may speak of a *Bouteloua gracilis* (Blue grama grass) association. Higher level groupings of associations are called *alliances* (Barbour et al., 1987; Bourgeron et al., 1995).

The physiognomic and floristic aspects of vegetation's classification often are combined under one system for particular applications, with physiognomy used to differentiate more generalized higher level taxa and floristics incorporated into lower taxa (FGDC, 1997; Grossman et al., 1994). For example, Kuchler's potential natural vegetation (PNV) map of the U.S. contains map units that describe both the life form (physiognomy) and dominant plant taxa (floristics) of the *presumed vegetation* resulting from plant succession (Kuchler, 1964). Mapping of *existing vegetation* also relies on combined physiognomic/floristic components for classification hierarchies (Bourgeron et al., 1995; Grossman et al., 1994).

Extent and Broad-Scale Distribution

By applying both land cover/use and vegetation classifications generalized at a high level, we develop a broad-scale picture of the extent, distribution, and ecological status of U.S. grazing lands. To characterize extent and distribution, we used several indices of existing land cover/use and the implicit vegetative land cover: the land cover/use categories of the NRI (NRCS, 1992); the land use categories from compilations of U.S. land use published by the Economic Research Service (ERS) (Daugherty, 1995); and the Anderson et al. (1976) land use land cover system as applied in the North American land cover characteristics (LCC) database (Loveland et al., 1991; USGS, 1998).

Private grazing land

Table 2.2 shows the extent of private U.S. grazing land, according to the NRI. The tabulated acreage does not include Alaska, which has a combined land and water area of about 158 Mha. The NRI is a scientifically based, statistically valid, periodic inventory of the state of private U.S. land, documenting changes in landscape characteristics such as land cover, land use, conservation treatment needs, soil erosion rates, etc. (Nusser and Goebel, 1997; NRCS, 1992). Data in Table 2.2 are in units of hectares, for the sake of consistency throughout this volume, although the reporting format for the NRI is acres.

For this tabulation, we considered grazing land to consist of the two NRI broad land cover/use categories of *rangeland* and *pastureland and native pasture* (NRCS, 1992). These categories represent a fundamental ecological distinction within grazing lands.

Rangeland is defined as "land on which the historic climax plant community (HCPC) (i.e., the late successional stage vegetation present in much of North America prior to massive European settlement) or the potential vegetative cover (i.e., potential natural vegetation or PNV) is principally native grasses, grasslike plants, forbs, or shrubs suitable for grazing and browsing" (NRCS, 1997; NRCS, 1992). Rangeland typically is managed as a natural ecosystem whose productivity and vegetative character reflect the biotic and abiotic habitat or site characteris-

Table 2.2. Extent of private grazing land in the U.S., according to the National Resource Inventory (NRCS, 1994).*

Land Cover/Use	1992		1987		1982	
	Mha	%[1]	Mha	%[1]	Mha	%[1]
Rangeland	161.5	20.6	163.0	20.7	165.5	21.1
Pastureland	51.0	6.5	51.7	6.6	53.4	6.8
Total private grazing land	212.5	27.1	214.7	27.3	219.0	27.9
U.S. land + water area[2]	785.4	—	785.4	—	785.4	—

[1]*As percentage of total U.S. land + water area (excluding Alaska).*
[2]*Does not include Alaska land + water area but does include Hawaii and Caribbean.*
*NRCS. 1994. Summary report 1992 National Resources Inventory. USDA-NRCS. Washington, DC.

tics. The vegetative land cover of rangeland often approximates the HCPC or PNV characteristic of a region (Driscoll et al., 1984; Passey et al., 1982).

Pastureland is considered "land used primarily for production of introduced or native forage plants for livestock grazing" (NRCS, 1992). Unlike rangeland, pasture is maintained in an ecological state in which the existing vegetative land cover usually differs considerably from the HCPC or the PNV, to maximize the land's resource value related to forage production (Joyce, 1989). Therefore, it often contains introduced forage plant species rather than native plant species. Pasture usually requires and receives more intensive management and production inputs than rangeland.

Figures 2.1 and 2.2 show the spatial distribution of private U.S. rangeland and pasture. During the three NRI inventory years, federal land area (excluding Alaska) totalled about 164.4 Mha, approximately 21% of the combined U.S. land and water area (Figs. 2.1 and 2.2; Table 2.2). Much of the federal land is in the western U.S. and consists of forest and rangeland. The NRI does not inventory the area of federal grazing land.

Since, by definition, rangeland is confined to land where the HCPC or PNV is grass, grasslike plants, or shrubs (NRCS, 1992), it tends to predominate in the western U.S. where bioclimatic conditions favor such vegetation (Fig. 2.1). Smaller areas of rangeland occur in the coastal southeastern U.S., often as land categorized as "marshland" under the NRI (NRCS, 1992). Pasture, on the other hand, is dominant in the eastern half of the U.S., on land where the HCPC or PNV is forest (Fig. 2.2). This emphasizes that, from an ecological standpoint, pasture is quite different from rangeland.

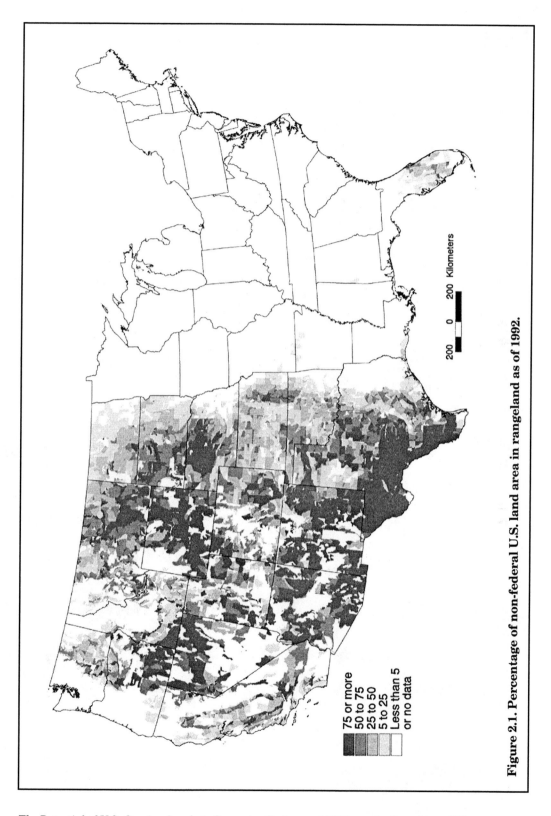

Figure 2.1. Percentage of non-federal U.S. land area in rangeland as of 1992.

75 or more
50 to 75
25 to 50
5 to 25
Less than 5
or no data

200   0   200 Kilometers

The Potential of U.S. Grazing Lands to Sequester Carbon and Mitigate the Greenhouse Effect

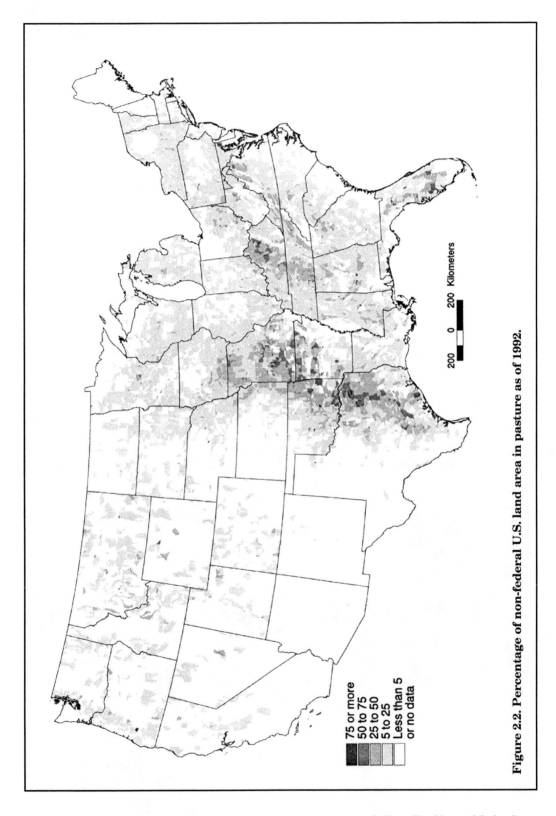

Figure 2.2. Percentage of non-federal U.S. land area in pasture as of 1992.

Total grazing land

We can get a general idea of the extent of all U.S. grazing land, privately and publicly owned, from other sources of data. Economic Research Service reports detailing major U.S. land use take into account all of the nation's land, private, state, and federal. The data come from a variety of sources, including Census of Agriculture, Bureau of Land Management and U.S. Forest Service, the NRI, and other federal agencies (Daugherty, 1995; Frey and Hexem, 1985). These reports, however, use different categories of land use than the NRI, preventing direct comparison with NRI figures for private land.

For our purposes, we considered grazing land to include the category of *grassland pasture and range* and specifically to exclude two other land use categories, *cropland pasture* and *forested land grazed* (Frey and Hexem, 1985). *Grassland pasture and range* is defined as "open land used primarily for pasture and grazing" and includes land with "vegetation of shrub and brush, native and tame grasses, legumes, and other forages" (Frey and Hexem, 1985). This category represents a diversity of vegetation and does not differentiate between rangeland and pastureland the way the NRI does. *Cropland pasture* is considered a category of cropland, not pastureland, and represents land used intermittently for grazing as part of a long-term crop rotation.

Table 2.3 shows the area of grassland pasture and range as a percentage of all U.S. land. Alaska was excluded to be consistent with the NRI land base (Table 2.2). According to Daugherty (1995), 468,016 ha, or less than 1% of Alaska's approximately 158 Mha, was in the "grassland pasture and range" category in 1992. Both because the definitions of grazing land differ from those the NRI uses and because some grazing land is included as miscellaneous areas under "special use" and "other land" categories in these compilations (Daugherty, 1995), area figures are not easily comparable to those from the NRI.

In general, however, a little less than 1/3 of the total U.S. land area (excluding Alaska) may be considered grazing land. By way of comparison, the category "forest land grazed" accounted for 58.9 Mha, or about 7.5% of U.S. land + water, in 1992. Corresponding figures for "forest land grazed" for 1987 and 1982 were 62.7 Mha (8.0%) and 63.9 Mha (8.1%), respectively.

Table 2.3. Extent of grazing land in the U.S., compiled from a variety of sources.[1]

Land Cover/Use	1992		1987		1982	
	Mha	%[2]	Mha	%[2]	Mha	%[2]
Grassland Pasture and Range	238.9	30.4%	238.8	30.4	241.6	30.8

[1] Area of grazing land in state of Alaska excluded to be comparable to land base used with the NRI data (see Table 2.2).
[2] As percentage of U. S. land + water area (excluding Alaska) (see Table 2.2).

About 58.7 Mha of the *grassland pasture and range* category were in federal ownership, about 30 Mha in state or other public ownership, and about 150.6 Mha in private holdings as of 1992 (Daugherty, 1995). Contrast the private figure of 150.6 Mha with the 1992 NRI data for total private grazing land (i.e., NRI *rangeland* and *pastureland*) of about 212.6 Mha (Table 2.2) and one sees the difficulty of comparing the two data sets. However, though a significant portion of U.S. grazing land is in public ownership, the *majority* of U.S. grazing lands is private, and it represents a considerable portion of the total U.S. land base.

Ecological Status

The preceding information on the general extent and distribution of U.S. grazing lands, while useful, does not provide the kind of ecological information needed to assess the functional potential of grazing lands to provide a variety of resource values (Joyce, 1989; NRCS, 1997), such as their ability to buffer increases in global atmospheric CO_2 levels. This is a global issue, but the regional and national character of these lands may bear on their functional ability to sequester atmospheric CO_2-C.

At a broad scale, we are forced to deal with the reality that U.S. grazing lands represent a complex, heterogeneous combination of land use and vegetative land cover types. To attempt to systematize some of this complexity, we might go back to a major distinction in the general category of grazing lands noted earlier: the difference in ecological status of rangeland versus pastureland. Other numerous distinctions also exist. We try to describe some of them by:

1. comparing the distribution of U.S. grazing land to the distribution of U.S. biophysical environments, and
2. comparing the distribution of generalized existing vegetative land cover types to a reference ecological state as a means of assessing the ecological status of U.S. grazing land.

Methodology

We used the distribution of U.S. ecoregions (Bailey, 1997) at several categorical levels to describe the biophysical environmental framework in which U.S. grazing lands occur. We chose to use generalized categories from Kuchler's PNV map of the U.S. (Kuchler, 1964) as an ecological reference state against which to compare existing vegetative land cover of U.S. grazing lands. We inferred the existing vegetative land cover from data in the North American LCC database (USGS, 1998). We completed the manipulation of all geospatial data layer and subsequent spatial analyses using a GIS, dominantly the ARC/INFO-GRID® package.

Other workers have used Kuchler's PNV as a reference potential vegetative land cover layer in geospatial land cover/use analyses, mainly because it is the only system available with national scale coverage yet reasonable detail (Klopatek et al., 1979). Potential natural vegetation differs from the HCPC, in that PNV assesses what the existing vegetation in an area (that is observable today) *would* look like if plant succession had proceeded in absence of human influence (Kuchler, 1964), as opposed to what the vegetation *looked* like prior to European settlement of North America (NRCS, 1997). The PNV provides a reference ecological state with which we can compare existing vegetation in different locations in order to assess that locale's ecological status or potential. The existing vegetative land cover in an area may be at a late-successional stage, so that it approximates the PNV characteristic of the bioclimatic environment of that location, or it may be quite different, due to natural and anthropogenic disturbances that interrupt or re-initiate plant succession.

The Kuchler map has a high degree of cartographic detail, considering its small scale (1:3,168,000). The map legend also is detailed categorically. Therefore, one can cartographically generalize for broad-scale analyses by categorically generalizing the taxa that name the map units. Upon categorical generalization, one gets less cartographic detail that is more appropriate when attempting to map PNV patterns for a broad-scale national assessment.

We used the Anderson land use and land cover classification (Anderson et al., 1976), as taken from the North American LCC database (USGS, 1998), as an index of existing vegetative land cover, in order to convey a broad-scale national picture of grazing land's ecological status. In addition to the Anderson land use and land cover classification, the LCC database contains a number of land cover *themes*, developed from digital analysis of seasonal AVHRR (Advanced Very High Resolution Radiometer) satellite imagery (Loveland et al., 1991).

The Anderson system, having fewer categories and being somewhat more generalized than some of the others in the LCC database, was more useful in outlining very general relationships describing the ecological status of U.S. grazing lands at a broad scale. We used the more detailed Olson global ecosystems legend (Olson, 1970) in the LCC to assess vegetative land cover ecological status in relation to the C sequestration potential of U.S. grazing lands (see the later section, "Broad-Scale Carbon Relationships").

The biophysical environment

U.S. grazing lands occur in a wide variety of biophysical environments. Bailey's hierarchical ecoregions offer a concise way to display the U.S. in a number of distinct ecological regions. *Ecoregions* are broad ecosystems of regional extent (Bailey, 1997), differentiated on the basis of land surface form, macroclimate, vegetation, soils, and fauna (Bailey, 1978). They represent the geospatial pattern of interaction of coarsely scaled biophysical driving gradients, such as climate, geology, topography, etc. (Bourgeron et al., 1994; Franklin, 1995), that constrain biotic processes and dictate the general character of the landscape. Ecoregions taken at a particular hierarchical level describe, at a particular scale, the biophysical setting in which grazing lands exist and function.

By comparing Figures 2.1 and 2.2 with Figure 2.3a, one can, at a very broad scale, see a major ecological difference between pasture and rangeland. Rangeland predominates in the dryer western U.S., while pasture tends to predominate in the humid eastern U.S. Within this broad pattern, however, is a more finely scaled biophysical landscape structure, upon which grazing lands are superimposed. Looking at ecoregions at the division level (Bailey, 1978) (Fig. 2.3b), we begin to appreciate the diversity of environments in which U.S. grazing lands, particularly rangelands, occur.

Present rangeland is most extensive in the dry steppe (semiarid grassland/shrub) and desert regions of the western U.S., with climatic regimes that range from temperate to subtropical (Fig. 2.3b). Isolated areas of rangeland do occur in the humid climate of the subtropical savanna (grassland/woodland ecotone) division of southern Florida (Fig. 2.3b). Of particular note is the relatively low proportion of grazing land in the upper central U.S. that was once a vast prairie or steppe ecosystem (Fig. 2.3b) but is now predominantly cropland (Fig. 2.4). Pasture in the eastern half of the U.S. (Fig. 2.2) is superimposed over a regional ecological structure more coarsely patterned than the western rangelands, but climatic conditions vary widely from warm continental to subtropical (Fig. 2.3b).

Figure 2.5 shows Kuchler's PNV generalized at a very broad life-form (biome) level, roughly equivalent to the UNESCO physiognomic *formation class* (Table 2.1) (UNESCO, 1973). Such categorical generalization results in considerable cartographic generalization, in that it does not differentiate shrublands and grasslands in the western U.S. However, it shows another aspect of the fundamental ecological distinction between rangeland and pasture — much of the pasture in the more humid eastern U.S. (Figs. 2.2 and 2.3a) has both an HCPC and PNV of forest (Fig. 2.5).

In contrast, rangelands generally occupy the drier regions (Figs. 2.1 and 2.3a), where the HCPC and PNV are shrub and grasslands (Fig. 2.5). The two notable exceptions to this generalization are the rangeland in the subtropical savanna and marshlands of southern Florida (Figs. 2.1, 2.3b, and 2.5) and the localized areas of pasture predominantly in high-elevation alpine meadows or in irrigated basins in the western U.S. (Fig. 2.2).

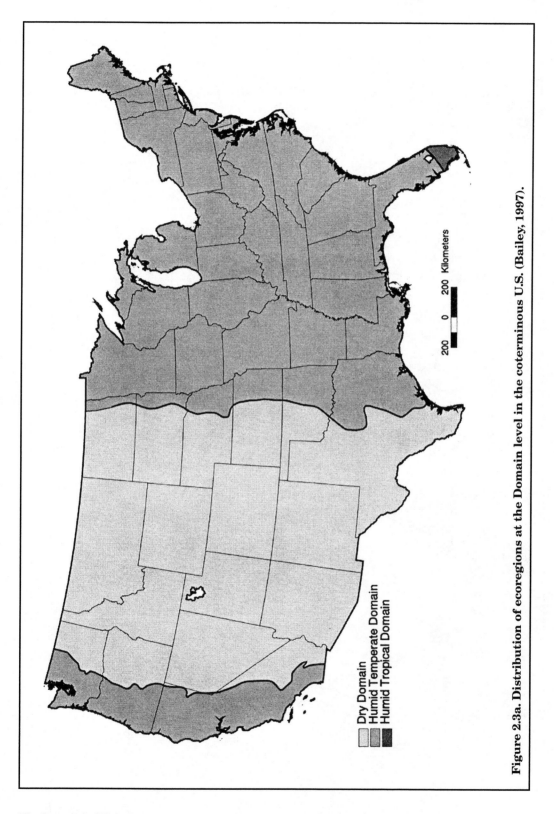

Figure 2.3a. Distribution of ecoregions at the Domain level in the coterminous U.S. (Bailey, 1997).

Dry Domain
Humid Temperate Domain
Humid Tropical Domain

200   0   200   Kilometers

The Potential of U.S. Grazing Lands to Sequester Carbon and Mitigate the Greenhouse Effect

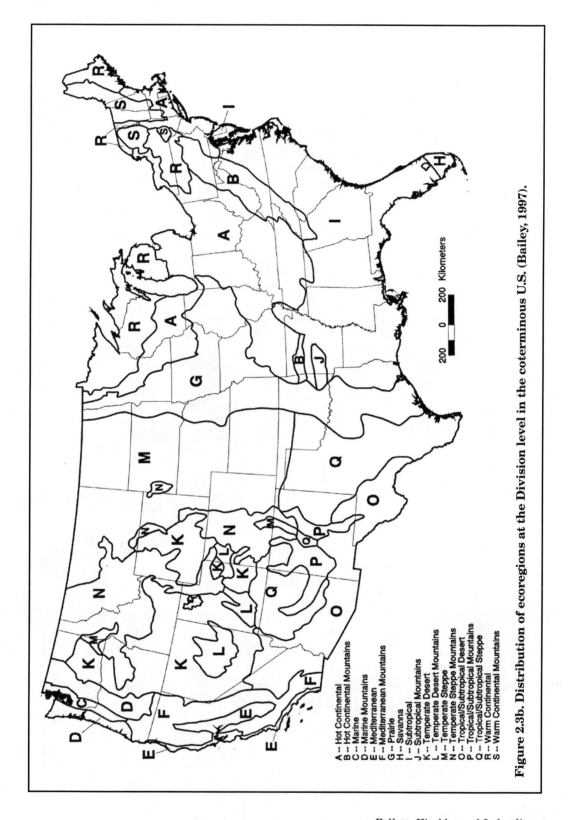

A -- Hot Continental
B -- Hot Continental Mountains
C -- Marine
D -- Marine Mountains
E -- Mediterranean
F -- Mediterranean Mountains
G -- Prairie
H -- Savanna
I -- Subtropical
J -- Subtropical Mountains
K -- Temperate Desert
L -- Temperate Desert Mountains
M -- Temperate Steppe
N -- Temperate Steppe Mountains
O -- Tropical/Subtropical Desert
P -- Tropical/Subtropical Mountains
Q -- Tropical/Subtropical Steppe
R -- Warm Continental
S -- Warm Continental Mountains

Figure 2.3b. Distribution of ecoregions at the Division level in the coterminous U.S. (Bailey, 1997).

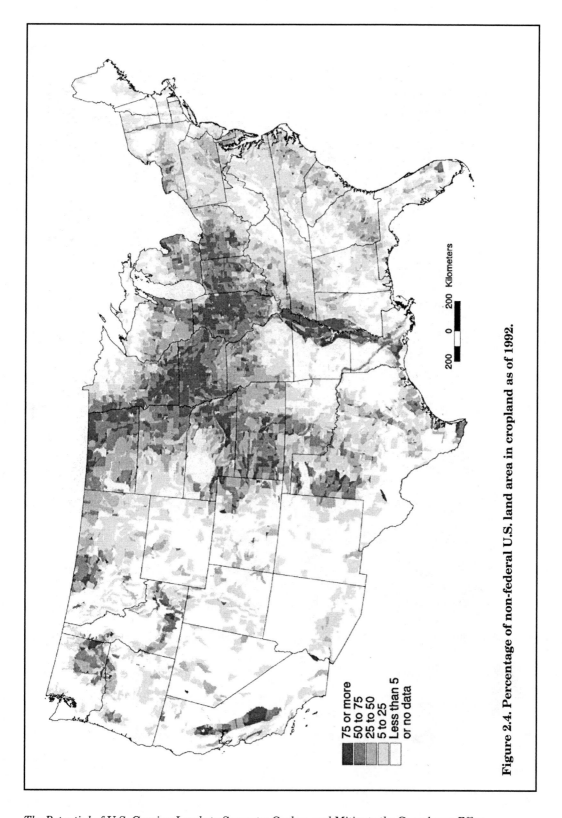

Figure 2.4. Percentage of non-federal U.S. land area in cropland as of 1992.

The Potential of U.S. Grazing Lands to Sequester Carbon and Mitigate the Greenhouse Effect

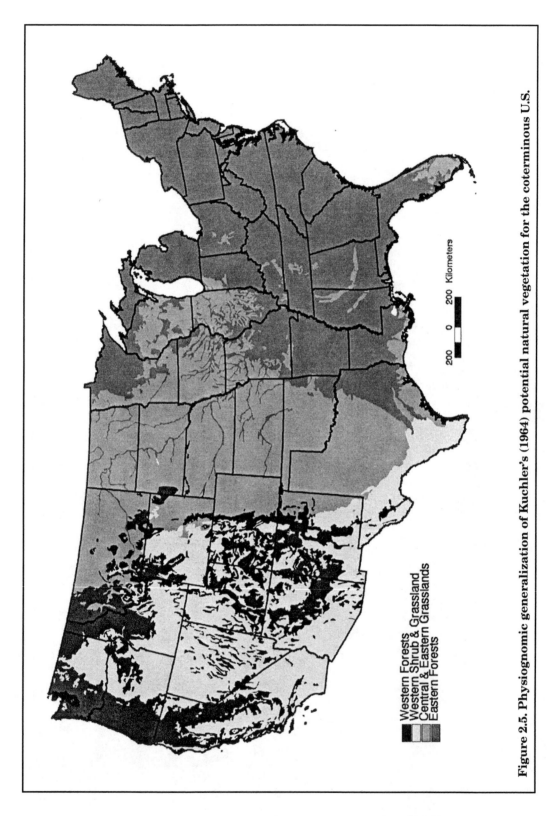

Figure 2.5. Physiognomic generalization of Kuchler's (1964) potential natural vegetation for the coterminous U.S.

Western Forests
Western Shrub & Grassland
Central & Eastern Grasslands
Eastern Forests

200   0   200   Kilometers

Existing versus potential natural vegetation

Applying at a broad scale the concept of transition states (NRCS, 1997), we can qualitatively assess the ecological status of grazing land with a simple comparison of acreage in rangeland and pasture today versus the generalized PNV (Fig. 2.5) as a reference transition state. Approximately 390.7 Mha of land (approximately 50% of the U.S. land area, excluding Alaska) have a PNV conceivably capable of supporting some degree of livestock or wildlife grazing. This estimate comes from combining the acreage of Figure 2.5's western shrub and grassland and central and eastern grassland categories. Contrast this acreage with that of existing grazing land from the NRI or ERS data of between 214.6 and 238.9 Mha, respectively (Tables 2.2 and 2.3).

This suggests that ecosystems in considerable areas of the U.S. depart significantly from an ecological state that effectively would support grazing land use. This is often because they are managed to realize a variety of resource values from the land in addition to or instead of those from grazing use alone. In the eastern U.S., this departure includes forest maintained as pasture, cropland, and urban land (Figs. 2.2 and 2.5). In the central and, to a lesser extent, the western U.S., much land has a PNV capable of supporting grazing but is maintained as cropland, urban land, or other land uses, to realize the associated resource values (Figs. 2.1, 2.4, and 2.5).

We can appreciate the complexity within today's grazing lands by looking at existing and potential vegetation at a finer scale and in more categorical detail. For example, within western rangelands, shifts in the ecological status of the vegetative land cover have converted the grassland HCPC to shrub-dominated or mixed grassland/shrub landscapes. Such changes have been associated with anthropogenic disturbances of grazing and fire patterns, or with periodic extended drought (Branson, 1985; Gibbens et al., 1983).

These changes reflect rather long-term (10s or 100s of years) trends, however. A subsample of NRI data on about 1.62 Mha, in 19 western and plains states, that have been in rangeland for each of the three inventory years of 1982, 1987, and 1992, indicates very little change in the overall distribution of woody canopy cover over that 10-year period. About 39% had <5% woody cover, about 29% had 5 to 20% woody cover, about 21% had 20% to 40% woody cover, and about 11% had >40% woody cover in each of the three years. This suggests that gradual change in vegetative cover of a region, or catastrophic change interspersed with long periods of relative stability (excepting outright anthropogenic change in land cover/use), may play the critical role in C sequestration on rangelands.

Compare the distribution of existing land use and land cover from the global land characteristics database (USGS, 1998), using the Anderson system (Anderson et al., 1976) (Fig. 2.6), with a more categorically detailed Kuchler PNV map that allows greater cartographic detail but is still based on general physiognomy (Fig. 2.7). Dif-

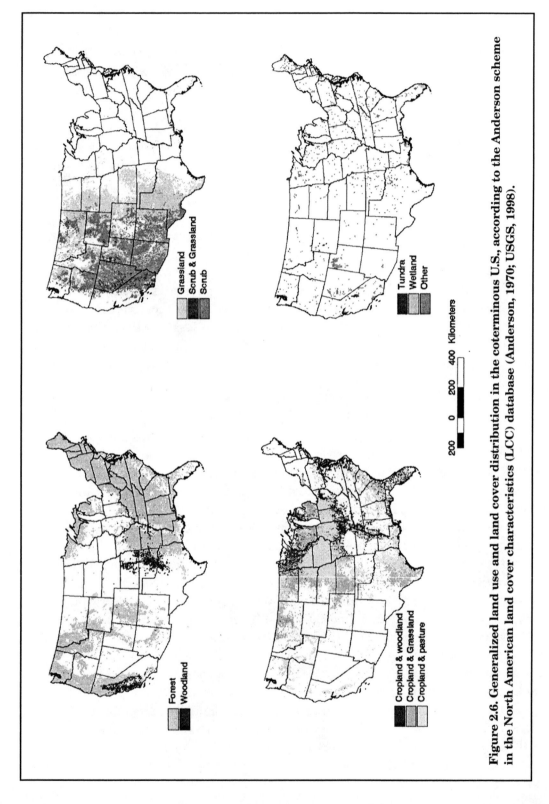

Figure 2.6. Generalized land use and land cover distribution in the coterminous U.S., according to the Anderson scheme in the North American land cover characteristics (LCC) database (Anderson, 1970; USGS, 1998).

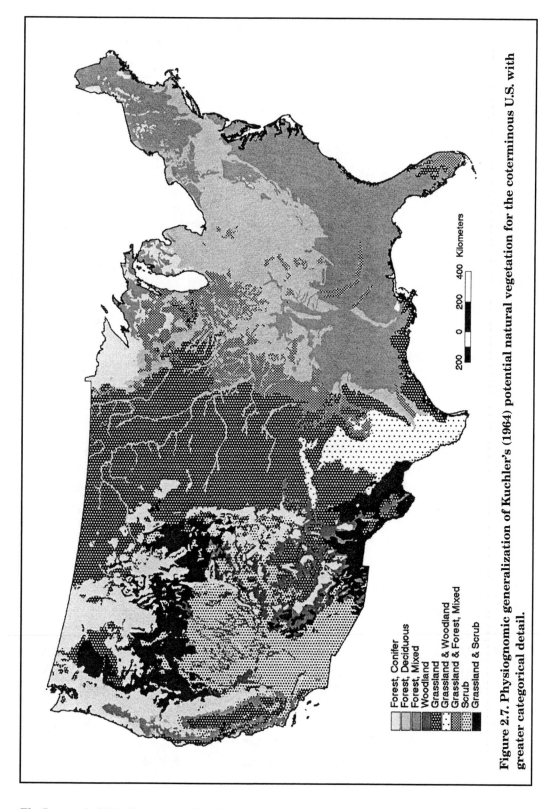

Figure 2.7. Physiognomic generalization of Kuchler's (1964) potential natural vegetation for the coterminous U.S. with greater categorical detail.

Forest, Conifer
Forest, Deciduous
Forest, Mixed
Woodland
Grassland
Grassland & Woodland
Grassland & Forest, Mixed
Scrub
Grassland & Scrub

ferences between the two in the pattern of generalized vegetation suggest that present grazing land exhibits a landscape structure (Forman and Godron, 1986) at a regional level that is different from what consideration of PNV might suggest, as the next section describes in detail.

Regional landscape structure

Landscape structure refers to the spatial distribution in the landscape of distinctive components of ecosystems. The *landscape* itself is defined as a spatially heterogeneous area characterized by structure (spatial configuration of components, including energy, matter, and species), function (interactions between the components in terms of flow of energy, matter, and species), and change over time (Forman and Godron, 1986; Turner and Gardner, 1991). Landscape structure and function are interwoven; thus there is much effort under way to assess landscape function via analysis of landscape patterns or structure.

Certain regional patterns of distribution of grazing lands or of lands formerly suited to grazing, in relation to the biophysical environment and potential vegetation, stand out, painting a broad-scale national picture of the structure of U.S. grazing land landscapes. These distribution patterns include:

1. extensive areas of cropland and pasture (Figs. 2.2, 2.4, and 2.6) in areas of forest PNV in the Piedmont province of the southeastern U.S. and in areas of forest and grassland PNV in the Coastal Plain province (Fenneman, 1938) along the Gulf coast (Fig. 2.7)

2. extensive areas of the till plains in the Central Lowlands province (Fenneman, 1938) of the upper Midwest in cropland and, to a lesser extent, pasture (Figs. 2, 4, and 6), that have a PNV of forest and, further to the west, grassland (Fig. 2.7)

3. large areas of the Great Plains province, particularly in the north, with a PNV of grassland (Fig. 2.7), that are predominantly cropland and pasture (Figs. 2.2, 2.4, and 2.6)

4. equally extensive areas of the Great Plains to the south, with a patchwork of existing grassland and woodland (Fig. 2.7), that is dominantly rangeland (Fig. 2.1), that to some extent approximates the PNV (Fig. 2.7)

5. a heterogeneous patchwork of existing shrub and grassland in the western U.S. (Fig. 2.6) that also approximates in a broad sense the PNV (Fig. 2.7) but seems more dominated by shrubland (scrub) than the PNV

6. extensive cropland or cropland and pasture and adjacent woodland in the Central Valley of California (California Trough, Fenneman, 1931) (Figs. 2.4 and 2.6) where PNV is dominantly grassland (Fig. 2.7)

7. PNV of grassland or scrub and grassland in the Palouse region (Fig. 2.7), where existing land cover is predominantly cropland (Figs. 2.4 and 2.6).

Such departures from the PNV reference state, in a variety of biophysical environments, show the considerable ecological and geographic complexity of U.S. grazing land, even when viewed on a national or regional scale. That complexity indicates considerable potential for change in the patterns of vegetation and ecological status. We can use this variation from PNV and potential for change in ecological status to assess the C sequestration potential associated with different landscapes, based on their existing and potential vegetative land cover.

Broad-Scale Carbon Relationships

Three potential reservoirs or pools of C exist on grazing lands: the standing biomass C, the soil organic C (SOC), and the soil inorganic C (carbonates). Other chapters go into detail on the dynamics of these C pools in grazing lands, and Monger et al. (Ch. 4) specifically discuss processes of inorganic C's cycling in grazing lands. Our discussion is of a more general nature and focuses only on the organic C pools: standing biomass C and SOC. We illustrate how broad-scale changes in the ecological status of the vegetative land cover of U.S. grazing lands can affect these pools.

Simplified steady-state model

A simplified, steady-state, C cycling model provides a functional, broad-scale framework for assessing C sequestration potential on U.S. grazing lands. From a steady-state perspective, we can view standing vegetative land cover biomass, both above- (plant shoots, stems, leaves, etc.) and belowground (plant roots), and SOC as terrestrial C repositories of C cycling in the global atmospheric-ocean system (Leith, 1975; Parton et al., 1987).

Via photosynthesis, the vegetative land cover captures atmospheric CO_2-C and "fixes" it as standing biomass. The size of this standing biomass C pool represents a balance between input, via fixation of atmospheric CO_2-C, and output of C, due to die off and decomposition of plant biomass (plant residue decomposition). We can quantify the net amount of C fixed, usually expressed on an annual basis relative to that lost due to plant residue decomposition, as net primary production (NPP) (Lieth, 1975), specifically as the equivalent amount of C fixed as a result of net primary production (NPP-C).

A portion of the C fixed in the vegetative land cover biomass ultimately ends up as the C of the soil organic matter (OM), SOC, because of the decomposition of plant residues in the plant-soil system (Lal et al., 1998). In a rough steady-state approximation, about $\frac{1}{2}$ of the plant residue C makes it into the SOC pool and balances SOC losses back to the atmosphere as CO_2-C that result from decomposition of the soil organic matter (SOM) hosting the SOC pool (Alexander, 1977; Parton et al., 1987; Paul and Clark, 1989).

Thus, the size of the SOC pool for any given state of the plant-soil system represents the difference between input of the fixed plant C from decomposed plant residues and other organic C sources and output as oxidative decomposition of SOM. Things in reality are more complex, the SOC pool consisting of multiple, kinetically differentiated pools, whose magnitude varies with the composition of the plant community and plant residues and the characteristics of the soil material, such as texture (Paul and Clark, 1989; Motavalli et al., 1994; Parton et al., 1987). For our purposes, however, the above, simple, steady-state model will suffice.

In this model, the character of the land's vegetative cover and how the vegetative cover changes from one state to another, due to natural processes and human land use, assumes critical importance relative to the land's C sequestration potential. This is because the atmospheric CO_2-C initially is captured or "fixed" by the vegetation and ultimately sequestered in the soil component of a highly integrated plant-soil system. A steady-state approach implies that perturbation of the state of the plant-soil system is the only way to realize *net* C sequestered or lost by grazing lands relative to the global atmospheric C pool.

Perturbation of the steady state is a necessary but not sufficient condition, however. Perturbation of the vegetation (i.e., change in its ecological status) may not result in net atmospheric CO_2-C sequestration or loss but may merely shift the dynamics between the standing biomass and SOC pools. This is due to the interplay of other important landscape process, such as soil erosion induced by vegetative changes, or changes in the root/aboveground biomass ratios that are also a function of the vegetative cover of the landscape (Connin et al., 1997; Jackson et al., 1996).

These complications aside, we can delineate areas of land that have the potential for perturbation and thus for net C gain or loss by identifying areas with an inherent potential for change in the ecological status of existing vegetative land cover (i.e., potential for change in the landscape structure). Therefore, we can map potential for change in grazing land landscape function (Forman and Godron, 1986), relative to C storage.

Potential for carbon sequestration — standing biomass pool

Perturbation of the steady state is possible if the grazing land's vegetation can move toward a vegetative transition state (NRCS, 1997). In our generalized broad-scale context, the chosen transition state is the PNV reference state. We use the PNV as a broad-scale equivalent of the ecological site, which describes the landscape unit's PNV at a very fine scale (Passey et al., 1982; NRCS, 1997).

Spatially distributed differences between PNV and existing vegetation (i.e., areas where the existing vegetative land cover is not equal to the PNV for the area) represent areas with a potential gradient for change in grazing land vegetation's ecological status (toward the PNV state in this case). Since both standing biomass and NPP (and the associated amounts of organic C) differ among broad vegetation types (Lieth, 1975), differences between existing vegetation and PNV in an area imply a "natural" potential for net C sequestration or loss first by the vegetation's standing biomass and, ultimately, by the soils underlying that vegetation.

The potential is realized if the landscape's vegetative cover changes toward the reference PNV state. Of course, the existing vegetation might move toward some other transition state (NRCS, 1997) and perturb C dynamics accordingly. However, these other transition states are not established well for many plant communities characteristic of U.S. grazing lands, nor are they available as a geospatial data set across the entire U.S.

Our approach merely makes use of a consistent reference transition state, PNV, that we can apply on a broad national scale, using existing geospatial data. It allows relative comparison of sequestration potential of different areas or vegetative land cover types, and it enables sequestration potential estimates, relative to the global C cycle, that consider grazing land's ecological complexity.

Methodology

To assess this potential for change in C storage, we compared existing vegetative cover of grazing land to the respective PNV reference state (Klopatek et al., 1979), in a way similar to what we did for land cover and land use (i.e., we assessed its ecological status), using GIS techniques. We associated values for standing biomass and net primary productivity (NPP) with each of the generalized existing and potential natural vegetation types, using data from several widely cited compilations (Bazilevich, 1967; Lieth, 1975) (Table 2.4). This allowed us to:

1. map potential for change in standing biomass and NPP as a function of ecological status of grazing land's vegetative land cover
2. quantitatively assess change in the size of the standing biomass C pool associated with various vegetative transitions on grazing lands, and
3. compare the size and distribution of the standing biomass and SOC pools on a broad scale.

Our goal was not to build detailed C budgets for all U.S. grazing land but to show that the ecological status of grazing land, both rangeland and pasture, is a key component of its capacity to store or release atmospheric CO_2-C on a scale

Table 2.4. Estimated representative values for standing biomass and net primary productivity as dry matter and the equivalent organic carbon content, for several generalized vegetative land cover types.[1]

Vegetation Type (based on Olson ecosystem legend)	Aboveground Mature Biomass[2] (MT/ha)	Organic Carbon[3] (MT/ha)	Belowground Biomass[6] (MT/ha)	Organic Carbon (MT/ha)	Net Primary Productivity (MT/ha/yr)	Organic Carbon Fixed[3] (MT/ha/yr)
Forest, Coniferous	350[2]	157.5	44.00	19.80	12.00[4]	5.40
Forest, Broadleaf (UNESCO Cold-Deciduous)	350	157.50	42.00	18.90	10.00	4.50
Forest, Mixed	285[5]	128.25	43.00	19.35	11.00[7]	4.95
Woodland	110	49.50	43.00	19.35	6.00	2.70
Grassland (temperate)	30	13.50	14.00	6.30	5.00	2.25
Shrub/Scrub (UNESCO Scrub)	20	9.00	48.00	21.60	.90	.40
Tundra/Desert	0	0	9.33	4.20	.01	.005
Cropland (annual crops)	35	15.75	1.50	.67	6.50	2.92

[1]Values taken from Lieth (1975) except where noted.

[2]Data quite variable, but suggests standing biomass can equal or exceed that of temperate broadleaf forest (McGuire et al., 1992), so assumed value equal to Forest, Broadleaf.

[3]Assumed 0.45% organic carbon in dry biomass (Lieth, 1975).

[4]Estimated from McGuire et al. (1992) data and Barbour (1987).

[5]Average of values from Lieth (1975) and McGuire et al. (1992).

[6]From Jackson et al. (1996). Forest, Mixed was computed as average of Forest, Coniferous and Forest, Broadleaf; Woodland was taken as equivalent to Forest, Mixed; and Tundra/Desert was taken as average of Jackson et al. (1996) values for Tundra, Cold and Warm Desert.

[7]Taken as average of Forest, Coniferous and Forest, Broadleaf.

relative to the global C cycle. We also show that the spatial pattern or "patchiness" (Turner et al., 1991) of grazing land's ecological status is an important landscape structural attribute as far as its C storage potential is concerned.

We used Olson's global ecosystems legend (Olson, 1970) from the LCC database (Loveland et al., 1991; USGS, 1998) as the index of existing vegetative land cover for this analysis. The relatively detailed Olson map units could be generalized easily according to the UNESCO vegetation classification (Ellenberg and Mueller-Dombois, 1967; UNESCO, 1973) (Table 2.1) at a level suitable for a national analysis, yet they retain vegetative distinctions, important at finer scales, that reflect the regional differences in landscape structure of grazing lands.

In addition, others have used the Olson global ecosytems to aggregate SOC data on a broad scale (Kern, 1994), and one of our objectives is to relate the ecological status of grazing land's existing vegetative land cover to the SOC pool. Figure 2.8 is a map of Olson's global ecosystems, categorically generalized to be comparable to generalized Kuchler PNV units of Figure 2.7, at a level equivalent to a com-

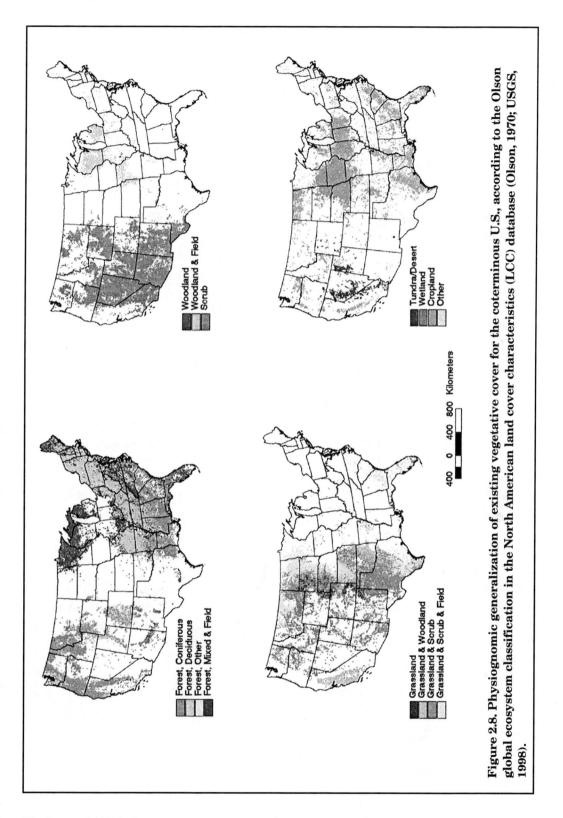

Figure 2.8. Physiognomic generalization of existing vegetative cover for the coterminous U.S., according to the Olson global ecosystem classification in the North American land cover characteristics (LCC) database (Olson, 1970; USGS, 1998).

bination of UNESCO Formation Class and Subclass hierarchical levels (Ellenberg and Mueller-Dombois, 1967; UNESCO, 1973).

We associated standing biomass and NPP for different generalized world vegetation types (Bazilevich, 1967; Lieth, 1975; McGuire et al., 1992; Olson, 1970) with physiognomically generalized vegetation types that we could relate to generalized units of both Olson global ecosystems and Kuchler PNV (Table 2.4). We added above- and belowground biomass from Table 2.4 for total standing biomass, which we used in the calculations.

Though values of standing biomass and NPP range considerably within vegetative types (Lieth, 1975), significant differences in mean values occur, so that vegetative differences affect broad patterns of C distribution in the landscape. For purposes of relative comparison, we used a C content of 45% of dry plant biomass (Lieth, 1975; Table 2.4) to compute standing biomass C and NPP-C. This enabled direct estimation of C gain or loss associated with a given vegetative transition to a PNV reference state.

Ecological status and the standing biomass carbon pool

Figure 2.9a displays areas of generalized *existing* vegetation with potential to support grazing land use (i.e., grasses, grasslike plants, shrubs, etc.) that depart in some way from the generalized PNV for the same area. It shows the fairly complex spatial pattern of ecological status of grazing land vegetation even at a broad scale. The mapped area covers 328,690,661 ha. It is greater than the figures given earlier for grazing land extent (Tables 2.2 and 2.3) because it includes areas mapped as complexes of grazing land vegetation plus forest, woodland, or "field" components.

Figure 2.9b shows one category of existing vegetation, grassland and scrub, and the categories of PNV found within its geographic distribution. It gives a good indication of the heterogeneity of grazing land's ecological status.

Table 2.5 shows the net change (+ or –) in standing biomass C and NPP-C of existing types of vegetative land cover (only those capable of supporting grazing land use), if they were to transition to the PNV state. The data are expressed on both a per unit area basis and a total basis (by taking into account the aerial extent of the given existing vegetative PNV units which Table 2.5 shows, some of which Figures 2.9a and b display). For change in standing biomass C, + implies a net potential for storage of C by the plant-soil system if it transitions to the PNV state, and – implies a net potential to lose C by transitioning to the PNV state. For NPP-C, + implies an increase in the annual *rate* of net C fixation by transitioning to the PNV state, and – implies a decrease in the *rate* of net C fixation for such a transition.

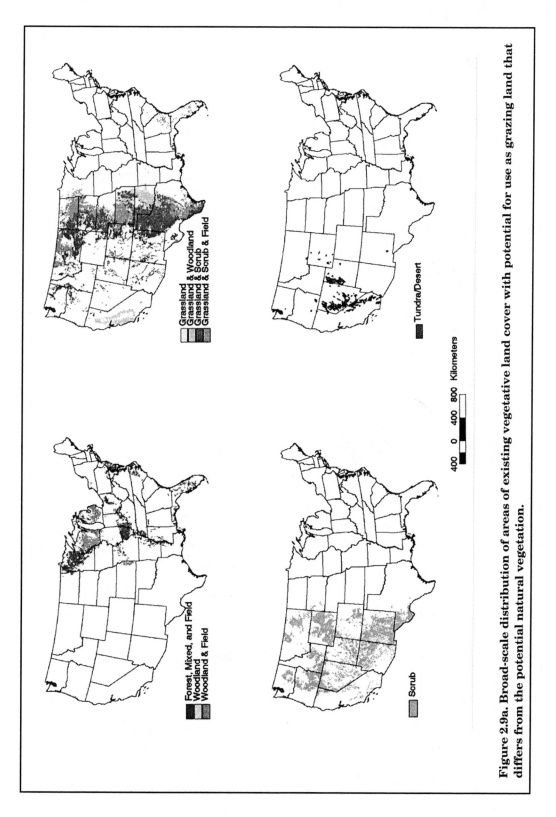

Figure 2.9a. Broad-scale distribution of areas of existing vegetative land cover with potential for use as grazing land that differs from the potential natural vegetation.

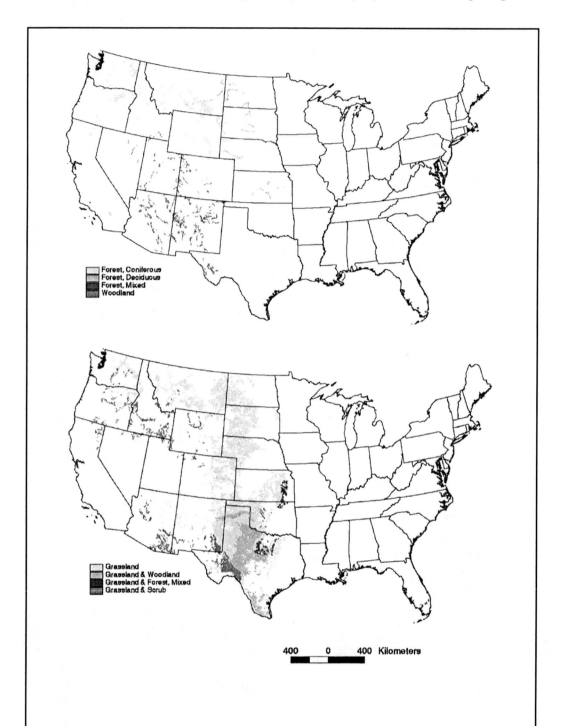

Figure 2.9b. Broad-scale distribution of areas of existing vegetative land cover of grassland and scrub in the U.S., and associated types of potential natural vegetation.

Table 2.5. Change (Δ) in estimated carbon associated with both standing biomass and net primary productivity for areas where existing grazing land vegetation[1] differs from potential natural vegetation.

Existing Vegetative Land Cover[2]	Potential Natural Vegetation[3]	Δ Standing Biomass C[4] (MT/ha)	Δ NPP C[3] (MT/ha/yr)	Area (Mha)	Δ Total Standing Biomass C (MMT)	Δ Total NPP C (MMT/yr)	Δ SOC[5] (MMT)
Forest, Mixed & Field	Forest, Coniferous	95.29	1.47	3.92	373.74	5.77	-138.47
	Forest, Deciduous	94.39	0.57	9.36	883.71	5.34	-36.95
	Forest, Mixed	65.59	1.02	6.88	451.30	7.02	21.92
	Grassland	-62.21	-1.68	1.57	-97.84	-2.64	-122.74
	Grassland & Forest, Mixed	1.69	-0.33	2.84	4.80	-0.94	—
					1615.71	**14.55**	
Woodland	Forest, Coniferous	108.45	2.70	.13	14.24	0.35	-268.33
	Grassland & Scrub	-43.65	-1.37	.04	-1.56	-0.05	-79.95
					12.68	**0.30**	
Woodland & Field	Forest, Coniferous	165.38	2.59	1.34	221.98	3.48	-104.17
	Forest, Deciduous	134.67	1.69	4.33	583.53	7.32	-87.61
	Forest, Mixed	104.97	2.14	.35	36.34	0.74	-18.15
	Grassland	-22.83	-0.56	.30	-6.74	-0.16	-59.49
	Grassland & Forest, Mixed	41.07	0.79	6.18	253.68	4.88	
					1088.79	**12.78**	
Grassland	Forest, Coniferous	157.50	3.15	.08	12.55	0.25	31.63
	Forest, Deciduous	156.60	2.25	.04	6.00	0.09	-2.59
	Woodland	49.05	0.45	.09	4.22	0.04	—
	Grassland	0	0.00	5.41	0.00	0.00	0.00
	Grassland & Woodland	24.52	0.22	.01	0.27	<0.01	10.89
					23.04	**0.38**	
Grassland & Woodland	Forest, Deciduous	132.08	2.03	.27	35.09	0.54	-29.81
	Forest, Mixed	103.28	2.48	3.49	360.85	8.66	83.63
	Woodland	24.53	0.23	.02	0.56	<0.01	—
	Grassland	-24.52	-0.22	4.78	-117.20	-1.05	-15.74
	Grassland & Woodland	0.00	0.00	.04	0.00	0.00	0.00
	Grassland & Forest, Mixed	39.38	1.13	4.99	196.62	5.64	—
	Scrub	-13.72	-2.07	.49	-6.75	-1.02	-8.82
					469.17	**12.77**	
Grassland & Scrub	Forest, Coniferous	152.10	4.07	5.43	825.91	22.10	13.93
	Forest, Deciduous	151.20	3.17	1.43	216.45	4.54	-10.35
	Forest, Mixed	122.40	3.62	.07	9.00	0.27	59.04
	Woodland	43.65	1.37	5.99	261.31	8.20	—
	Grassland	-5.40	0.92	54.27	-293.05	49.93	-13.61
	Grassland & Woodland	19.12	1.14	14.67	280.54	16.73	—
	Grassland & Forest, Mixed	58.50	2.27	2.55	149.05	5.78	—
	Scrub	5.40	-0.93	2.45	13.22	-2.28	-29.96
	Grassland & Scrub	0.00	0.00	12.05	0.00	0.00	0.00
					1462.43	**105.27**	

(continues)

Table 2.5., cont.

Existing Vegetative Land Cover[2]	Potential Natural Vegetation[3]	Δ Standing Biomass C[4] (MT/ha)	Δ NPP C[3] (MT/ha/yr)	Area (Mha)	Δ Total Standing Biomass C (MMT)	Δ Total NPP C (MMT/yr)	Δ SOC[5] (MMT)
Grassland & Scrub & Field	Forest, Coniferous	156.49	2.79	.19	29.39	0.52	-9.24
	Forest, Deciduous	155.59	1.89	1.59	247.19	3.00	-60.95
	Forest, Mixed	126.79	2.34	1.13	143.53	2.65	30.89
	Grassland	-1.01	-0.36	27.53	-27.81	-9.91	-32.62
	Grassland & Woodland	23.51	-0.14	6.84	160.83	-0.96	—
	Grassland & Forest, Mixed	62.89	0.99	.41	25.70	-0.40	—
	Scrub	9.79	-2.21	.07	0.72	-0.16	-54.26
	Grassland & Scrub	4.39	-1.28	.02	0.10	-0.03	-16.23
					579.65	**-5.81**	
Scrub	Forest, Coniferous	146.70	5.00	3.17	465.50	15.87	27.64
	Forest, Deciduous	145.80	4.10	.08	11.70	0.33	13.72
	Forest, Mixed	117.00	4.55	.23	27.31	1.06	83.43
	Woodland	38.25	2.30	19.07	729.36	43.86	—
	Grassland	-10.80	1.85	22.63	-244.37	41.86	13.52
	Grassland & Woodland	13.72	2.07	.23	3.16	0.48	—
	Scrub	0.00	0.00	44.12	0.00	0.00	0.00
	Grassland & Scrub	-5.40	0.93	38.22	-206.38	35.54	11.44
					786.28	**139.00**	
Tundra/Desert	Forest, Coniferous	173.10	5.39	.11	19.10	0.59	34.82
	Woodland	64.65	2.69	.13	8.28	0.34	—
	Grassland	15.60	2.24	.22	3.45	0.49	30.21
	Scrub	26.40	0.39	6.38	168.36	2.49	3.76
	Grassland & Scrub	21.00	1.32	.45	9.51	0.60	25.61
					208.70	**4.51**	
	Total Net Change				**6246.45**	**283.75**	

[1]*Forest land excluded, cropland included because of potential to revert to natural vegetation.*

[2]*Generalized vegetative land cover type based on the Olson global ecosystems legend (Loveland et al., 1991; Olson, 1970) (see Fig. 2.8).*

[3]*Generalized Kuchler potential natural vegetation units (Kuchler, 1964) within the area of the respective existing vegetation unit (Fig. 2.9b).*

[4]*Potential Natural Vegetation (PNV) – Existing Vegetation (ENV). For standing biomass, + implies a net potential for storage of C by the plant-soil system by moving toward the PNV state, – implies a net potential to lose carbon by moving towards the PNV state. For net primary productivity, + implies an increase in the rate of carbon fixation by moving toward the PNV state, – implies a decrease in the rate of carbon fixation by moving toward the PNV state.*

[5]*Value of 0.0 for units where existing vegetative land cover equals or approximates PNV; — implies that no areas with existing vegetation equal to or approximating PNV could be identified at the scales analyzed, making it impossible to calculate a value for the indicated existing vegetation-PNV transition.*

The Olson existing vegetative units, which Table 2.5 includes, for the most part exclude existing forestland, large contiguous areas of cropland, and other land unavailable for potential use as grazing land (i.e., urban and built up, salt flats, ice, extensive water, wetlands, etc.).

Standing biomass carbon pool and the global carbon balance

We can compare estimated changes in standing biomass C, which result from the sum total of all the stated transitions to PNV in Table 2.5, to the absolute size of various components of the global C budget as an index of the C sequestration potential of U.S. grazing lands. Nationally (coterminous U.S.), +6,246 million metric tons (MMT) more C (+6.3 Pg C) would exist in the terrestrial standing biomass C pool as a result of transitioning to PNV from any existing vegetative land cover capable of supporting grazing (Table 2.5). Compared to either the average size of the global atmospheric CO_2-C pool, which is ~720 x 10^3 MMT of C (~720 Pg C), or to the estimated global terrestrial biomass C pool, which is ~550 x 10^3 MMT of C (~550 Pg C) (Lal et al., 1998), this is a rather small proportion, <0.009% and <0.010% of the respective global C pools.

A more meaningful comparison is between the estimated change in NPP-C that results from the sum total of all the stated transitions to PNV, as Table 2.5 shows, and the global *annual NPP* figures and *transfer rates* between global C pools, rather than *absolute size* of C pools at any particular time. The NPP-C (i.e., net rate of atmospheric C fixation) *increases* by about 284 MMT of C/yr (~0.3 Pg C/yr) for the sum total of all the stated transitions to PNV in Table 2.5. This change in C fixation rate is minuscule in comparison with the annual exchange of C between the atmosphere and the ocean, <0.003% of that *annual rate*, which is about 105 x 10^3 MMT of C/yr (~105 Pg C/yr) (Lal et al., 1998).

It comprises a somewhat larger proportion of the global annual terrestrial NPP-C (estimates range from 45 to 76 x 10^3 MMT of C/yr [~45 to 76 Pg C/yr]), however, about 0.5% of that value, and is about 1% of annual NPP-C in the ocean (~25 x 10^3 MMT of C/yr [~25 Pg C/yr]) (Lieth, 1975). Compared to the estimated annual C input into the atmospheric CO_2-C pool from destruction of terrestrial vegetation worldwide, which is about 2000 MMT of C/yr (~2 Pg C/yr) (Lal et al., 1998), or compared to that input due to other anthropogenic sources, which is about 5000 MMT of C/yr (~5 Pg C/yr), the increase in NPP-C on U.S. grazing lands would make up a more significant proportion, 14.2 and 5.7%, respectively.

Perhaps the best estimate of the impact of changes in the vegetative landcover of U.S. grazing land on global C dynamics, as measured by the NPP-C increase of 284 MMT of C/yr (~0.3 Pg C/yr), is to compare that figure with the estimated annual increase in atmospheric CO_2-C, which is 3.3 Pg C/yr (Follett et al., Ch. 16;

Lal et al., 1998). The NPP-C due to vegetative transition on U.S. grazing lands represents about 9% of this latter figure. This is a fairly direct index of the potential of grazing land's vegetative changes in the coterminous U.S. to offset atmospheric CO_2 buildup.

Capacity versus intensity of carbon sequestration

The above calculations do not explicitly consider time, since they are based on steady-state assumptions (i.e., a change in state of the system is required to achieve net C sequestration or release). However, one can assume that a finite time period is needed to reach a relatively steady-state biomass and NPP level associated with grazing land's vegetative transitions at any one location.

Changes in vegetative land cover's biomass C storage, due to vegetative transitions, represent one component of the C sequestration *capacity* of U.S. grazing lands, the other component being storage as SOC. One might compare the *change* in the standing biomass C pool to vegetative transitions, +6,296 MMTC (+6.3 Pg C), a measure of C sequestration capacity on U.S. grazing lands, to the intensity with which atmospheric CO_2 is fixed globally, which usually is exressed as an annual rate. This gives a more meaningful picture of the potential of U.S. grazing lands to affect global C cycles. Contrasted in this way, U.S. grazing land C sequestration capacity that results from biomass changes associated with vegetative transitions represents about 9.5% of the global annual terrestrial NPP-C and about 25% of ocean annual NPP-C.

Pasturelands

Pasturelands present complications in the preceding assessment for two reasons. First, their more complex pattern in the landscape is of a finer grain than the resolution of the PNV and LCC data (Loveland et al., 1991; USGS, 1998), which prevents us from treating them as geographically extensive, contiguous areas. Rangeland tends to occur as extensive, spatially contiguous areas on a national scale and thus fits more easily into a broad-scale analysis with the LCC database.

Similarly, it was relatively easy to exclude from this analysis the extensive contiguous areas in the Midwest mapped as "cropland" (Fig. 2.8). However, the patchwork of pasturelands in the eastern U.S. cannot be resolved (Fig. 2.8), making it impossible to deal with them as a distinct geographic entity. We must deal with pasture as a component occurring in complex with other vegetative land cover types, given the grain or resolution (Turner and Gardner, 1991) of the available LCC data (USGS, 1998).

For example, Figure 2.6 shows considerable area in the southeastern U.S. that, in terms of the Anderson land use and land cover classification, consists of a complex unit of combined "cropland and pasture." Considering these data and also the distribution of pastureland Figure 2.2 shows, a considerable portion of parts of the southeastern U.S. mapped as a pure "cropland" unit in the Olson map of existing vegetation (Fig. 2.8) probably represents a map unit that is a complex of pastureland and cropland. This is in addition to other areas shown explicitly on the Olson map as units of forest, mixed & field (Fig. 2.8), where the field component probably represents pasture to some degree.

Second, unlike rangelands, pasturelands usually are not managed as natural ecosystems. Pastures are not as constrained by natural plant community processes such as succession or other transition state dynamics that figure heavily into rangeland management. This makes the concept of a PNV reference state less relevant than for rangelands.

To overcome the lack of spatial resolution of pastureland in the LCC data yet still make some estimates of potential sequestration in the context of vegetation's ecological status, we considered a representative pasture system on a common forested soil type, typical of those used for pasture in the southeastern U.S. We then compared maintaining the ecological status of pasture as permanent pasture with changing from pasture (introduced or native forage plants) to the PNV of forest.

An annual dry matter production of 12.1 MT/ha of coastal bermuda grass (*Cynodon dactylon*) is typical on a Bowie fine sandy loam soil type (Typic Paleudult), representative of the upland coastal plain so extensive in the southeastern U.S. (Hatherly, 1993). This annual production, 12.1 MT/ha, expressed as NPP-C, is 5.4 MTC/ha/yr, which compares favorably with the NPP-C values for forest, which range from about 4.5 to 5.4 MTC/ha/yr (Table 2.4). From this standpoint alone, productive pasture vegetation has about the same *annual* C fixation potential as forest.

We estimated the size of the pastureland's steady-state, standing biomass C pool and compared it to that for forest:

1. Annual pasture production (as NPP-C) was multiplied by the approximate time necessary to accumulate the steady-state biomass C level of forest, coniferous PNV (i.e., forest standing biomass C/forest NPP-C). This was necessary in order to place the two vegetative land cover types on equivalent temporal scales. Thus, 5.4 MTC/yr x 32.9 years (Table 2.4) yields an estimated equivalent steady-state standing biomass C amount for permanent pasture of 177 MTC/ha.

2. Compare the above standing biomass C equivalent of 177 MTC/ha with the change in standing biomass C that would be realized if we allowed a vegetative change from existing forest, mixed & field (the only available unit mapped in the southeastern U.S. that is not either "cropland" or "forest, coniferous") (Fig. 2.8) to a PNV state of forest, coniferous. This later number is 95 MTC/ha (Table 2.5).

This analysis suggests that maintaining a productive pasture system potentially stores more C as terrestrial standing biomass than converting an equivalent unit of forest, mixed & field (assumed cropland for the field component, with 2.9 MTC/ha/yr NPP-C) to PNV of, in this case, forest, coniferous. These numbers would vary with the productivity of the pasture or cropland system (i.e., the field part), the fate of the removed biomass C up the food chain, and the vegetative patterns in the pastured landscape. However, they do suggest, in terms of standing biomass's potential to store C on eastern grazing lands, the rough equivalence of steady-state pasture systems with transitions to forest PNV.

Implications for U.S. grazing lands

Though the estimated net changes in size of the standing biomass C due to transitions toward PNV on U.S. grazing lands appears to have a relatively small impact on the absolute size of the various global C pools (atmospheric, oceanic, and terrestrial), and the total capacity is a relatively small proportion of global annual atmospheric CO_2-C fixation, the change in NPP-C is more significant in terms of potential to offset the global rate of annual atmospheric CO_2-C increase.

The down side is that rather extensive and significant changes in the ecological status of the existing grazing land's (especially rangeland's, not pasture's) vegetative land cover are required. Most of this potential would have to be realized by converting extensive areas of this land's vegetative land cover to woodland and forest. This is clear in Table 2.5, where such conversions to forest or woodland PNV account for the largest part of the positive change in standing biomass C and NPP-C in each existing vegetation category. Others have noted the necessity for such drastic changes in land cover and land use patterns to affect the global C budget (Lal et al., 1998).

In addition, converting pasturelands to PNV may not have much impact at all, relative to maintaining the land cover/use as well managed, productive, permanent pasture. Conversely, a significant net loss of C stored as standing biomass occurs with transitions from existing vegetation with a woody component to grassland or, to a lesser extent, to grassland and scrub PNV states (Table 2.5). Much of this is due to the "loss" of the considerable belowground biomass associated with dominantly woody vegetative land cover types (Table 2.4).

According to ecosystem hierarchy theory, broad-scale ecological changes should occur at relatively slow rates (O'Neill et al., 1986). Thus, in assessing impacts of the management of vegetation on the overall potential for C sequestration on grazing lands, whether globally or locally, one should consider actions that may affect the plant-soil system in the landscape over extended periods of time (10s or 100s of years). Such periods are on the order of magnitude of historical, post-European settlement changes in the vegetative land cover of U.S. grazing lands (Branson, 1983).

One of the limitations of the preceding broad-scale analyses is that it is difficult to capture more finely scaled landscape patterns and their associated processes that often dictate grazing land management strategies. As we have suggested, the landscape pattern for potential C storage or loss by grazing land vegetation appears to be quite complex. It depends heavily on the composition of the existing vegetative cover in a particular area and the ecological site potentials there (NRCS, 1997; Passey et al., 1982).

Potential for Carbon Sequestration —
Soil Organic Carbon Pool

The other component of long-term C storage in grazing lands, in addition to standing biomass, is the SOC pool. Figure 2.10 shows the U.S. distribution of SOC for the soil profile down to 1 m deep, expressed as unit area, derived from STATSGO level soils data and adjusted along physiographic province boundary lines (Fenneman, 1931, 1938). The state-factor model of soil formation (Buol et al., 1973) implies that a spatial pattern of SOC content like that in Figure 2.10 would, in part, manifest steady-state C cycling between the landscape's vegetative land cover and the underlying soil cover.

If one could identify areas where the vegetative land cover has been relatively stable (i.e., late successional stage) and unaffected by catastrophic anthropogenic influences, and thus still reflects the prevailing bioclimatic setting and site conditions, one could consider the organic C content of the underlying soil to represent the steady state with that particular vegetative land cover type. In the vegetative transition model we have been using, these "stable" areas would be areas where the existing vegetation equals or approximates the PNV.

In many areas, dramatic changes in land cover/use have occurred associated with settlement of the U.S., particularly conversion of large areas of rangeland and forest to cropland, and they have destroyed the HCPC or have prevented attainment of the PNV state. A map such as Figure 2.10, made from soil survey data collected mostly within the last 30 years, also is likely to bear the imprint of recent changes in SOC levels due to widespread cultivation.

In some areas, however, the generalized existing vegetative land cover type does approximate the PNV. For vegetative land cover types that could support grazing (i.e., those with components of grass, grasslike plants, and shrubs), four such units exist. They are those units in Table 2.5 where the entry for existing vegetation has an equivalent entry in the potential natural vegetation column. Listed as existing vegetation-PNV, they are: grassland-PNV grassland; grassland & woodland-PNV grassland & woodland; grassland & scrub-PNV grassland & scrub; and scrub-PNV scrub. Figure 2.9b shows the variety of PNV that exists within the current geographic distribution of one of these vegetative units, grass-

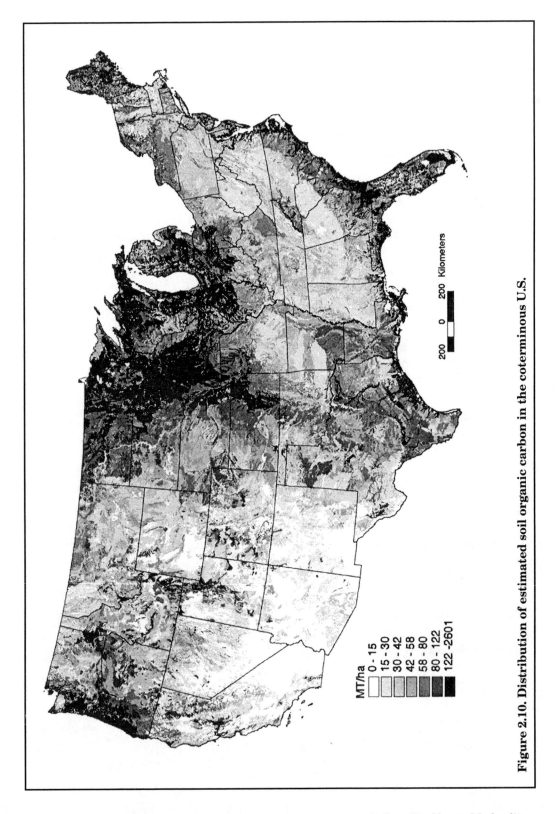

Figure 2.10. Distribution of estimated soil organic carbon in the coterminous U.S.

land & scrub. Table 2.5 does not show the dominantly forested units where existing vegetative land cover approximated PNV (Fig. 2.11).

Method

We estimated the representative SOC content under the four "stable" vegetative land cover types capable of supporting grazing and under the stable forested land cover types by computing the area-weighted mean SOC. We also computed area-weighted mean SOC contents of land areas where the existing vegetative land cover differs from the PNV. Using these computed SOC values for each of the existing vegetation-PNV units, and the areal extent of those respective units (shown in Table 2.5), we estimated the *change* in size of the SOC pool that would result from transitions from grazing land's existing vegetative land cover to PNV (Table 2.5). We could do this only for existing vegetation-PNV units for which we could estimate the SOC value representative of the PNV state.

Ecological status and the soil organic carbon pool

The computed SOC contents (whole soil profile to 1 m) of the existing vegetation-PNV units in Table 2.5 ranged from a low of 10.23 MTC/ha for the tundra/desert-PNV woodland unit to a high of 323.75 MTC/ha in the woodland-PNV forest, coniferous unit, though most of the forested units were below 150 MTC/ha. These values are of the same order of magnitude as the standing biomass C content expressed as per unit area. These later values range from 4.2 MTC/ha for the tundra/desert vegetative land cover type, 30.6 MTC/ha for scrub, to 177 MTC/ha for forest, coniferous, with other types intermediate (Table 2.4).

Grassland existing vegetation-PNV grassland and grassland & woodland existing vegetation-PNV grassland & woodland had relatively low coefficients of variation (CVs) (39 and 14%, respectively), so we assumed the steady-state SOC contents estimated for these units, 41.4 and 45.6 MTC/ha, respectively, to represent reasonably certain values for these PNV types at the scale of this analysis. The CVs for the other units were generally >50%, with many >100%. For example, the CVs for other units where existing vegetation equaled or approximated the PNV, particularly forested units, ranged from 76 to 138%.

A graphic example of this compares forested areas, where the generalized existing vegetative land cover equaled PNV and the variability in the SOC contents (compare Figs. 2.11 and 2.10). Such a large amount of variability in SOC within Olson-based ecosystem units has been documented before (Kern, 1994). This makes estimates of the steady-state SOC for those PNV types more uncertain and therefore imparts a greater degree of uncertainty to the estimated changes in size of the SOC pools for those transitions (Table 2.5).

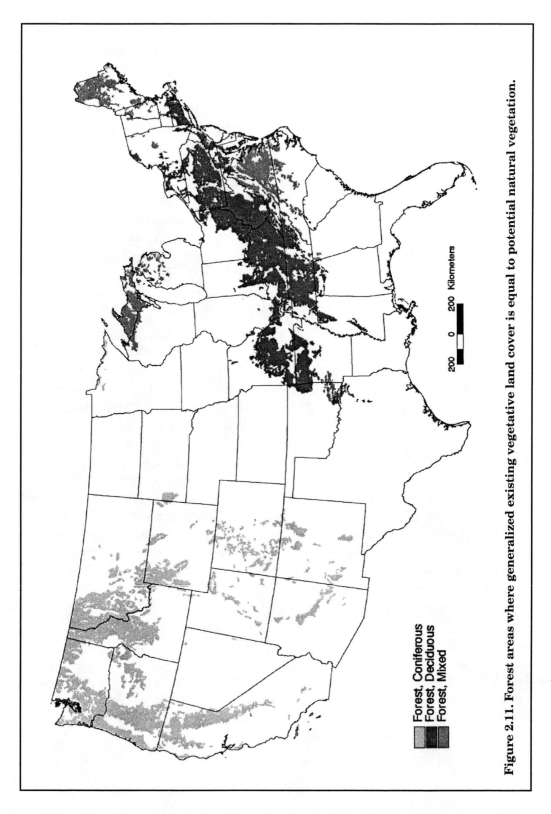

Figure 2.11. Forest areas where generalized existing vegetative land cover is equal to potential natural vegetation.

Forest, Coniferous
Forest, Deciduous
Forest, Mixed

Implications for U.S. grazing lands

Some general trends are apparent from the above analysis but, due to the variability and uncertainty of the results, we must interpret them with caution.

Transitions from grassland-existing vegetative land cover to grassland & woodland PNV increases the SOC pool, while transitions from grassland & woodland to grassland PNV reduces the SOC pool. The magnitude of reduction is of the same order of magnitude and in the same direction as the change in size of the standing biomass C pool (Table 2.5). All transitions from scrub to various PNV states are associated with a net increase of the SOC pool. Transitions from grassland & scrub to scrub PNV are associated with a net decrease in the SOC pool while transitions from scrub to grassland PNV or grassland & scrub PNV are associated with a net increase.

Interpreted in light of the state-factor model of soil formation, this suggests that, on a broad scale, the size of the SOC pool increases with transitions from lower standing biomass, vegetative land cover to higher standing biomass types with woody components, and decreases with transitions from higher biomass, dominantly woody, vegetative land cover to lower biomass grassland or scrub PNV states (Tables 2.4 and 2.5). In transitions involving grassland and scrub, a grassland component in either the existing vegetative land cover or PNV state seems beneficial in either maintaining the size of the SOC pool (relative to the scrub PNV state) or increasing it in transitions to a PNV with a grassland component.

The large amount of variability in SOC contents, and the apparent scale incongruity between SOC content patterns and those of vegetative land cover, may result from the limitations this analysis induced by arbitrarily generalizing existing vegetation and PNV at relatively high categorical levels. This was necessary to accomplishing a national analysis conveying broad ecological relationships of grazing land's vegetative land cover.

A more fine-grained analysis, though unwieldy for the broad-scale analysis presented here, might be possible using less categorically generalized LCC vegetation data. Such a more finely scaled analysis, perhaps stratified within existing vegetation-PNV units by bioclimatic zones or, as data in Kern (1994) suggest, by soil type itself, may allow establishment of more reliable approximations of steady-state SOC levels for the other areas where existing vegetative land cover equals PNV. This would allow more reliable estimates of changes in size of the SOC pool associated with those vegetative transitions.

Conclusion

Grazing lands cover approximately 1/3 of the land area of the U.S., excluding Alaska. They represent a very heterogeneous land cover/use category, comprising a variety of vegetative land cover types and occurring in climatic regimes ranging from arid to humic tropical. As rangeland, they approximate the potential natural vegetation in many areas of the western U.S. As pasturelands, however, they can represent a considerable departure from the potential natural vegetation of a region, having been converted from former forest and grassland areas in much of the central and eastern U.S.

Thus, the ecological status of grazing lands' vegetation forms a heterogeneous patchwork across the U.S. landscape, even on a national scale. This pattern is a major aspect of C storage potential in U.S. grazing lands' landscapes, particularly in relation to storing C in the standing biomass of vegetative land cover, relative to global C pools. C storage relationships in the SOC pool are confounded by considerable variability, but this broad-scale analysis suggests some increase in the SOC pool with conversion to woody vegetative components of higher standing biomass.

Potential for significant change in C storage, relative to the global C budget, seems limited without major outright change in land cover/use of grazing lands. However, the ecological status of grazing lands and their heterogeneous patterns have important implications for management for local and regional resource values in terms of soil quality and overall grazing land productivity.

The small-scale patterns evaluated here suggest that effectively addressing grazing lands relative to global C dynamics may involve long-term strategies capable of encompassing rates of change of extensive areas of the landscape on the order of tens or hundreds of years. Analyzing more finely scaled patterns in the landscape may be critical to making optimum assessments for particular situations when considering the other resource values of the land.

Acknowledgments

The authors extend their appreciation to Sharon Waltman of the USDA-Natural Resource Conservation Service Soil Survey Center for providing the digital soil C base map which some of the analyses use.

References

Alexander, M. 1977. *Introduction to Soil Microbiology,* 2nd ed. John Wiley & Sons. New York, NY.

Anderson, J.R., E.E. Hardy, J.T. Roach, and R.E. Witmer. 1976. *Land use and land cover classification system for use with remote sensor data. U.S. Geological Survey Professional Paper 964.* U.S. Government Printing Office. Washington, DC.

Bailey, R.G. 1978. *Description of the ecoregions of the United States.* USDA-FS Intermountain Region. Ogden, UT.

Bailey, R.G. 1997. *Ecoregions Map of North America:* explanatory note. Misc. Publ. No. 1548. USDA-FS. Washington, DC.

Barbour, M.G., J.H. Burk, and W.D. Pitts. 1987. *Terrestrial Plant Ecology,* 2nd edition. The Benjamin/Cummings Publishing Company, Inc. Menlo Park, CA.

Bazilevich, N.I. 1967. Cartoschemes of the productivity and biological cycle of the chief types of terrestrial vegetation of the earth. *Izv. Vses. Geograf. Obs.* 99 (3):190-194.

Bourgeron, P.S., L.D. Engelking, H.C. Humphries, E. Muldavin, and W.H. Moir. 1995. Assessing the conservation value of the Gray Ranch: rarity, diversity, and representativness. *Desert Plants* 11(2-3):2-68.

Branson, F.A. 1985. *Vegetation Changes on Western Rangelands.* Society for Range Management. Denver, CO.

Buol, S.W., F.D. Hole, and R.J. McCracken. 1973. *Soil Genesis and Classification.* Iowa State University Press. Ames, IA.

Connin, S.L., R.A. Virginia, and C.P. Chamberlain. 1997. Carbon isotopes reveal soil organic matter dynamics following arid land shrub expansion. *Occologia* 110:374-386.

Daugherty, A.B. 1989. *U.S. Grazing Lands: 1950-82. SB-771.* USDA-NRED-ERS. Washington, DC.

Daugherty, A.B. 1995. *Major Uses of Land in the United States. 1992. Agric. Econ. Report No. 723.* USDA-NRED-ERS. Washington, DC.

Driscoll, R.S., D.L. Merkel, D.L. Radloff, D.E. Snyder, and J.S. Hagihara. 1984. *An ecological land classification framework for the United States. Misc. Publ. No. 1439.* USDA-FS. Washington, DC.

Ellenberg, H., and D. Mueller-Dombois. 1967. Tentative physiognomic-ecological classification of plant formations of the earth. *Ber. Geobot. Inst. ETH Stiftig. Rubel, Zurich* 37:21-55.

Fenneman, N.M. 1938. *Physiography of the Eastern United States.* McGraw-Hill Book Co. Inc. New York.

Fenneman, N.M. 1931. *Physiography of the Western United States.* McGraw-Hill Book Co., Inc. New York.

FGDC. 1997. *National spatial data infrastructure: vegetation classification standard.* Federal Geographic Data Committee. Washington, DC.

Forman, R.T.T., and M. Godron. 1986. *Landscape Ecology.* John Wiley & Sons. New York.

Franklin, J. 1995. Predictive vegetation mapping: geographic modelling of biospatioal patterns in relation to environmental gradients. *Prog. Phys. Geog.* 19(4):474-499.

Frey, H.T. and R.W. Hexem. 1985. *Major Uses of Land in the United States. 1982. Agric. Econ. Report No. 535.* USDA-NRED-ERS. Washington, DC.

Gibbens, R.P., J.M. Tromble, J.T. Hennessy, and M. Cardenas. 1983. Soil movement in mesquite dunelands and former grasslands of southern New Mexico from 1933 to 1980. *J. Range. Manage.* 36:145-148.

Grossman, D., Goodin, K.L., D. Faber-Langendoen, and M. Anderson. 1994. *The USGS-NPS vegetation mapping program.* USGS. Reston, VA.

Hatherly, D.T. 1993. *Soil survey of Smith County, Texas.* USDA-SCS. Washington, DC.

Jackson, R.B., J. Canadell, J.R. Ehleringer, H.A. Mooney, O.E. Sala, and E.D. Schulze. 1996. A global analysis of root distributions for terrestrial biomes. *Occologia* 108:389-411.

Joyce, L.A. 1989. *An analysis of the range forage situation in the United States: 1989-2040. A technical document supporting the 1989 USDA Forest Service RPA assessment. USDA Forest Service General Technical Report RM-180.* Rocky Mountain Forest and Range Experiment Station. Fort Collins, CO.

Kern, J.S. 1994. Spatial patterns of soil organic carbon in the contiguous United States. *Soil Sci. Soc. Am. J.* 58:439-455.

Klopatek, J.M., R.J. Olson, C.J. Emerson, and J.L. Joness. 1979. Land-use conflicts with natural vegetation in the United States. *Environ. Conserv.* 6(3):191-199.

Kuchler, A.W. 1964. *Potential natural vegetation of the conterminous United States. Spec. Publ. No. 36.* American Geographical Society. New York.

Lal, R., J. Kimble, and R.F. Follett. 1998. Pedospheric processes and the carbon cycle. *In* R. Lal, J. M. Kimble, R.F. Follett, and B.A. Stewart (eds.), *Soil Processes and the Carbon Cycle*, CRC Press, Boca Raton, FL, pp. 1-8.

Lieth, H. 1975. Primary production of major vegetation units of the world. *In* H. Lieth and R.H. Whittaker (eds), *Primary Productivity of the Biosphere*, Springer-Verlag. New York. p. 203-215.

Lillesand, T.M., and R.W. Kiefer. 1987. *Remote Sensing and Image Interpretation.* John Wiley & Sons. New York.

Loveland, T.R., J.W. Merchant, D.O. Ohlen, and J.F. Brown. 1991. Development of a land-cover characteritics database for the conterminous U.S. *Photogram. Eng. Rem. Sens.* 57:1453-1463.

McGuire, A.D., J.M. Melillo, L.A. Joyce, D.W. Kicklighter, A.L. Grace, B. Moore III, and C.J. Vorosmarty. 1992. Interactions between carbon and nitrogen dynamics in estimating net primary productivity for potential vegetation in North America. *Global Biogeochem. Cycles* 6:101-124.

Motavalli, P.P., C.A. Palm, W.J. Parton, E.T. Elliott, and S.D. Frey. 1994. Comparison of laboratory and modeling simulation methods for estimating soil carbon pools in tropical forest soils. *Soil Biol. Biochem.* 26:935-944.

NRCS. 1992. (formerly SCS. 1992.) *Instructions for collecting 1992 National Resources Inventory Sample Data.* USDA-NRCS. Washington, DC.

NRCS. 1994. *Summary report 1992 National Resources Inventory*. USDA-NRCS. Washington, DC.

NRCS. 1997. *National Range and Pasture Handbook*. Grazing Lands Technology Institute, USDA-NRCS. Washington, DC.

Nusser, S.M., and J.J. Goebel. 1997. The National Resources Inventory: a long-term multiresource monitoring programme. *Environ. and Ecol. Stats*. 4(3):181- 204.

Olson, J.S. 1970. Geographic index of world ecosystems. *In* P.E. Reichle (ed), *Analysis of Temperate Forest Ecosystems, Ecological Studies*, vol. 1, Springer-Verlag, New York, pp. 297-304.

O'Neill, R.V., D.L. DeAngelis, J.B. Waide, and T.F.H. Allen. 1986. *An Hierarchical Concept of Ecosystems*. Princeton University Press. Princeton, NJ.

Parton, W.J., D.S. Schimel, C.V. Cole, and D.S. Ojima. 1987. Analysis of factors controlling soil organic matter levels in Great Plains grasslands. *Soil Sci. Soc. Am. J.* 51:1173-1179.

Passey, H.B., V.K. Hughie, E.W. Williams, and D.E. Ball. 1982. *Relationships between soil, plant community, and climate on rangelands of the intermountain west. Technical Bulletin No. 1669*. USDA-SCS. Washington, DC.

Paul, E.A., and F.E. Clark. 1989. *Soil Microbiology and Biochemistry*. Academic Press, Inc. San Diego, CA.

Turner, M.G., and R.H. Gardner. 1991. Quantitative methods in landscape ecology: an introduction. *In* M.G. Turner and R.H. Gardner (eds), *Quantitative Methods in Landscape Ecology*, Springer-Verlag, New York, pp. 3-14.

Turner, S.J., R.V. O'Neill, W. Conley, M.R. Conley, and H.C. Humphries. 1991. Pattern and scale: statistics for landscape ecology. *In* M.G. Turner and R.H. Gardner (eds), *Quantitative Methods in Landscape Ecology*, Springer-Verlag, New York, pp. 17-49.

UNESCO. 1973. *International classification and mapping of vegetation. Ecology and Conservation No. 6*. United Nations Educational, Scientific, and Cultural Orgnaization. Paris.

USGS. 1998. *North America land cover characteristics database. ver. 1.2*. USGS. Reston, VA.

CHAPTER **3**

Organic Carbon Pools in Grazing Land Soils

R.F. Follett[1]

Introduction

Private grazing lands in the U.S. occupy about 212 Mha. If both public and private lands are considered, grazing lands occupy about 336 Mha, or a little less than a third of the U.S. (Sobecki et al., Ch. 2).

We determined amounts of soil organic C (SOC) in soil profiles ~2 m deep for 14 native grassland sites in 9 states in the historic grasslands of the U.S. They contained an average of 123 ± 48 MTC/ha with an average of $24 \pm 7\%$ of the SOC located in the top 10 cm of depth (Follett et al., unpublished data). Assuming conservative average estimates, 100 and 20 MTC/ha contained in the top 200 and 10 cm, respectively, then private U.S. grazing lands alone contain 21,000 MMTC in the surface 200 cm and 4,200 MMTC in the top 10 cm of soil.

In comparison, the MMTC emissions for U.S. agriculture recently reported are about 43 MMTC (Lal et al., 1998), and the MMTC of CO_2 from all U.S. sources is reported to be 1442 (DOE/EIA, 1996) to 1488 (USEPA, 1999). These numbers imply that an increase (or loss) of only 1% of the SOC in the top 10 cm of these grazing land soils is equivalent to the total C emissions from all U.S. cropland agriculture.

Very small changes in the amounts of SOC sequestered in grazing land soils become extremely important because of the great land areas involved. The vast amount of CO_2 previously removed from the atmosphere by photosynthesis and sequestered into grazing land soils as SOC underscores the importance of grazing lands to potentially help mitigate global climate change resulting from increasing atmospheric levels of CO_2 (OSTP, 1997).

---

[1]Supervisory Soil Scientist, USDA-ARS, P.O. Box E, Fort Collins, CO 80522.

Burke et al. (1997) proposed a conceptual framework for the biogeochemical responses of grasslands to management practices. The framework focuses on disturbances' potential to alter the distribution and flux of biologically active elements in grasslands of the central grasslands region of North America. Disturbances are most likely to alter nutrient storage and cycling if they affect large biological/chemical pools. Small pools, with short turnover times, may change easily but are likely to recover rapidly once the disturbance ceases. Although other grassland sites may vary significantly in pool sizes, the relative proportions of C in pools and the turnover times Burke et al. (1997) proposed for the central U.S. grasslands are likely to represent grasslands in other regions (Fig. 3.1).

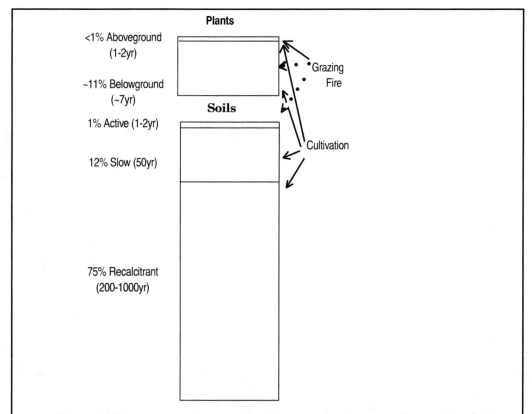

Figure 3.1. Distribution of organic C in a short grass steppe ecosystem, showing the direct (solid lines) and indirect (dashed lines) influences of management on the component pools. Values are in percent of total aboveground plus belowground C. Values in parentheses are turnover times of subpools.

The main points from Figure 3.1 are:

- The aboveground, plant biomass C pool is less than 1% of the total above- and belowground mass of C and turns over in ~1 to 2 years; its mass of C may be only about 10% of that in the belowground plant biomass C pool. The belowground, plant biomass C pool is about 10% of total mass of C and turns over more slowly (~7 years). Thus, short-term perturbations of aboveground biomass alone are not likely to cause large changes in SOC storage; recovery from such perturbations is likely to be rapid.
- Most of the SOC in grassland soils is recalcitrant and has a slow rate of turnover, is well protected from natural disturbances, and generally resists change.
- Most input of SOC into grazing land soils is from belowground organs (i.e., plant roots, etc.). Consequently, the effects of aboveground perturbations (such as from grazing, fire, or other) on belowground SOC are indirect or delayed.
- The exception, for large perturbations on belowground SOC, occurs when removal of all or part of the aboveground biomass results in significant soil erosion or when management practices (i.e., cultivation) disrupt the soil and associated SOC pools. Such disturbances enhance decomposition, cause physical loss of SOC, and reduce the ecosystem's ability to maintain the size and activity of its critical nutrient cycles.

The extent of grazing lands and total amount of C sequestered in their soils make understanding C-storage dynamics into various SOC pools, and mechanisms whereby C is gained or lost, very important. This chapter discusses a number of these pools and their relationships. The objective is to encourage a deeper understanding of the various SOC pools and their potential roles, especially with improved grazing land management, in enhancing of SOC sequestration in grazing land soils.

The Connection between Atmospheric Carbon Dioxide and Soil Organic Carbon

Figure 3.2 shows the cycle of agriculture (CAST, 1992) in producing food and fiber. At the top of the figure are plants that capture atmospheric CO_2 through photosynthesis and convert it into useful C-products, and herbivores (animals) that eat the plants. Below plants is the soil with its nutrient ions, plant residues, and microbes. The arrows indicate the movement of matter, such as when nutrients are added or plant residues and animal manures cycle into the soil. The respiration of plants, animals, or microbes in turn can release the CO_2 back into the

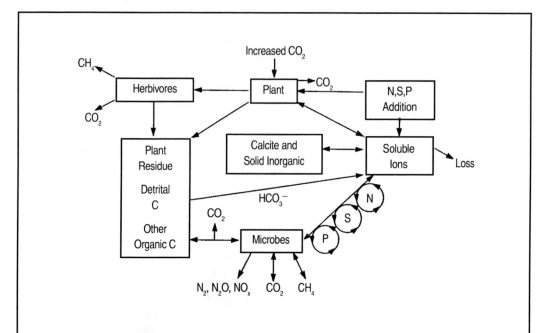

Figure 3.2. The cycle of agriculture with production of plants and animals for food, fiber, and timber; with the cycling of nutrients in the soil and its microbes; and with the leakage of some greenhouse gases (Follett, 1991).

atmosphere during the decomposition of organic materials. Understanding the processes that convert and store atmospheric CO_2 in SOC may lead to practical ways that soils can help offset overall U.S. CO_2 emissions.

Although not covered here, other greenhouse gases (CH_4, N_2O, and NO_x) and N_2 gas are emitted also as a result of microbial activities in soil or, in the case of CH_4, from the rumen of cattle. Native grassland soils also are a major sink for atmospheric CH_4 (Mosier et al., 1997).

In the cycle of agriculture with its interacting processes (Fig. 3.2), changing one process may alter the flow of C and other nutrients and thus may cause either increased or decreased rates of C sequestration in soil. Understanding the processes involved may make it possible to enhance C sequestration in grazing land soils.

Biological Pools of Carbon in the Plant-Soil System

Plant-derived materials

Plant-derived materials are the primary source of SOC, and researchers make constant efforts to separate them from the soil, primarily based on their size or density. Various authors may or may not use similar laboratory methods and often use different terms to describe similar fractions of plant-derived materials separated from the soil.

Litter (sometimes called macroorganic matter) generally is identified as originating from plants and being larger than 2 mm in size. *Light fraction* is smaller than litter, generally considered to range from 0.25 mm to 2 mm.

Another size category used in the literature is *particulate organic matter* (POM), as Cambardella and Elliott describe (1992). Paul and Clark (1996) indicate that the size of POM is from 0.050 mm to 2 mm, thus overlapping the size range of the light fraction. More recently, Amelung et al. (1998) measured SOC in sizes of coarse sand (0.025 mm to 2 mm) and fine sand (0.002 mm to 0.025 mm) fractions separated from soils obtained across temperature and precipitation transects from Canada to Texas.

Regardless of their names, these small fractions are highly important in the overall processes whereby plant-derived C is sequestered into SOC. However, the methods to separate the various fractions (i.e., density flotation vs. chemical dispersion vs. sonication and wet sieving, respectively) do provide fractions with different properties (Elliott et al., 1996; Six et al., 1998).

Litter

Other than living plant material, litter is the most familiar form of plant C in the plant-soil system. Litter is the macroorganic matter on or near the top of the soil surface and is of considerable importance to nutrient cycling, especially for nitrogen (N). Herbaceous litter on grazing lands can include that derived from grasses, shrubs, or trees. The amount of aboveground litter on grazing land depends on such factors as season of the year, climatic conditions, land use (grazing intensity, recentness of grazing, etc.), and management (use of fertilizer, shrub and brush control, the introduction of improved species, etc.).

Plant species have an important effect on litter quality. For example, legumes or plants with a low C/N ratio produce litter that decomposes more rapidly than that from woody plants that have high C/N ratios. The chemical composition of litter, such as lignin content, also influences the rate of decomposition.

Research by Naeth et al. (1991) illustrates the potential effects of grazing on species composition and on plant-root dry matter (DM). They observed more large POM, particularly standing litter, on control treatments than on grazed treatments. More medium and small particle OM occurred on grazed treatments than on ungrazed controls, and the vegetation probably decomposed more rapidly because the grazing animals trampled it and broke it down.

Light fraction

We probably should standardize the lower limit for litter. This lower limit would then become the upper limit for "light fraction" organic matter. No agreed division seems to exist, although 2 mm has been suggested for the upper limit for the light fraction and 250 µm has been suggested for the lower limit (Stevenson and Elliott, 1989).

The light fraction generally is separated from mineral particles by flotation using dense liquids. Densities for separating light fractions have ranged from 1.6 to 2.0 Mg/m^3. Following separation from mineral particles, the light fraction includes plant debris and fauna, appreciable amounts of plant root material, and low amounts of inorganic material. The very fine plant root material that separates with the light fraction can be difficult to separate more from the rest of the light fraction. However, Jastrow et al. (1998) recently published data including that of fine fibrous roots (0.2 to 1 mm dia.), very fine fibrous roots (<0.2 mm dia.), and external mycorrhizal hyphae.

Plant root material

Dormaar et al. (1977) reported that a change in root-mass DM in the top 15 cm of soil accompanied the change in aboveground vegetation with increased grazing intensity. At their Mayberries, Alberta, study site, root DM averaged 15.0 MT/ha on the ungrazed site and 24 MT/ha on the heavily grazed site. At their fescue grassland location near Stavely, Alberta, *Festuca scabrella* dominated on the ungrazed site but essentially was eliminated from the heavily grazed study site. Secondary plants the authors identified as "other grasses" and "other forbs and shrubs" were more abundant under heavy grazing than on nongrazed land. A change in root DM in the top 15 cm of soil accompanied the change in vegetation. Root DM averaged 17.2 MT/ha on the ungrazed site and 25.3 MT/ha on the heavily grazed site.

Accumulation of SOC relates to mass of root material (including exudates, etc.) and litter and aboveground plant-derived materials. Dormaar et al. (1977) did not measure mass of SOC; but average concentrations of SOC in the A1 soil horizon at the mixed prairie site near Mayberries, Alberta, were 18.7 and 13.6 g/kg for the heavily grazed and ungrazed sites, respectively. Concentrations of SOC for heavily grazed and ungrazed sites at the fescue grassland location near Stavely, Alberta, were 95.1 and 90.3 g/kg, respectively. Additional comparisons

(Dormaar et al., 1995) were that native mixed prairies had about 7.6 times more root mass than did seeded crested wheatgrass (*Agropyron cristatum* (L.) Gaertn.) or Russian wildrye (*Elymus junceus* Fisch.) pastures in the 0- to 7.5-cm depth, with about the same amounts of root mass in the 7.5- to 40-cm depth.

Particulate organic matter (POM)

Particle size separates have been used widely to distinguish pools of different soil organic matter (SOM) quality and turnover rates (Christensen, 1992). The ability to estimate SOM fractions quantitatively is especially important for understanding the dynamics of SOM in intensively managed systems.

Cambardella and Elliott (1992) developed a simple method that combined chemical dispersion of soil with a subsequent, easily standardized, physical separation, based on particle size. They suggested their method isolated a POM fraction from soil that closely matches the characteristics of slow (Parton et al. 1987), decomposable (van Veen and Paul, 1981), or stabilized (Paul, 1984) OM pools conceptualized in simulation models. They further suggested that POM accounts for the majority of the SOM initially lost as a result of cultivation of grassland soils. Thus, POM would be the most rapidly lost fraction from continuously cultivated soil (Tiessen and Stewart, 1983; Dalal and Mayer, 1986).

Paul and Clark (1996) identify that sieving to separate the residues in the 50- to 2000-m sand fraction yields POM which generally makes up 5 to 15% of the C of cultivated top soils and much more of uncultivated soils. They considered POM highly related to the C and N mineralization potential and thus to soil quality.

For long-term uncultivated grazing land soils, Cambardella and Elliott (1992) suggest that the POM fraction is less important for providing mineralized N than are the microbially produced decomposition products that physically occlude within soil aggregates. Jastrow (1996) noted the suggestion that most of the labile pool within microaggregates is either light-fraction POM or relatively low density, mineral associated OM, probably of microbial origin.

Free particulate organic matter

In a model which Golchin et al. (1998) proposed for a fraction they referred to as *free POM*, they assumed that POM entering the soil from root or plant debris becomes colonized by rhizosphere microbes and at the same time adsorbs mineral materials. The mineral phase and close contact between plant residues, microbes, and microbial mucilages facilitate this adsorption. However, the degree of association between primary particles and POM is still low, and free POM can still be floated out of the soil with gentle shaking and by separation using a high density ($\sim$1.6 Mg/m^3) solution.

Fine and coarse sand sized fractions

Recently, Amelung et al. (1998) reported research from soils collected from 21 native sites in the North American Great Plains, with a range of mean annual temperatures (MATs) from 0.9 to 23.4°C and mean annual precipitation (MAP) from 343 to 1308 mm. Soils were separated into particle sand sizes of fine (20 to 250 µm) and coarse (250 to 2000 µm), thus partitioning into two fractions the size range of POM (50 to 2000 µm) which Paul and Clark (1996) identified.

To describe their observations, Amelung et al. (1998) used an enrichment ratio (E_{SOC}) equal to the concentration of SOC in a particular size of fraction divided by the concentration of SOC in the bulk soil. They observed that the C/N ratio of their fine sand sized fraction was significantly lower than that of the coarse sand sized fraction. This observation plus electron microscopy reveals that SOM in their coarse sand sized fraction was comprised mainly of undecomposed plant residue. The SOM in their fine sand sized fraction was comprised mainly of altered and decomposed organic debris and fine root particles. For these grassland soils, POM rapidly degraded, and E_{SOC} for the fine sand sized fraction decreased as MAT increased. The coarse sand sized fraction made up a separate SOM pool that was not related to climate when the input data were varied.

Amelung (1997) earlier observed more pronounced lignin oxidation and higher ratios of microbe-derived to plant-derived sugars in the OM of the fine sand sized fraction than in the coarse sand sized fraction, confirming that microbes altered the latter less. In the later study of 21 grassland sites along temperature and precipitation transect from Saskatchewan to Texas, Amelung et al. (1998) observed that, for both wet and dry sites, E_{SOC} of the clay fraction (<2 m) increased with increasing MAT. The SOC and total N concentrations in the clay fraction, relative to those in the bulk soil, increased significantly across sites with increasing annual temperature, decreasing annual precipitation, and decreasing clay content; the concentration of SOM in the fine sand fraction showed the opposite trend. Thus, with a higher MAT, clay seems to stabilize more SOM, preferentially.

Principal axis component analyses confirmed that both clay and fine sand sized fractions comprised sensitive SOC and N pools related to climate. The results suggest SOM, with increasing temperature, decays preferentially from pools of the fine sand sized fraction, resulting in a relative enrichment of SOC stabilized on clay.

Total SOC

Many interacting and complex factors control SOC content in soil. A regional study by Burke et al. (1989) evaluated the relative importance of these factors and how they control SOC content in soil. They obtained pedon and climate data for 500 rangeland and 300 cultivated soils in the U.S. Central Great Plains grasslands and statistically analyzed relationships among SOC, soil texture, and climate.

That analysis indicated that SOC increased with precipitation and clay content and decreased with temperature. This supports earlier work by Parton et al. (1987), who predicted regional trends in SOM, using temperature, moisture, soil texture, and plant lignin content. Regional studies that help provide insights about the dynamics and pools of SOC in plant-soil systems provide an illuminating context for this chapter's discussion of the relationships of various SOC pools to the processes that control C sequestration in soil.

Soil fauna

The overall mineralization of OM by soil fauna is generally small compared to that by bacteria and fungi. As Paul and Clark (1996) describe, microflora constitute the primary feeders, with fauna in secondary and tertiary roles. The microflora participate most in N mineralization by attacking plant residues that have C:N ratios as high as 100:1 and processing it down to a biomass with a C:N ratio of about 10:1. In laboratory and other studies, nutrient release accelerates when a faunal component is present. Hunt et al. (1987) diagrammed and described the detrital food web in a short grass prairie. The faunal grazers feed on the microbial biomass.

Paul and Clark (1996) provide relative sizes of these pools from pine forest studies. The densities of the faunal, bacterial, and fungal biomass were 1.7, 3.9, and 120 g/m^2, respectively. Estimates of faunal consumption of bacteria were 30 to 60% of microbial production in the pine forest, 38% for a prairie soil in Colorado, and 19 to 38% for grass leys and arable soils in Sweden.

Larger soil fauna (animalia), such as earthworms, affect both soil physical properties and rates of plant residue decomposition. Their channeling activities increase the transport of material, soil pore volume, water intake and flow, and soil aeration, while decreasing water runoff and soil erodibility (Linden and Clapp, 1998).

Earthworms increase cycling of plant residues into SOM by ingesting plant residues, mixing them into the soil, and transporting material from its point of ingestion to its point of egestion. Most nutrients undergo some degree of change during gut transit. Microbial activity and biochemical breakdown occur in the gut and in the egested cast. Aged casts can form stable aggregates and may protect the organic residues trapped within them from rapid breakdown. The release of CO_2 and nutrients from both the soil and fresh residue material of the ingested mix primarily results from microbial respiration within the earthworm gut and fecal casts.

Earthworms' ingestion of plant and soil materials and its deposition as fecal pellets sequesters considerable C. Linden and Clapp (1998) showed that an immediate encapsulation of fresh plant residue could occur. Because of the magnitude of C encapsulation, its isolation within the cast from environmental changes, and the deposition of casts at depth by some earthworm species, the likely result from their activity is a gain in C sequestration.

Soil microbial biomass

The availability of C substrate limits the growth and functioning of soil microbial biomass (SMB). When flushes of substrate C enter the soil, the size of the SMB increases and then returns to a quasi-equilibrium level. The energy flux through the SMB drives the decomposition of organic residues (Smith and Paul, 1990) and SOM.

The SMB is generally in a resting state, with periodic flushes of activity and growth. Even though the SMB is very active for nutrient cycling, SMB levels are fairly stable for a particular soil/land use. Much of the yearly throughput of energy is used for SMB maintenance. If the available C in soil becomes increasingly depleted, then the size of the SMB pool itself will decrease but continues to subsist for very extended periods of time (Follett, 1998).

The chloroform fumigation-incubation (CFI) method that Jenkinson and Powlson (1976) introduced is used widely. Measured amounts of SMB can deviate considerably among various methods, but Kaiser et al. (1992) observed that the rankings for the quantities of SMB remained the same in the soils that they tested with any single method. Data in Table 3.1 (from a 2-year-old poplar stand) illustrate the size and type of microbial biomass and the relationships of its size to that of its distribution with depth, plant roots, soil C and N, and microbial biomass C and N (Paul and Clark, 1996). I could not locate a similarly thorough data set for a grassland site.

All of the components Table 3.1 measures decreased with increasing depth in the soil profile. Microbial biomass calculated from microscopic counts was about 70% of that determined by chloroform fumigation incubation (CFI), since the counts do not include all of the biota that would be sensitive to the CFI treatment. Similar relationships to those Table 3.1 shows are expected under grazing lands, although the magnitude and relative ratios of the numbers may be different. Microbial biomass C was observed to decrease fairly rapidly with depth for grassland soils in CO, NE, and IA (Follett, 1998). However, for the soils developed under grasslands, Follett showed larger weights of biomass C and SOC per unit of near-surface soil depth and also larger ratios of biomass C to SOC (from 1.4% to >5.0%) than Table 3.1 shows.

Table 3.1. The effect of soil depth on microbiota C, N, and activity; soil and root C and N content; and soil C dynamics in a 2-year-old poplar plantation (adapted from Paul and Clark, 1996).

Soil Profile	Depth (cm)		
	0 to 25	25 to 60	60 to 100
SOC (kg/ha)	30,000	20,000	17,500
Soil N (kg/ha)	2,250	1,800	1,650
Root C (kg/ha)	1,240	560	400
Fungal C (kg/ha)	210	181	155
Bacterial C (kg/ha)	74	24	25
Total microbial C (kg/ha)	284	205	180
Biomass C (kg/ha)	410	347	327
Biomass N (kg/ha)	75	54	51
Fungal C:Bacterial C	2.8	7	6.3
Biomass C:N	5.5	6.5	6.4
Biomass C:SOC, %	1.3	1.9	1.8
Biomass N:soil N_T, %	3.2	3.3	3.2

Chemical Pools of Organic Carbon and Their Properties

The need for realistic methods to assess the quantity and quality of SOM, including occluded organic C, is continuing and urgent. For example, Golchin et al. (1995, 1998) report that the stabilities of occluded organic materials relate inversely to O-alkyl C content but directly to their aromatic C content. Some fractionation of the complex and dynamic mixture of organic material in soils is necessary. Desirably, the fractions separated should relate to the response of SOM to its natural environment, management, sources of organic C inputs, and observed biological and physical C pools.

Much of the SOM in soils consists of acidic, yellow to black macromolecules referred to as "humic" substances. These humic substances represent a complex mixture of molecules of various sizes and shapes. Besides humic substances, the "nonhumic" substances in SOM include carbohydrates, lipids, and proteins. Generally, humic and nonhumic substances are associated closely with and not readily separated from each other (Stevenson, 1986).

Many procedures exist for fractioning SOM. A recently reported method (Ping et al., 1999) allows for the separation and study of a wide range of organic materials from SOM, including lipids, cell wall components, fibrous materials, and humin.

Humic acid, fulvic acid, and humin

SOM commonly is extracted by repeated steps with caustic alkali (usually 0.1 to 0.5 M NaOH) to obtain maximum recovery. More recently, milder reagents, such as neutral $Na_4P_2O_7$, are being used. The fraction that is soluble in alkali but insoluble in acid commonly is defined as *humic acid* (HA), while the fraction that is soluble in both alkali and acid is referred to as *fulvic acid* (FA). The fraction that is insoluble in both alkali and acid is referred to as *humin*. The solubility of HA in alkali is believed to be caused by disruption of bonds binding the molecules to clay and polyvalent cations.

Substantial amounts of C are lost during these procedures. Humic acids extracted with alkali contain variable amounts of inorganic materials, much of which can be removed by repeated dissolution and precipitation and by passage through an ion exchange resin column (Ping et al., 2000).

Campbell et al. (1967) conducted research using ^{14}C isotope to determine the mean residence time of soil humic fractions. Given the natural ^{14}C activities of the OM fractions, humin was determined to be more stable than HA and more resistant to decomposition than FA. Later, this concept (Paul et al., 1997) was extended to the use of acid hydrolysis of soil in hot, 6 M HCl followed by ^{14}C dating of the residue to determine both the size and mean residence time of the old resistant SOC fraction in soil (Leavitt et al., 1996).

Separation of the acid hydrolysis fraction is used in an approach Paul and Clark (1996) described to interpret SOM pool sizes and fluxes relative to ecosystem dynamics. Also, Leinweber and Schulten's recent research (1998) indicates that 50 to 80% of nonhydrolyzable N is extracted together with pedogenic oxides. It therefore was considered possible that these oxides bind or occlude N-containing molecules and prevent them from hydrolyzing with 6 M HCl. The same processes should apply to the C-containing molecules.

Identifying the acid soluble fraction as FA requires recognizing that this fraction invariably contains organic substances belonging to well known classes of organic compounds. The term *FA* should be reserved as a generic name for the pigmented components of the acid soluble fraction. Consequently, the soluble material remaining after removal of HA should be referred to as the *FA fraction* (Stevenson, 1982).

The humin fraction may include one or more of the following: (a) Humic acids that are intimately bound to mineral matter, (b) highly condensed humic matter with a high concentration of C and thereby matter insoluble in alkali, (c) fungal melanin, (d) paraffinic structures, (e) undecomposed plant material, if not removed prior to alkali extraction, and (f) charcoal that has been left behind by burning plant residues (Stevenson and Elliott, 1989). While studying nonhydrolyzable organic N in soil, Leinweber and Schulten (1998) identified organo-mineral bonds as responsible for the nonhydrolyzability of perhaps 8 to 30% of the N in soil. Repeated acid and alkali extraction, followed by treatment of the final soil residue with an acid solution containing HF, can measure concentration of humin.

Soluble carbon

Dissolved organic C (DOC) is the OC that passes through a filter of 0.45 μm pores. It consists of a wide range of chemicals, ranging from simple acids and sugars to complex humic substances with large molecular weights. Estimates of the release of DOC from vegetation and soil organic horizons range from 10 to 500 kg DOC/ha/yr. However, much of the terrestrially produced DOC probably is consumed microbially, photo-degraded, or precipitated in soils (Moore, 1998).

When comparing no-tilled and conventionally tilled soils from a long-term tillage experiment of 10 years of continuous barley, Arshad et al. (1990) found the no-till soil higher in carbohydrates and amino acids. The no-till soil was also higher in aliphatic C and contained less aromatic C than the conventionally tilled soil. Although the data were not from grazing lands, they provide important insights concerning compounds likely to be contained in DOC and their relative availability for microbial consumption.

Soils containing large amounts of Fe and Al and small amounts of organic C are reported to absorb large quantities of DOC, whereas soils rich in organic C and low in Fe and Al allow DOC to pass through the subsoil with little DOC sorption. DOC sorption is also likely to be strong in soils containing calcium carbonate. Moore (1998) also reports that the origin and chemistry of DOC affects DOC dynamics, although the DOC fraction which causes this variation has not been identified.

Physically Protected Organic Carbon Pools in Soil

Soil structure

Soil flora and fauna attack plant and other fresh organic materials to begin the process of decomposition. At this early stage, free OC has little association with the mineral phase and can be separated from the rest of the soil visually or on the basis of density as a light or POM fraction.

At a later stage, ultrasonic dispersion of the aggregates, based on density or other separation methods, is required to allow separation of occluded OC from the mineral fraction (Golchin et al., 1995). The occluded OC in the light (POM) fraction may contribute more to the water held by soil than does the free OC, because of occluded C's stronger association with mineral material. Also, as Kay (1998) discussed, the occluded fraction may have important effects on structural stability because fine roots and fungal hyphae are included in this fraction and these materials are particularly important in strengthening micro-aggregate binding.

The structure of soils can be considered in terms of its form, stability, resiliency, and vulnerability. Spatial distribution of failure zones of different strengths leads to the creation of aggregates (Kay, 1998). Failure zones are located between aggregates. They occur where bonding between elementary particles is weak and must therefore include pores as well as space containing less cementing material than the surrounding matrix. The cementing materials may include amorphous and crystalline inorganic materials as well as various organic materials. Because of the failure zones, aggregates result with characteristic size and pore space distributions between them.

Soil texture has a major influence on soil structure as well as its response to weather, biological factors, and management. Sands can be considered in the simplest case as single grains, and all characteristics of structure are determined by the distribution of grain sizes. Clay or silt sized materials then fill the interstices or exist as coatings between the sand grains. Organic materials, together with fine clays and amorphous and crystalline inorganic materials, provide the cement that exists between grains.

However, as clay content increases, characteristics of the soil matrix increasingly exhibit the characteristics of the clay, the nature and quantities of cementing materials, and the composition of the pore fluid (Kay, 1998). Roots, fungal hyphae, and polysaccharides associated intimately with the mineral fraction appear to fill a particularly important role in stabilizing soil aggregates. For example, Wright (1997) identified arbuscular mycorrhizal fungi as producing copious amounts of a highly stable glycoprotein named *glomalin* and observed a linear relationship between glomalin concentrations and percent aggregate stability within the 5 to 80% range. In the 80 to <97% stability range, a hyper-accumulation of glomalin (4 to 15 mg/g) appeared.

The soil aggregation process

Micro-aggregates

Adsorption of organic material onto the reactive surfaces of clay minerals in soils and other clay particle surfaces is an important mechanism for binding soil particles together. Among the organic materials are extracellular polysaccharides which soil microorganisms produce as a metabolic product while decomposing various organic substances.

Polysaccharide molecules can be adsorbed to clay surfaces strongly through polyvalent bridging and because of the net negative charge that results from the presence of glucuronic, galacturonic, mannuronic, pyruvic, or succinic acids and other acid groups (Chenu, 1995). Uncharged polysaccharides may strongly adsorb to mineral surfaces because of hydrogen bonding or van der Waals forces (Golchin et al., 1998). Extracellular polysaccharides are effective building agents because of their strong affinity for clay surfaces and the surfaces of other soil minerals and because of their ability to form intermolecular linkages.

In the aggregation model Golchin et al. (1998) proposed, adsorbed organic molecules such as microbially derived polysaccharides are important for binding the soil particles together into "packets" of clays and other small primary particles. These packets eventually result in small micro-aggregates. Some micro-aggregates have been shown resistant to the dispersive effect of ultrasonic energy; they remain as silt and clay sized aggregations of primary soil particles.

Because of the high amounts of roots and hyphae in restored prairie soils (Cook et al., 1988; Miller and Jastrow, 1992), formation of micro-aggregates that are stable enough to withstand slaking and wet sieving is rapid. At the same time, SOC continues to increase with time but at a much slower rate of accumulation than macro-aggregates (Jastrow, 1996). Jastrow et al. (1998) also observed that, at least in the tall grass prairie region and with parent materials with adequate amounts of clay minerals and polyvalent cations, a significant proportion of OM inputs rapidly become associated with mineral matter. This process promotes the formation of very stable micro-aggregates while also serving to protect the OM physically.

Macroaggregates

Angers and Chenu (1998) demonstrated that recently incorporated POM participates in aggregate initiation. The POM acts as a substrate for the fungi and bacteria which are vital to the aggregation process because of their associated hyphae and mucilages, which promote physical attachments of mineral and other particles.

In plant-residue, mediated aggregation, microorganisms are the active binding agents, through the physical enmeshment and adhesion of mineral particles to microorganism surfaces and through their associated by-products. Fresh plant residues also may exhibit some adhesive and binding properties, such as by the plant-root mucilages (Morel et al., 1991). Production of metabolic binding agents from microbial colonization of free POM, and metabolism of POM C and nutrients, strengthens interactions between POM and the surrounding minerals and increases the encrustation of the POM particles with silt and clay. The presence of fresh POM provides a readily available source of C that, in turn, allows increased

microbial and fungal activity, further encrustation by silt and clay, and the formation of the macro-aggregates.

Growth of roots and fungal hyphae facilitates the accrual of OM. Continued
decomposition and the production of various binding agents makes the POM a
center of intense biological activity around which mineral particles stabilize. In-
tramacro-aggregate POM also may be important for binding micro-aggregates
into macro-aggregates because it provides nucleating sites for growth of fungal
hyphae and for other microbial activities which cause deposits of extracellular
polysaccharides.

The size of the stabilized macro-aggregates is a function of the size, geometry,
and mode of decomposition of the POM (Golchin et al., 1998). The formation of
macro-aggregates stable enough to withstand slaking can be rapid.

Jastrow et al. (1998) used path analysis (Table 3.2) to consider the observed
correlations between the percentage of water-stable aggregates (>212 μm) and
partitioned measurements of various organic binding agents into direct, indirect,
and total causal effects. They considered the differences between the correlations
and the total effect to be due to conceptually noncausal relationships between
variables. Fine roots (0.2 to 1 mm dia.) probably exert considerable indirect effect
because of their strong influence on external hyphae and SMB. Very fine roots
(<0.2 mm dia.) contributed more to SOC than fine roots, but their overall indirect
contribution to aggregation was minimal. Jastrow et al.'s (1998) further path
analyses, beyond those that Table 3.2 shows for three size classes of macro-aggregates, support the hypothesis that the effectiveness of various binding mechanisms depends on the physical dimensions of the binding agents relative to the
spatial scale of the aggregate planes of weakness being bridged.

Macro-aggregate stabilization by POM is a brief process, and maintaining a
given level requires continued addition of POM to the system. The macroaggregates break into smaller pieces as POM decomposition continues and the organic
cores holding the macro-aggregates together break into smaller pieces. The fragmentation is likely to result from physical disruption of the soil, soil fauna, and
microbial action (Golchin et al., 1998).

The micro-aggregates which the fragmentation process releases are likely to
consist of small fragments of partially degraded plant debris, bound in a matrix of
mucilages and mineral matter. When the POM cores of micro-aggregates degrade
completely, cavities within the micro-aggregates remain, as Waters and Oades
(1991) showed, using electron microscopy. Such aggregates accumulate in the
>2.0 MT/m^3 density fraction (Golchin et al., 1998).

Table 3.2. Observed correlations between the percent water-stable aggregates (>212 μm) and measurements of selected organic binding agents, with the effects partitioned into direct, indirect, and total causal effects on the basis of path analysis (n = 49) (Jastrow and Miller, 1998).

Measured Parameter	Correlation (r)	Direct Effect	Indirect Effect	Total Effect
Fine root length	0.91	0.25	0.47	0.72
Very fine root length	0.85	0.26	−0.04	0.22
External hyphal length	0.89	0.38	0	0.38
Soil organic C (SOC)	0.43	0	0.09	0.09
Microbial biomass C	0.65	0.14	0.03	0.17
Hot-water sol. carbohydrate C	0.55	0.05	0	0.05

Stability of soil aggregates

Because of the physical protection soil structure affords, significant interactions exist between SOC dynamics and the formation, stabilization, and degradation of soil aggregates. The distinction between macro- and micro-aggregates is based on both size and susceptibility to slaking, i.e., rapid wetting. Macro-aggregates, subjected to slaking, break down into micro-aggregates rather than primary particles. Stabilized micro-aggregates are not disrupted by rapid wetting and mechanical disturbance, including cultivation (Tisdale and Oades, 1980, 1982).

The stability and nature of aggregates in a given soil depends on the relative amounts and strengths of various organo-mineral associations that function as binding and stabilizing agents. At the same time, the nature of these organo-mineral associations and their spatial locations within the aggregate hierarchy determine the degree to which SOC physically is protected. All pools of organic binding agents are subject to simultaneous effects of loss through decomposition and inputs from the creation of new materials. However, the long-term existence of persistent agents may be due more to protection from decomposition by their physicochemical associations with inorganic soil minerals than to any inherent biochemical inertness (Kay, 1990).

Conclusions

Very small changes in the amounts of SOC sequestered in grazing land soils become extremely important because of the vast land areas involved. The potential of grazing lands to help mitigate global climate change resulting from increasing atmospheric levels of CO_2 thus must be recognized (OSTP, 1997). For example, an increase (or loss) of only 1% of the SOC in the top 10 cm of soil is equivalent to the C emissions from all U.S. cropland agriculture.

Several pools that store C in soils, and their sizes, relate either directly or indirectly to plants, species of plants, and types of grazing land management. The sequestration of C as SOC begins with the entrance into the soil of plant-derived material, such as litter and plant root material, as variously measured particulate fractions. The quantity and properties of the plant materials entering the soil are important. Plant properties such as size (degree of fragmentation), plant type (C_3, C_4, or woody species), lignin content, and C:N ratio of the plant material are important, and these properties need better definition as they relate to C-sequestration.

Important soil properties include the effects of soil texture; Fe, Al, and $CaCO_3$ content; bulk density; SOM content; exchange capacity; and perhaps other properties. Environmental conditions including MAT and MAP also affect whether there is an enrichment of the SOC in the clay sized fraction versus other size fractions. The combined roles of soil fauna and soil microbial biomass, as they use plant-derived material, contribute both to degradation and to soil aggregation processes that lead to the physical protection of SOC.

An understanding of how to enhance the physical protection of SOC, including its relation to biological and chemical processes, through grazing land management must be considered one of the higher priority areas for research and possibly increased C sequestration in grazing land soils. Finally, we must carefully guard against abusive grazing or other practices that may decrease the efficiency of the biological and chemical processes that enhance C sequestration or that may increase soil erosion or other losses of physically protected SOC.

References

Amelung, W. 1997. *Zum Klimaeinflub auf die organische Substanz nordamerikanischer Prarieboden.* Bayruether Bodenkundlich Berichte 53. Lersstuhl fur Bodendunde und Bodengeographie. Univsitat Bayreuth.

Amelung, W., W. Zech, X. Zhang, R.F. Follett, H. Tiessen, E. Knox, and K.W. Flach. 1998. Carbon, nitrogen, and sulfur pools in particle sized fractions as influenced by climate. *Soil Sci. Soc. Am. J.* 62:172-181.

Angers, D.A., and C. Chenu. 1998. Dynamics of soil aggregation and C sequestration. *In* R. Lal, J.M. Kimble, R.F. Follett, and B.A. Stewart (eds), *Soil Processes and the Carbon Cycle*, CRC Press, Boca Raton, FL, pp. 199-206.

Arshad, M.A., M. Schnitzer, D.A. Angers, and J.A. Ripmeester. 1990. Effect of till vs no-till on the quality of soil organic matter. *Soil Biol. Biochem.* 22:595-599.

Burke, I.C., W.K. Lauenroth, and D.G. Milchunas. 1997. Biogeochemistry of managed grasslands in central North America. *In* E.A. Paul, K. Paustian, E.T. Elliott, and C.V. Cole (eds), *Soil Organic Matter in Temperate Agroecosystems: Long-Term Experiments in North America,* CRC Press, Boca Raton, FL, pp. 85-102.

Burke, I.C., C.M. Yonker, W.J. Pardon, C.V. Cole, K. Flach, and D.S. Schimel. 1989. Texture, climate, and cultivation effects on soil organic matter content in U.S. grassland soils. *Soil Sci. Soc. Am. J.* 53:800-805.

Cambardella, C.A. and E.T. Elliott. 1992. Particulate soil organic-matter changes across a grassland cultivation sequence. *Soil Sci. Soc. Am. J.* 52:777-783.

Campbell, C.A., E.A. Paul, D.A. Rennie, and K.J. MCCallum. 1967. Applicability of the carbon-dating method of analysis to soil humus studies. *Soil Sci.* 104:217-224.

Cook, B.D., J.D. Jastrow, and R.M. Miller. 1988. Root and mycorrhizal endophyte development in a chronosequence of restored tallgrass prairie. *New Phytol.* 110:355-362.

CAST. 1992. *Preparing U.S. Agriculture for Global Climate Change. Task Force Report No. 119.* Council for Agricultural Science and Technology. Washington, DC.

Chenu, C. 1995. Extracellular polysaccharides: An interface between microorganisms and soil constituents. *In* P.M. Huang, J. Berthelin, J.M. Bollag, W.B. McGill, and A.L. Page (eds), *Environmental Impact of Soil Component Interactions, Natural and Anthropogenic Organics*, CRC Press, Boca Raton, FL, pp. 217-233.

Christensen, B.T., 1992. Physical fractionation of soil and organic matter in primary particle size and density separates. *Adv. Soil Sci.* 20:1-90.

Dalal, R.C., and R.J. Mayer. 1986. Long-term trends in fertility of soils under continuous cultivation and cereal cropping in southern Queensland. III. Distribution and kinetics of soil organic carbon in particle size fractions. *Aust. J. Soil Res.* 24:393-300.

DOE/EIA. 1996. *Emission of greenhouse gases in the United States 1995.* Energy Information Administration, U.S. Dept. of Energy. Washington, DC.

Dormaar, J.F., A. Johnston, and S. Smoliak. 1977. Seasonal variation in chemical characteristics of soil organic matter of grazed and ungrazed mixed prairie and fescue grassland. *J. Range Manage.* 30:195-198.

Elliott, E.T., K. Paustian, and S.D. Frey. 1996. Modeling the measurable or measuring the modelable: a hierarchical approach to isolating meaningful soil organic matter fractionations. *In* D.S. Powlson, P. Smith, and J.U.Smith (eds), *Evaluation of Soil Organic Matter Models, NATO ASI Series, Vol. I 38*, Springer-Verlag. Berlin, pp. 161-179.

Follett, R.F. 1991. *Strategic Plan for Global Climate Change Research by the Agricultural Research Service: Biogeochemical Dynamics.* USDA. Washington, DC.

Follett, R.F. 1998. CRP and microbial biomass dynamics in temperate climates. *In* R.Lal, J.M. Kimble, R.F. Follett, and B.A. Stewart (eds), *Management of Carbon Sequestration in Soil*, CRC Press, Boca Raton, FL, pp. 305-322.

Golchin, A., J.A. Baldock, and J.M. Oades. 1998. A model linking organic matter decomposition, chemistry, and aggregate dynamics. *In* R. Lal, J.M. Kimble, R.F. Follett, and B.A. Stewart (eds), *Soil Processes and the Carbon Cycle*, CRC Press, Boca Raton, FL, pp. 245-266.

Golchin, A., P. Clarke, J.M. Oades, and J.O. Skjemstad. 1995. The effects of cultivation on the composition of organic matter and structural stability of soils. *Aust. J. Soil Res. J.* 33:975-933.

Hunt, H.W., D.C. Coleman, R.E. Ingham, E.T. Elliott, J.C. Moore, S.L. Rose, C.P. Reid, and C.K. Morely. 1987. The detrital food web in a shortgrass prairie. *Biol. Fert. Soils* 3:17-68.

Jastrow, J.D. 1996. Soil aggregate formation and the accrual of particulate and mineral-associated organic matter. *Soil Biol. Biochem.* 28:665-676.

Jastrow, J.D. and R.M. Miller. 1998. Soil aggregate stabilization and carbon sequestration: Feedbacks through organomineral associations. *In* R. Lal, J.M. Kimble, R.F. Follett, and B.A. Stewart (eds), *Soil Processes and the Carbon Cycle*, CRC Press, Boca Raton, FL, pp. 207-223.

Jastrow, J.D., R.M. Miller, and J. Lussenhop. 1998. Interactions of biological mechanisms contributing to soil aggregate stabilization in restored prairie. *Soil Biol. Biochem.* 30:905-916.

Jenkinson, D.S., and D.S. Powlson. 1976. The effects of biocidal treatments on metabolism in soil. V. A method for measuring soil biomass. *Biochem.* 8:209-213.

Kaiser, E.A., T. Mueller, R.G. Joergensen, H. Insam, and O. Heinemeyer. 1992. Evaluation of methods to estimate the soil microbial biomass and the relationship with sil texture and organic matter. *Soil Biol. Biochem.* 24:675-683.

Kay, B.D. 1990. Rates of change of soil structure under different cropping systems. *Adv. Soil Sci.* 12:1-52.

Kay, B.D. 1998. Soil structure and organic carbon: a review. *In* R. Lal, J.M. Kimble, R.F. Follett, and B.A. Stewart (eds), *Soil Processes and the Carbon Cycle*, CRC Press, Boca Raton, FL, pp. 169-197.

Lal, R., J.M. Kimble, R.F. Follett, and C.V. Cole. 1998. *The Potential of U.S. Cropland to Sequester Carbon and Mitigate the Greenhouse Effect.* Ann Arbor Press. Chelsea, MI. 128 pp.

Leavitt, S.W., R.F. Follett, and E.A. Paul. 1996. Estimation of slow- and fast-cycling soil organic carbon pools from 6N HCl hydrolysis. *Radiocarbon* 38:231-239.

Leinweber, P., and H.R. Schulten. 1998. Nonhydrolyzable organic nitrogen in soil size separates from long-term agricultural experiments. *Soil Sci. Soc. Am. J.* 62:383-393.

Linden, D.R., and C.E. Clapp. 1998. Effect of corn and soybean residues on earthworm cast carbon content and natural abundance isotope signature. *In* R. Lal, J.M. Kimble, R.F. Follett, and B.A. Stewart (eds), *Soil Processes and the Carbon Cycle*, CRC Press, Boca Raton, FL, pp. 345-351.

Miller, R.M., and J.D. Jastrow. 1992. Extraradical hyphal development of visicular-arbuscular mycorrhizal fungi in a chronosequence of prairie restroations. *In* D.J. Read, D.H. Lewis, A.H. Filter, and I.J. Alexander (eds), *Mycorrhizas in Ecosystems*, C.A.B. International, Wallingford, Oxon, U.K., pp. 171-176.

Moore, T.R. 1998. Dissolved organic carbon: sources sinks, and fluxes and role in the soil carbon cycle. *In* R. Lal, J.M. Kimble, R.F. Follett, and B.A. Stewart (eds), *Soil Processes and the Carbon Cycle*, CRC Press, Boca Raton, FL, pp. 282-292.

Morel, J.L., L. Habib, S. Planturaux, and A. Guckert. 1991. Influence of maize root muscilage on soil aggregate stability. *Plant Soil* 136:111-119.

Mosier, A.R., W.J. Parton, D.W. Valentine, D.S. Ojima, D.S. Schimel, and O. Heinemeyer. 1997. CH$_4$ and N$_2$O fluxes in the Colorado shortgrass steppe. 2. Long-term impact of land use change. *Global Biogeochem. Cycles* 11:29-42.

Naeth, M.A., A.W. Bailey, D.J. Pluth, D.S. Chanasyk, and R.T. Hardin. 1991. Grazing impacts on litter and soil organic matter in mixed prairie and fescue grassland ecosystems in Alberta. *J. Range Manage.* 44:7-12.

OSTP. 1997. *Climate Change: State of Knowledge.* Office of Science Technology Policy. Washington, DC.

Parton, W.J., D.S. Schimel, C.V. Cole, and D.S. Ojima. 1987. Analysis of factors controlling soil organic matter levels in Great Plains grasslands. *Soil Sci. Soc. Am. J.* 51:1173-1179.

Paul, E.A. 1984. Dynamics of organic matter in soils. *Plant Soil* 76:275-285.

Paul, E.A., and F.E. Clark. 1996. *Soil Microbiology and Biochemistry* (2nd ed). Academic Press. New York. 340 pp.

Paul, E.A., R.F. Follett, S.W. Leavitt, A. Halvorson, G.A. Peterson, and D.J. Lyon. 1997. Radiocarbon dating for determination of soil organic matter pool sizes and dynamics. *Soil Sci. Soc. Am. J.* 61:1058-1067.

Ping, C.L., G.J. Michaelson, X.Y. Dai, and R.J. Candler. 1999. Characterization of soil organic matter. *In*. J.M. Kimble, R. Lal, and R.F. Follett (eds), *Methods of Assessment of Soil Carbon*, CRC Press., Boca Raton, FL, (in press).

Six, J., E.T. Elliott, K. Paustian, and J.W. Doran. 1998. Aggregation and soil organic matter accumulation in cultivated and native grassland soils. *Soil Sci. Soc. Ame. J.* (in press).

Smith, J.L., and E.A. Paul. 1990. The significance of soil microbial biomass estimation. *In* J. Bollag and G. Stotzky (eds), *Soil Biochemistry* (Vol 6), Marcel Dekker, New York, pp. 357-396.

Stevenson, F.J. 1982. *Humus Chemistry: Genesis, Composition, Reactions.* Wiley-Interscience. New York.

Stevenson, F.J. 1986. *Cycles of Soil: Carbon, Nitrogen, Phosphorus, Sulfur, Micronutrients.* Wiley-Interscience. New York.

Stevenson, F.J., and E.T. Elliott. 1989. Methodologies for assessing the quantity and quality of soil organic matter. *In* D.C. Coleman, J.M. Oades, and G. Uehara (eds), *Dynamics of Soil Organic Matter in Tropical Ecosystems*, University of Hawaii Press, Honolulu, pp. 173-199.

Tiessen, H., and J. Stewart. 1983. Particle-size fractions and their us in studies of soil organic matter: II. Cultivation effects on organic matter composition in size fractions. *Soil Sci. Soc. Am. J.* 47:509-514.

Tisdale, J.M., and J.M. Oades. 1980. The effect of crop rotation in a red-brown earth. *Aust. J. Soil Res.* 18:423-433.

Tisdale, J.M., and J.M. Oades. 1982. Organic matter and water stable aggregates in soils. *J. Soil Sci.* 33:141-163

USEPA. 1999. *Inventory of U.S. greenhouse gas emissions and sinks: 1990-97.* USEPA. Washington, DC.

van Veen, J.A., and E.A. Paul. 1981. Organic dynamics in grassland soils. 1. Background information and computer simulation. *Can. J. Soil Sci.* 61:185-201.

Waters, A.G., and J.M. Oades. 1991. Organic matter in water stable aggregates. *In* W.S. Wilson (ed), *Advances in Soil Organic Matter Research, The Impact on Agriculture and Environment,* Roy. Soc. Chem., Cambridge, pp. 163-175.

Wright, S.F. 1997. Acid soils optimize aggregate stabilization by a glycoprotein from arbuscular mycorrhizal hyphae. *Agron. Abstr.* 89:189.

CHAPTER **4**

Inorganic Carbon Sequestration in Grazing Lands

H.C. Monger[1] and J.J. Martinez-Rios[2]

Introduction

Carbonate C as $CaCO_3$ (also termed soil inorganic C, or SIC) is a principal component of many arid and semiarid soils throughout the world (Dregne, 1976). In reference to soil profiles in southern New Mexico, for example, Ruhe (1967, p. 55) said that the subsoil horizon of calcium carbonate is so prominent that it is the first feature to catch an observer's eye. The amount of carbonate that forms in an arid or semiarid soil depends highly on soil age (Gile et al., 1966; Hawley et al., 1976). In young soils of Holocene age, only enough carbonate has formed to make filaments or coatings composed of silt and clay sized $CaCO_3$ crystals. With increasing age, the amount of carbonate crystals progressively accumulates, plugging soil pores, engulfing other soil particles, and eventually forming calcic and petrocalcic horizons (Soil Survey Staff, 1998).

In terms of C sequestration from the atmosphere, soil carbonate C is important because it is the third largest pool of C, containing approximately 750 to 950 Pg of C (Schlesinger, 1985; Eswaran et al., 1999). Only the oceanic (38,000 Pg C) and soil organic (1,550 Pg C) pools are larger (Schlesinger, 1997, p. 359).

Many uncertainties exist, however, about the role of soil carbonate C in the global C cycle. First, is the formation of soil $CaCO_3$ a sink for atmospheric CO_2 or does it only represent a lateral transfer of C from one location to another (Grossman et al., 1995)? Second, if soil $CaCO_3$ does sequester CO_2, would the rate of sequestration be rapid enough to play an important role in mitigating the greenhouse effect? Third, instead of being a CO_2 sink, is the carbonate C pool more im-

[1] Assoc. Prof., Agronomy and Horticulture, New Mexico State Univ., Las Cruces, NM 88003; e-mail: cmonger@nmsu.edu.

[2] Asst. Prof., Universidad Juarez del Estado de Durango; Av. Fco. Sarabia No. 998 Sur, Col. 20 de Noviembre, Cd. Lerdo, Durango CP 35155, Mexico; e-mail: martinez@taipan.nmsu.edu.

portant as a CO_2 source when exhumed calcic and petrocalcic horizons are exposed to the dissolving effects of acidic rain and microbiotic crusts?

Although grazing occurs on various patches of land across the U.S., this chapter focuses on carbonate C in grazing lands of the drylands regions of the West and Midwest — that is, soils with aridic, ustic, and xeric moisture regimes (Soil Survey Staff, 1998). In this sense, our use of the term *grazing lands* is synonymous with *rangelands* (Holechek et al., 1989), except we exclude forests of the West that are used for grazing and browsing. We assess carbonate C in these western grazing lands in three categories: woodlands, shrublands, and grasslands. Within this context, this chapter:

1. reviews processes of soil $CaCO_3$ formation
2. estimates the amount of soil carbonate C in the woodlands, shrublands, and grasslands of the U.S.
3. estimates the annual sequestration rates of carbonate C
4. discusses the potential for promoting C sequestration as soil carbonate C in grazing lands.

Types of Inorganic Carbon

Inorganic C refers to the chemical species involved in the carbonic acid system (Morse and Mackenzie, 1990). These species include gaseous CO_2 ($CO_{2(g)}$), dissolved CO_2 ($CO_{2(aq)}$), carbonic acid ($H_2CO_{3(aq)}$), bicarbonate ($HCO_{3\ (aq)}^-$), and the carbonate ion ($CO_3^{2-}{}_{(aq)}$). We can represent the relationships among these chemical species with these reactions:

$$CO_{2(g)} \rightleftharpoons CO_{2(aq)} \qquad\qquad \text{(Reaction 1)}$$

$$CO_{2(aq)} + H_2O_{(l)} \rightleftharpoons H_2CO_{3(aq)} \qquad\qquad \text{(Reaction 2)}$$

$$H_2CO_{3(aq)} \rightleftharpoons HCO_{3\ (aq)}^- + H^+{}_{(aq)} \qquad\qquad \text{(Reaction 3)}$$

$$HCO_{3\ (aq)}^- \rightleftharpoons CO_3^{2-}{}_{(aq)} + H^+{}_{(aq)} \qquad\qquad \text{(Reaction 4)}$$

In the soil solution, as with other natural waters, we can illustrate the interaction of atmospheric CO_2 with the carbonic acid system by combining the above reactions into this one:

$$CO_{2(g)} + H_2O_{(l)} \rightleftharpoons HCO_{3\ (aq)}^- + H^+{}_{(aq)} \rightleftharpoons CO_3^{2-}{}_{(aq)} + H^+{}_{(aq)} \qquad \text{(Reaction 5)}$$

Increasing $CO_{2(g)}$ drives the reaction to the right. In contrast, the addition of acid ($H^+{}_{(aq)}$) drives the reaction to the left.

Another important part of the inorganic C system is its mineral component. In soil, the overwhelming carbonate mineral is calcium carbonate ($CaCO_3$), occurring mainly as calcite (Doner and Lynn, 1989). In fact, in terms of C sequestration, the phrase *soil inorganic carbon* progressively has become synonymous with *soil carbonate* (e.g., Lal et al., 1999).

We can represent the interaction of soil carbonate minerals with the carbonic acid system in Reaction 5 in two ways. First, Reaction 6 represents the dissolution of preexisting soil carbonates by carbonic acid and its reprecipitation (Krauskopf, 1967):

$$CO_2 + H_2O$$

$$\updownarrow$$

$$CaCO_3 + H_2CO_3 \rightleftharpoons Ca^{2+} + 2HCO_3^- \qquad \text{(Reaction 6)}$$

Second, Reaction 7 represents the precipitation of soil carbonate in a system supplied with Ca released by chemical weathering of non-carbonate minerals, such as Ca feldspars, and the production of bicarbonate from root and microbial respiration (Schlesinger, 1997, p. 115):

$$Ca^{2+} + 2HCO_3^- \rightleftharpoons CaCO_3 \downarrow + H_2O + CO_2 \qquad \text{(Reaction 7)}$$

In many arid and semiarid regions, large amounts of pedogenic carbonate have been produced in soils formed in parent materials without preexisting $CaCO_3$. For these dryland soils, we can write a precipitation-dissolution reaction by combining the reaction representing the dissolution of carbonate (Reaction 6) with the reaction representing the precipitation of pedogenic carbonate (Reaction 7):

$$CO_2 + H_2O$$

$$\text{pedogenic} \quad \updownarrow$$

$$Ca^{2+} + 2HCO_3^- \rightleftharpoons CaCO_3 + H_2CO_3 \qquad \text{(Reaction 8)}$$

For example, supplying ample Ca^{2+} from weathering, from atmospheric additions, or from plant litter decomposition drives the reaction to the right, provided enough HCO_3^- is available. Likewise, providing ample HCO_3^- from respiration under suitable pH conditions drives the reaction to the right as long as enough Ca^{2+} is available. Conversely, depleting Ca^{2+} by leaching or plant uptake, or depleting HCO_3^- by leaching or acidification, drives the reaction to the left, causing $CaCO_3$ to dissolve. An increase in the amount of CO_2 and H_2O also drives the reaction to the left, also causing $CaCO_3$ to dissolve.

Much of what we know about carbonate equilibria as expressed in the above reactions has been developed for marine systems (e.g., Bathurst, 1975; Morse and Mackenzie, 1990). Nevertheless, these principles appear valid for the soil biogeochemical system even though it differs from the marine system in several important ways. Most notably, unlike the marine system, the soil system is alternately wet and dry. Moreover, it is a porous system, consisting largely of silicate minerals, organic particles, gases and solutions, and a vast variety of organisms, including roots, arthropods, nematodes, fungi, and bacteria. In this environment, CO_2 can reach concentrations hundreds of times greater than atmospheric CO_2.

In this environment, carbonate precipitates in the ephemeral aqueous zones of the soil, which are areas where the soil solution resides in pores and surrounds particles. Carbonate also precipitates in the aqueous environments associated with root and microbial tissue (Monger et al., 1991). Carbonate, because of its low solubility of 0.0014 g/100 mL (Weast, 1986), is one of the first minerals to precipitate when the soil dries, which consumes Ca^{2+} and HCO_3^- from the soil solution. Again, due to its low solubility in alkaline conditions, carbonate does not readily dissolve upon subsequent wettings but accumulates with time, progressively plugging the soil porosity and, eventually, pushing apart framework grains (Gile et al., 1966; Machette, 1985).

Soil carbon dioxide

The biological processes of root and microbial respiration produce soil CO_2. In addition, some atmospheric CO_2 may enter the upper soil profile if soil respiration rates are low (Cerling and Quade, 1993). The link between atmospheric CO_2 and soil CO_2 exists because plant C, which photosynthesis previously fixed from the atmosphere, generates respired soil CO_2.

Several methods, of which the most common are adsorption by alkali and soda lime, gas chromatography, and infrared, can measure soil CO_2 (Anderson, 1982; Cropper et al., 1985). These techniques reveal that soil CO_2 is seasonally dynamic. It generally reaches highest concentrations in the rooting zone during warm growing seasons. Afterwards, pulses of CO_2 propagate downward through the soil profile (Reardon et al., 1979).

In some cases, concentrations of soil CO_2 can reach levels of 8% in cropland soils (Buyanovsky and Wagner, 1983), which is over 200 times greater that the atmosphere concentration of about 0.036%. Lower concentrations are reported for desert soils. In stony desert soils of Nevada, for example, seasonal levels of soil CO_2 range from 0.03% (near atmospheric levels) in January to 0.25% in April (Terhune and Harden, 1991).

Bicarbonate

Bicarbonate (HCO_3^-) is produced when CO_2 dissolves in water, making carbonic acid, which subsequently dissociates into H^+ and HCO_3^- (Reactions 1, 2, and 3). Bicarbonate is one of the major anions in soil (Bohn et al., 1985) as well as one of the dominant anions in rivers (Schlesinger, 1997, p. 94). It is more common than CO_3^{2-} in natural waters except in solutions above pH 10.3 (Morse and Mackenzie, 1990, p. 8).

Measurements of HCO_3^- most commonly are based on its amount in saturated soil extracts determined by titration to a pH end point with a chemical indicator (Soil Survey Staff, 1996). Because this process destroys the three-dimensional architecture of the soil-root-microbial infrastructure, and because the amount of HCO_3^- in the saturation paste extract depends highly on the water content used to make the extract, it seems likely that these HCO_3^- data are only partially useful for understanding its exact role in $CaCO_3$ precipitation and C sequestration.

Origin and classification of soil carbonate ($CaCO_3$)

The pedologic literature has discussed the origin of soil carbonate for several decades (e.g., Dokuchaev, 1883). Depending on local conditions, soil carbonate can originate from one or more of these sources: (1) parent material, (2) atmospheric additions, (3) biogenic precipitation, and (4) groundwater.

In much of the midwestern U.S., for example, researchers largely have attributed the origin of soil carbonate to its loess and glacial till parent materials (Aandahl, 1982), and they commonly have viewed carbonate concentration in the subsoil of these parent materials as the result of its dissolution in the upper profile and its reprecipitation after downward leaching (Jenny and Leonard, 1934; Arkley, 1963). In many limestone terranes, the *in situ* dissolution and reprecipitation of limestone bedrock is another example of carbonate derived from parent material (Rabenhorst and Wilding, 1986).

However, research on soil carbonates that formed in non-calcareous parent materials made clearer the importance of an external source of carbonate (Bretz and Horberg, 1949; Gile et al., 1965, 1966). In this case, Ca in rain and calcareous dust were important sources of carbonate. Russian scientists, who observed that decomposed plant rootlets leave behind carbonate minerals, also offered biogenic precipitation of carbonate as a source of soil carbonate (Lobova, 1967). Moreover, lab studies have shown carbonate precipitation by soil microorganisms (Monger et al., 1991). Carbonate from calcareous groundwater, in contrast to other sources of soil carbonate, occurs in localized areas where water tables are shallow (Dregne, 1976).

Researchers have measured the amount of soil carbonate mainly by applying acid to soil and determining the amount of evolved CO_2 or the sample weight loss or measuring a titration end point (Soil Survey Staff, 1996). These procedures give the total amount of carbonate in soil, which would include limestone detritus plus secondary carbonate, if both were present.

For the purpose of understanding the role of soil carbonate in sequestering atmospheric CO_2, soil carbonate can be classified into three pools: *primary*, *pedo-lithogenic*, and *pedo-atmogenic* carbonate (Fig. 4.1). The *primary soil carbonate*, as defined here, is the mineral fraction composed of marine or lacustrine carbonate detritus, such as limestone and dolostone particles. It is *allogenic* and includes $CaCO_3$ dust. *Lithogenic carbonate*, as in Rabenhorst et al. (1984), West et al. (1988), and Nordt et al. (1998), is synonymous with our *primary carbonate*.

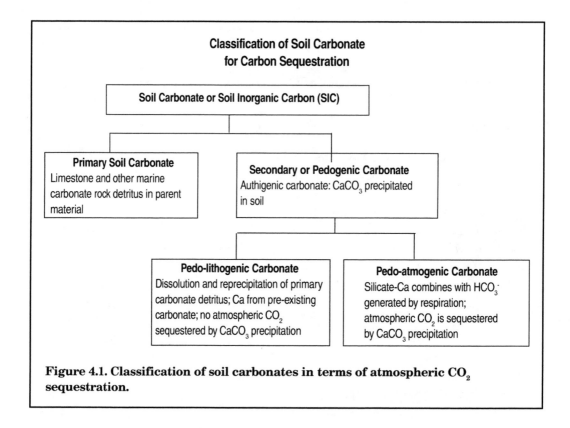

Figure 4.1. Classification of soil carbonates in terms of atmospheric CO$_2$ sequestration.

In contrast to primary carbonate, *secondary* or *pedogenic carbonate* is formed *in situ* and is *authigenic*. Pedogenic carbonate, in turn, can be subdivided into two groups: *pedo-lithogenic* and *pedo-atmogenic* (Fig. 4.1). Pedo-lithogenic carbonate is the result of dissolution and reprecipitation of primary carbonate. Pedo-lithogenic soil carbonate includes not only marine carbonate detritus that dissolved and reprecipitated, but also CaCO$_3$ dust that dissolved and reprecipitated, as well as carbonate formed from Ca in rain derived from CaCO$_3$ dust. Monger and Gallegos (1999) formerly termed pedo-lithogenic carbonate *lithogenic carbonate*.

With respect to C sequestration, pedo-lithogenic carbonate would not sequester additional atmospheric CO$_2$, given the premise that the amount of CO$_2$ consumed by dissolution of primary carbonate would be released by its reprecipitation as pedo-lithogenic carbonate (Reaction 10). This concept, however, depends on scale. Over long geologic time periods (millions of years), the C in marine carbonates also can be attributed to an atmospheric origin (Berner, 1999).

If the source of Ca during the precipitation of soil carbonate is non-carbonate minerals (e.g., Ca silicates), a second type of pedogenic carbonate forms. This form has been termed *atmogenic soil carbonate* (Monger and Gallegos, 1999), reflecting the sequestration of atmospheric CO$_2$. In this chapter, we have updated the term to *pedo-atmogenic carbonate* in order to emphasize its pedogenic origin and relation to pedo-lithogenic carbonate (Fig. 4.1). In addition to Ca from Ca silicate weathering in the soil profile, pedo-atmogenic carbonate forms when the source of Ca is gypsum or when rain or groundwater provides Ca from non-carbonate sources.

Sequestration of Atmospheric Carbon Dioxide by Soil Carbonates

Sequestration of atmospheric CO_2 depends on the timescale of the C cycle and parent material. The short-term timescale occurs over decades to centuries and involves the exchange of C in the surficial system, consisting of the atmosphere, biota, soils, and ocean. The long-term timescale operates over millions of years and involves the slow exchange of C between rocks and the surficial system (Berner, 1999). Parent material is important because of its role in supplying Ca. The two dominant parent materials for grazing land soils of the western U.S. are sediments from marine carbonates and igneous rock (cf., Hunt, 1986; Raisz, 1995).

Carbonate weathering and the consumption or release of atmospheric carbon dioxide

The following cases can illustrate whether or not the weathering of $CaCO_3$, either primary or pedogenic, consumes or releases atmospheric CO_2:

Case #1: Transport of bicarbonate from carbonate weathering via river water to the oceanic reservoir

$$CO_2 + H_2O$$
$$\downarrow$$

$$CaCO_3 + H_2CO_3 \rightarrow Ca^{2+} + 2HCO_3^- \qquad \text{(Reaction 9)}$$

$$\downarrow \text{ leached from soil and transported to ocean}$$

$$\text{oceanic } Ca^{2+} + 2HCO_3^- \rightarrow CaCO_3 + H_2O + CO_2 \uparrow \text{ release to atm.}$$

Adams (1993) argued that weathering of caliche ($CaCO_3$) would take up CO_2 from the atmosphere and contribute the resulting hydrogen carbonate (bicarbonate), via river water, to the oceanic reservoir. But as Case #1 (Reaction 9) illustrates, this process only temporarily consumes CO_2, because calcium carbonate precipitation in the oceans liberates CO_2 back into the atmosphere (Berner and Lasaga, 1989).

Also, during the journey of bicarbonate to the ocean, any drop in pH below about 6 would cause bicarbonate to release CO_2 via Reaction 5 (Morse and Mackenzie, 1990, p. 8). Moreover, this case would be most common to the weathering of limestone and soil carbonates in regions with enough rainfall to leach Ca and bicarbonate from soil profiles into phreatic water, but it would be of limited importance in arid regions. In terms of C sequestration, this long-term process operates at a timescale greater than the scale important for management designed to offset anthropogenic CO_2 emissions.

Case #2: Dissolution and reprecipitation of carbonate in aridland soil profiles

$$CO_2 + H_2O$$
$$\downarrow$$

$$CaCO_3 + H_2CO_3 \;\rightarrow\; Ca^{2+} + 2HCO_3^- \qquad\qquad\qquad \text{(Reaction 10)}$$

$$\downarrow \text{ illuviated} \downarrow \text{ illuviated}$$

$$Ca^{2+} + 2HCO_3^- \rightarrow CaCO_3 + H_2O + CO_2 \uparrow \text{ to soil atm. or atm.}$$

In Case #2, carbonate weathers, dissolves, and is illuviated from the upper profile to the subsoil. Yet, rainfall is insufficient to flush Ca and bicarbonate from the soil profile, so they reprecipitate. Note that, in this reaction (10), one of the two C moles in $2HCO_3^-$ is from atmospheric CO_2 and the other is from $CaCO_3$ (Krauskopf, 1967). Therefore, the CO_2 used to form carbonic acid is released with the precipitation of $CaCO_3$.

If the released CO_2 goes back into the atmosphere, no C has been sequestered. If the released CO_2 remains in the soil, it could be recycled to produce more carbonic acid and be re-released upon the reprecipitation of $CaCO_3$. Therefore, as in Case #1, the weathering of existing carbonate and its reprecipitation in soil is both a sink and source of C, and therefore neutral in terms of CO_2 sequestration. Unlike the first case, however, Case #2 would occur at the short-term C cycle scale.

Case #3: Congruent dissolution of carbonate

$$CO_2 + H_2O$$
$$\downarrow$$

$$CaCO_3 + H_2CO_3 \;\rightarrow\; Ca^{2+} + 2HCO_3^- \qquad\qquad\qquad \text{(Reaction 11)}$$

$$\downarrow \text{ acidic conditions (added H}^+\text{)}$$

$$2HCO_3^- + 2H^+ \rightarrow 2H_2O + 2CO_2 \uparrow \text{ to soil atm. or atm}$$

In Case #3, sufficiently low pH causes the complete (or congruent) dissolution of carbonate. As a result, the two moles of C in $2HCO_3^-$ of Reaction 11 are released as two moles of CO_2, following the path Reaction 5 illustrates. This is the reaction that would occur in acid soils that have received lime, and it is the reaction that would occur, to some extent, in the rhizosphere microenvironments in arid soils. In terms of greenhouse gases, this reaction would occur when exhumed soil carbonate is exposed to acids in rain and from microbiotic crust. In this case, weathering of $CaCO_3$ would be a source of CO_2 and may be important to managing grazing lands in the U.S. and elsewhere.

Consumption of atmospheric carbon dioxide by calcium silicate weathering

In contrast to carbonate weathering, the weathering of Ca silicate minerals is an unidirectional sink for atmospheric CO_2. Berner (1993) expressed the consumption of atmospheric CO_2, in general terms, by the weathering of Ca Mg silicates using these reactions:

$$CO_2 + CaSiO_3 \rightarrow CaCO_3 + SiO_2 \qquad \text{(Reaction 12)}$$
$$CO_2 + MgSiO_3 \rightarrow MgCO_3 + SiO_2 \qquad \text{(Reaction 13)}$$

A more detailed form of Reaction 12 can be written as (Berner, 1992):

Weathering

$$2CO_2 + 3H_2O + CaAl_2Si_2O_8 \rightarrow Al_2Si_2O_5(OH)_4 + Ca^{2+} + 2HCO_3^- \quad \text{(Reaction 14)}$$

Carbonate precipitation in oceans

$$2HCO_3^- + Ca^{2+} \rightarrow CaCO_3 + CO_2 + H_2O \qquad \text{(Reaction 15)}$$

Overall reaction

$$CO_2 + 2H_2O + CaAl_2Si_2O_8 \rightarrow Al_2Si_2O_5(OH)_4 + CaCO_3 \qquad \text{(Reaction 16)}$$

These same reactions of global weathering and marine carbonate precipitation also apply to the weathering of Ca silicates in arid and semiarid soils. In arid and semiarid soils, because of limited leaching, the Ca is not removed from the profile. Instead, Ca combines with bicarbonate generated by root and microbial respiration in the soil profile.

This reaction may be represented best in a general stoichiometric form, where various clay minerals and partially weathered Ca silicates are produced:

$$2CO_2 + 3H_2O \qquad \text{(Reaction 17)}$$
$$\downarrow$$

Ca silicates + $2H_2CO_3 + H_2O \rightarrow Ca^{2+} + 2HCO_3^-$ + clay minerals + partially weathered Ca silicates

$$\downarrow \text{ in soil} \downarrow \text{ in soil}$$
$$Ca^{2+} + 2HCO_3^- \rightarrow CaCO_3 \downarrow + H_2O + CO_2 \uparrow \text{to atm.}$$

Roots and microbes play a major role in this process for two reasons. First, they elevate the concentration of CO_2 by respiration. Second, they provide an aqueous environment of rhizosphere films and mucilaginous sheaths where carbonate precipitation can occur (e.g., Monger et al., 1991; Monger and Gallegos, 1999).

Model of carbon sequestration processes within limestone and igneous soils

Figure 4.2 illustrates potential sources of atmospheric additions that lead to soil carbonate formation and processes occurring internally in soil among the three carbonate pools — pedo-atmogenic, pedo-lithogenic, and primary carbonates. The block diagrams of limestone and igneous soils are hypothetical and referred to as soils at point B (Fig. 4.2). They illustrate the relationships between the carbonate pools, the atmosphere, parent material, and rainfall. In theory, the pedo-atmogenic carbonate pool would be small in the limestone soil but large in the igneous soil.

Limestone is used in this theoretical treatment, but other marine carbonate rocks also would pertain, such as dolostone, marl, and calcareous sandstones. Similarly, igneous rock is used in this model, but other rocks composed of Ca silicates would pertain equally, such as many metamorphic rocks, arkosic sandstones, and shales, as well as gypsiferous parent material containing Ca from igneous sources.

Limestone soil

The limestone soil, like all soils with vascular plants, contains CO_2 brought into the soils via a two-step process (see Fig. 4.2 and read it, beginning on the left side of the limestone soil diagram, left to right). First, photosynthesis removes CO_2 from the atmosphere to make plant biomass. Second, C in biomass is released into soil by respiration, either directly as root respiration or subsequently as microbial decomposition of plant material.

A small portion of soil CO_2 combines with water and dissociates to make HCO_3^-. This amount may be only a very small fraction of the total amount of CO_2 that fluxes through soil (Chadwick et al., 1994). If HCO_3^- encounters Ca^{2+} from a silicate source, as from an atmospheric addition of silicate Ca^{2+}, it precipitates as $CaCO_3$ and contributes to the pedo-atmogenic carbonate pool. On the other hand, if HCO_3^- encounters Ca^{2+} released by the dissolution of primary carbonate from limestone detritus or limestone dust, it precipitates as pedo-lithogenic carbonate and releases CO_2.

The third carbonate pool, the primary carbonate pool, is composed of limestone detritus and limestone dust (Fig. 4.2). The C in this pool was atmospheric during

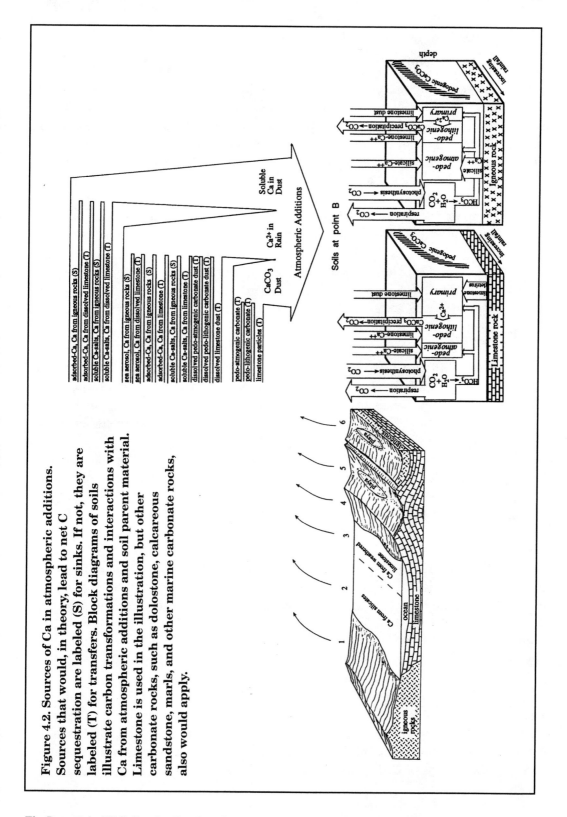

Figure 4.2. Sources of Ca in atmospheric additions. Sources that would, in theory, lead to net C sequestration are labeled (S) for sinks. If not, they are labeled (T) for transfers. Block diagrams of soils illustrate carbon transformations and interactions with Ca from atmospheric additions and soil parent material. Limestone is used in the illustration, but other carbonate rocks, such as dolostone, calcareous sandstone, marls, and other marine carbonate rocks, also would apply.

some previous geologic era (Berner, 1993), with respect to mitigating the greenhouse gases, it is considered here as a C reservoir.

The right side of the diagram illustrates the climatic effect of rainfall. As rainfall increases, the amount of pedogenic $CaCO_3$ (i.e., the combined pedo-atmogenic and pedo-lithogenic pools) decreases and becomes deeper in the profile (Jenny and Leonard, 1934). When pedogenic carbonate ceases to exist as the result of humid climates, the soils have changed from Pedocals to Pedalfers (Marbut, 1928). In humid regions, the effect of rainfall is so prevalent that primary carbonates are flushed from soils overlying limestone, leaving only insoluble residues dominated by quartz and clay minerals (Monger and Kelly, in press).

With respect to C sequestration, the pedo-atmogenic carbonate pool is the C sink, because one of the two moles of CO_2 required to weather Ca silicates is sequestered by $CaCO_3$ (Reaction 17). In contrast, the pedo-lithogenic carbonate pool does not sequester C during its formation. This is because one mole of CO_2 used to dissolve primary $CaCO_3$ is released during the precipitation of pedo-lithogenic $CaCO_3$ (Reaction 10).

Igneous soil

Soils formed in igneous parent material also could contain all three pools of soil carbonate (Fig. 4.2). In theory, the pedo-atmogenic pool would be the largest if chemical weathering released Ca^{2+} that precipitated with HCO_3^- generated by respiration. Any silicate Ca^{2+} supplied via atmospheric additions also would contribute to the pedo-atmogenic pool.

The primary and pedo-lithogenic carbonate pools in igneous soils could result from atmospheric additions. The primary carbonate pool would be composed of limestone dust or other airborne $CaCO_3$ particles. If such dust dissolved and reprecipitated, it would become pedo-lithogenic carbonate and release CO_2. Pedolithogenic carbonate also would form from any Ca^{2+}, dissolved in rain, that originated as limestone Ca^{2+}.

The effect of rainfall on pedogenic carbonate in igneous soils is similar to its effect in limestone soils. As rainfall increases, pedogenic carbonate deepens and diminishes. Unlike limestone soils, however, igneous soils lose pedogenic carbonate more readily in humid environments. This is apparent when tracing soils upslope from arid lowlands into forested udic moisture regimes of the mountainous Southwest. In these situations, limestone soils have calcic horizons extending farther into progressively wetter climates than do neighboring igneous soils.

In terms of C sequestration, the igneous soil Figure 4.2 illustrates contains the greater potential for sequestering atmospheric CO_2, provided chemical weathering makes a supply of Ca^{2+} available. To this end, the "mining" of Ca^{2+} from silicate minerals by plant roots and associated mycorrhizae may greatly hasten and increase the amount of Ca^{2+} available for $CaCO_3$ precipitation. In comparison to the size of the pedo-atmogenic pool, the size of pedo-lithogenic and primary pools theoretically would be small in igneous soils. An exception, however, is where igneous soils received abundant amounts of calcareous dust.

Atmospheric additions

Atmospheric additions can be a major source of Ca in pedogenic carbonate (Gile et al., 1981; Reheis et al., 1995). Understanding the source of Ca in atmospheric additions is important for knowing whether or not atmospheric CO_2 is sequestered as $CaCO_3$. As Figure 4.2 shows, "atmospheric additions" is a broad category composed of multiple subcategories. The three major subcategories are $CaCO_3$ dust, soluble Ca in dust, and Ca^{2+} in rain (Gile and Grossman, 1979).

For purposes of this hypothetical discussion, the same two soils at point B, representing the two major parent materials, are considered the recipients of atmospheric additions. Atmospheric additions are labeled transfers (T) if no C sequestration is expected, or sinks (S) if C sequestration is expected.

All of these cases depend on the timescale. We can differentiate transfers and sinks on the short-term C cycle timescale. On the long-term C cycle timescale, the C in limestone was originally a sink for atmospheric CO_2.

Calcium carbonate dust

In theory, $CaCO_3$ dust has at least three sources. The first source is limestone particles derived from limestone terrane. As Figure 4.2 illustrates, their transport by wind originated from source 4. If the limestone particles landed on the soil surfaces, were translocated in suspension, and accumulated in the soils, no net C sequestration would occur. Therefore, the soil carbonate is a transfer (Grossman et al., 1995). Similarly, if the limestone dust, upon landing on the soils at point B, dissolved and reprecipitated, as Reaction 10 illustrates, no sequestration would occur because reprecipitation would release the CO_2 dissolution consumed (Reaction 10).

The second source of $CaCO_3$ dust is pedo-lithogenic carbonate. The same reasoning also applies to pedo-lithogenic carbonate dust blown from soils at source 4 (Fig. 4.2). The third source of $CaCO_3$ dust is pedo-atmogenic carbonate. Pedo-atmogenic carbonate, although having sequestered C initially during its formation at source 1, for example (Fig. 4.2), also would represent a transfer.

Soluble calcium in dust

A second major category — soluble Ca in dust — is that fraction of Ca resolved from dust samples contained in water extracts (Gile and Grossman, 1979). The origin of this Ca could be (1) soluble Ca minerals or (2) water soluble Ca adsorbed on the cation exchange complex of the dust particles.

If the Ca in the Ca minerals came from limestone weathering and, for example, moved downslope to a playa as at source 5 (Fig. 4.2), then Reaction 10 is most applicable. However, the first part of this reaction would occur in the weathering of limestone at source 5, while the second part of Reaction 10 would occur in the soil at point B. Consequently, the CO_2 consumed during the weathering of limestone at source 5 would be released during the formation of $CaCO_3$ in the soils at point B and would be a transfer.

If, however, the Ca in soluble Ca minerals came from igneous weathering, as at source 6 (Fig. 4.2), the situation would be different. In this case the Ca, after travelling on windblown dust to the soils at point B, would combine with bicarbonate generated by respiration, and C sequestration would occur. Reaction 17 represents this case, where the formation of $CaCO_3$ sequesters one of the two moles of C required to dissolve Ca silicates.

Ca adsorbed on dust particles follows the same sequences. If the adsorbed Ca comes from limestone weathering, as at source 5 (Fig. 4.2), then Reaction 10 is most applicable, and no net C sequestration occurs from any subsequent $CaCO_3$ precipitation in soils at point B. In contrast, if adsorbed Ca is from the weathering of igneous rocks, as at source 6 (Fig. 4.2), C sequestration occurs with subsequent $CaCO_3$ precipitation.

Calcium in rain

A third, and probably the most important, source of atmospheric Ca is Ca dissolved in rain, represented in its cationic form — Ca^{2+} — in this discussion. For example, dust trap and rainwater analysis in southern New Mexico strongly suggests that Ca^{2+} in rain could produce roughly two to three times more soil $CaCO_3$ than would result from the combined amounts of $CaCO_3$ dust and soluble Ca in dust (Gile et al., 1981, p. 63).

The source of Ca^{2+} in rain largely comes from droplets containing dissolved Ca dust. Therefore, the potential sources of Ca^{2+} in rain are the same as those for dust and soluble Ca in dust. Whether or not the precipitation of $CaCO_3$ in soils at point B sequesters C depends on those Ca sources:

1. Ca^{2+} from dissolved limestone dust is a transfer.
2. Ca^{2+} from dissolved pedo-lithogenic soil carbonate is a transfer.
3. Ca^{2+} from dissolved pedo-atmogenic soil carbonate also is a transfer, but its initial formation was a sink.
4. Ca^{2+} from soluble Ca minerals derived from limestone is a transfer.
5. Ca^{2+} from soluble Ca minerals derived from igneous rocks is a sink.
6. Ca^{2+} from adsorbed Ca derived from limestone is a transfer.
7. Ca^{2+} from adsorbed Ca derived from igneous rocks is a sink.

In addition to these Ca^{2+} sources in rain, two other sources of Ca^{2+} are possible, and both are sea aerosols from ocean waters (Bohn et al., 1985). First, if the Ca^{2+}

in aerosols comes from the dissolution of terrestrial limestone, as at source 3 (Fig. 4.2), then the precipitation of $CaCO_3$ in soils at point B is a transfer. In this case, Reaction 9 is most applicable. Instead of $CaCO_3$ precipitating in a marine environment, however, $CaCO_3$ precipitates in soils at point B. In the second case, Ca^{2+} comes from the weathering of igneous rocks, as at source 2 (Fig. 4.2). In this situation, Reaction 17 is most applicable. Of the two moles of CO_2 used to weather Ca silicates, one is sequestered by $CaCO_3$ precipitation in soils at point B and therefore is a sink.

Differences between inorganic and organic carbon sequestration

C sequestered as carbonate C differs from C sequestered as OC, in at least three ways. SOC exists to some extent in most soils, but pedogenic $CaCO_3$ is limited to dryland soils. As annual rainfall climbs above 50 to 70+ cm per year, soil $CaCO_3$ progressively ceases to exist (Jenny and Leonard, 1934; Arkley, 1963). This boundary is the Pedocal-Pedalfer boundary (Marbut, 1928).

Figure 4.3 illustrates the processes occurring across this boundary. In the three ecosystems it shows, photosynthesis and subsequent root and microbial respiration bring atmospheric CO_2 into the soil. Pools of C storage include the aboveground C pool, root C pool, soil organic C (SOC) pool, and $CaCO_3$ pool. As rainfall increases and climate changes from dry to intermediately dry to humid, the relative sizes of the pools change. For example, the aboveground C pool is smallest in the desert scrub system and greatest in the forest system (e.g., Dick-Peddie, 1993). In contrast, the $CaCO_3$ pool is greatest in desert scrub systems (Schlesinger, 1982) and nonexistent in humid forest systems.

A second difference between organic and carbonate C sequestration is that $CaCO_3$ formation requires additional steps. Photosynthesis converts the C in organic matter (OM) from gaseous CO_2 to solid phase C, but the C in $CaCO_3$ passes from the solid phase OM to the gaseous CO_2 phase by respiration, then to the aqueous HCO_3^- phase by dissolution and dissociation, and then to the solid $CaCO_3$ phase by crystallization. This requires a dry climate, because Ca^{2+} and HCO_3^- progressively leach out with increasing rainfall (Fig. 4.3). It also requires suitable pH and wet-dry cycles for carbonate precipitation. Consequently, C sequestered by $CaCO_3$ captures only a small portion of photosynthetic C that passes through the soil (Chadwick et al., 1994).

A third difference is the temporal scales of the two processes. Plant photosynthesis brings OC into the soils, and microbial decomposition returns it to the atmosphere, at turnover rates ranging from a few to a few thousand years (Trumbore et al., 1996; Paul et al., 1997). In contrast, $CaCO_3$ may remain in the soil for hundreds of thousands of years (Gile et al., 1981). The annual movement of CO_2 from the atmosphere into the soil carbonate pool may be only 0.023 Pg C/yr, yielding a turnover time of 85,000 years (Schlesinger, 1997, p. 365).

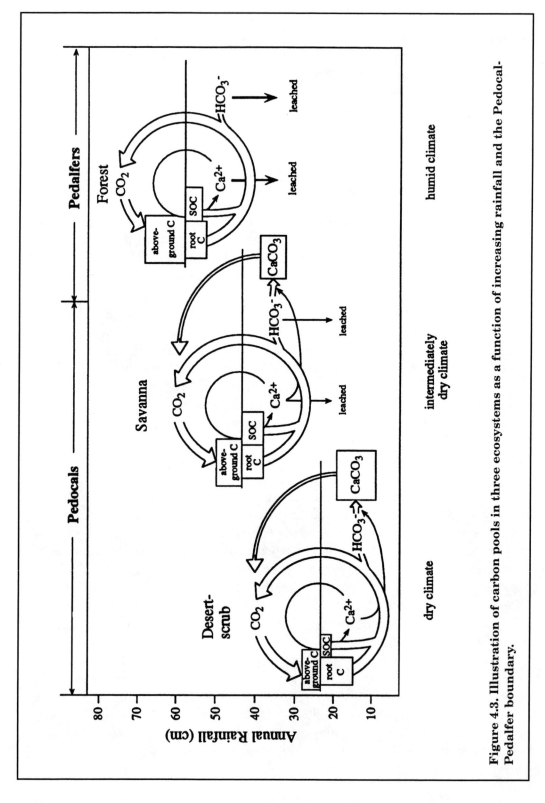

Figure 4.3. Illustration of carbon pools in three ecosystems as a function of increasing rainfall and the Pedocal-Pedalfer boundary.

However, although this may be true as a long-term global average, the possibility lingers that, like OC, carbonate may have both a recalcitrant pool (slow turnover) and a labile pool (fast turnover), the latter because of the involvement of roots and soil microbes in carbonate formation. In lab experiments, for example, soil bacteria and fungi can precipitate carbonate crystals in days to months (Monger et al., 1991). Furthermore, mesquite shrubs *(Prosopis glandulosa)* on the Jornada Experimental Range in New Mexico, known to be under 75 years old, have carbonate crystals on their roots (Gallegos, 1999).

Amount of Soil Carbonates in Grazing Lands

Estimating method

We estimated the amount of soil carbonate in grazing lands by calculating its individual amounts in woodlands, shrublands, and grasslands that occur within aridic, ustic, and xeric moisture regimes of the U.S. The ecoregion map of Bailey et al. (1994) provided these ecoregion categories, based on dominant species.

For example, their unit 315C of the rolling plains section of central Texas (mesquite-buffalo grass) was grouped into the woodland category because mesquite was listed first. In contrast, its neighboring unit to the west, 315B (grama-buffalo grass, shinnery oak), was grouped into the grassland unit because grama grass was listed first. Woodlands, as used in this study, are mainly areas dominated by pinyon and juniper, but they also include the central Texas savannas of medium grass and mesquite (Küchler, 1995).

We excluded ecoregions in the udic moisture regime because udic soils generally do not contain pedogenic carbonate except when the parent material contains high amounts of limestone or other marine carbonate parent material. For the same reason, and because of their small size, we excluded aquic moisture regimes.

After grouping the ecoregions into woodlands, shrublands, and grasslands, we digitized each ecoregion, using ARC/INFO GIS software as a vector layer. We then converted this layer to raster format (grid), using raster-based GIS software, assigning a size of 1,068,705.6 m^2 per grid cell (a square of approximately 1,000 m per side). Figure 4.4 presents the map of the three ecoregion categories and Table 4.1 presents their areas.

We estimated soil carbonate in the woodland, shrubland, and grassland categories in three ways. First, we digitized Machette's (1985) calcic soil map (Fig. 4.5), encoding the calcic soils and marginal areas in the vector layer, then converting it to raster format. Using random sampling for at least 25 sites per ecoregion on Machette's map, we determined the mean concentration of CaCO$_3$ (Table 4.1) based on the National Soil Survey Center database and the STATSGO database.

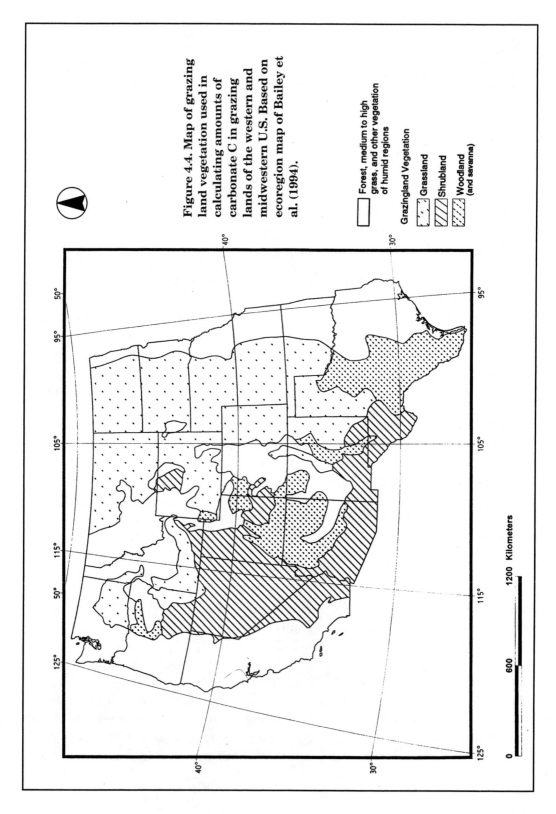

Figure 4.4. Map of grazing land vegetation used in calculating amounts of carbonate C in grazing lands of the western and midwestern U.S. Based on ecoregion map of Bailey et al. (1994).

Forest, medium to high grass, and other vegetation of humid regions

Grazingland Vegetation

Grassland

Shrubland

Woodland (and savanna)

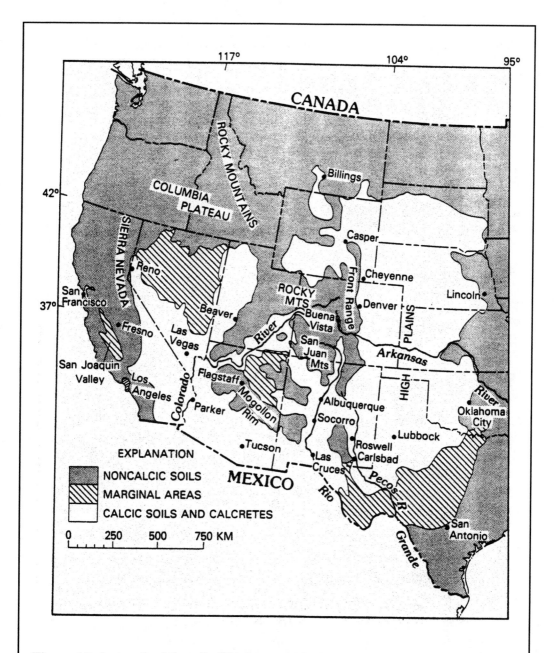

Figure 4.5. Areas of calcic soils (Machette,1985). Areas of discontinuous or poorly preserved calcic soils are designated as marginal areas.

Table 4.1. Estimated amounts of carbonate C in grazing land soils of ecoregions that lie within aridic, ustic, and xeric moisture regimes of the western and midwestern U.S.

Grazing Land Ecoregions	Carbonate C Area (Mha)	Carbonate Concentration (MTC/ha)	C (Pg)
Based on Machette's (1985) map and the Ecoregion map (1994)			
woodland	35.6	177	6.3
shrubland	73.9	225	16.6
grassland	<u>100.1</u>	273	<u>27.3</u>
	Total 209.6		Total 50.2
Based on Schlesinger's 1982 concentrations and Ecoregion map (1994)			
woodland	58.0	50	2.9
shrubland	89.0	310	27.6
grassland	<u>158.0</u>	223	<u>35.2</u>
	Total 305.0		Total 65.7
Based on STATSGO concentrations and Ecoregion map (1994)			
woodland	58.0	183	10.6
shrubland	89.0	207	18.4
grassland	<u>158.0</u>	145	<u>22.9</u>
	Total 305.0		Total 51.9

Second, we estimated soil carbonate C in the woodland, shrubland, and grassland ecoregions by multiplying the areas of each category (Fig. 4.4) by the concentrations of carbonate C that Schlesinger (1982) derived for Arizona soils.

Third, we determined average SIC amounts state by state, using STATSGO, and then multiplied them by the total areas of the three ecoregion categories (Fig. 4.4 and Table 4.1).

Sources of error

Several assumptions and simplifications may have introduced error into the estimated carbonate C amounts in grazing lands. First, we based the results on general maps that oversimplify vegetation inclusions and ecotones. For example, many isolated mountain ranges occur in the shrubland category. Grassland typically skirts these ranges, and woodland caps them. At this scale, however, these categories do not show.

Second, we greatly underestimated the vertical distribution of $CaCO_3$. The soil databases do not represent adequately the massive calcretes in many parts of the Southwest, such as those of the Llano Estacado in eastern New Mexico and west Texas. They also do not represent adequately the stacked sequences of buried soils containing pedogenic carbonate (Grossman et al., 1995). Neither do they account for groundwater carbonate, despite its importance in the terrestrial C cycle and its dependence, in part, on soil processes (L. Wilding, personal communication, 1999).

Third, listing $CaCO_3$ content by ecological categories oversimplifies the great variability that exists because of differences in soil age. For example, grassland soils of the middle Pleistocene Dona Ana geomorphic surface near Las Cruces have 453 kg $CaCO_3/m^2$ (Gile et al., 1981). In contrast, nearby grassland soils of the late Holocene Organ geomorphic surface have less than 1 kg $CaCO_3/m^2$ (cf., pedons 59-1 and 60-12 in Gile and Grossman, 1979).

Fourth, measuring areas represented on a two-dimensional map does not account for the topographical third dimension. This can be important for the mountainous regions of the arid Southwest. Without accounting for the greater surface area resulting from arid mountains, we will underestimate $CaCO_3$ by as much as 20% in some mountainous regions.

Fifth, many steep slopes have been mapped Entisols-Rockland with little significance given to the degree of pedogenic development, including $CaCO_3$ accumulation, that can occur there. In fact, many steep alluvial fans draping the sides of mesas in the west have well developed calcic and petrocalcic horizons, as do soils occurring as pockets in mountain bedrock outcrops. Therefore, we again underestimate $CaCO_3$.

Sixth, we excluded the central valley of California, which contains some grazing land areas, from either the grassland or woodland categories because of the dominance of agriculture. In contrast, we included large areas of dryland and irrigated agriculture in the High Plains in the grassland category.

Seventh, we grouped the sagebrush steppes that occur in parts of eastern Oregon, southern Idaho, and Wyoming (Bailey et al., 1994; Küchler, 1995) with the grasslands instead of the shrublands, for three reasons. First, they were classified as steppes, so we kept them grouped with the other steppes, which are grasslands. Second, they contain large amounts of Mollisols, which also fit best with grasslands. Third, they largely occur outside the desert boundary MacMahon defined (1985), which also fit better with the grasslands than the shrublands. We also grouped most of central Texas into the woodland category, based on Bailey et al. (1994), when much of that region, especially the western part, is considered grassland (Küchler, 1995).

Amounts

Table 4.1 presents three estimated amounts of C (as carbonate C) for the three grazing land ecoregions. Method #1 (the amounts derived using Machette's map) is considered least accurate of the three methods. This is because the calcic and marginal units of that map omit large areas of calcic soils in Montana, North Dakota, and Washington. As the result, the ecoregion areas are smaller than those used in the other two methods. Using this method, however, woodlands contain 6.3 Pg of carbonate C, shrublands contain 16.6 Pg of carbonate C, and grasslands contain 27.3 Pg of carbonate C, for a total of 50.2 Pg of carbonate C.

Method #2 (extrapolating Schlesinger's 1982 C concentrations) is more accurate because it does not omit large areas of calcic soils. Based on this extrapolation, woodlands in the U.S. contain about 2.9 Pg of carbonate C, shrublands contain about 27.6 Pg of carbonate C, and grasslands contain about 35.2 Pg carbonate C, for a total of 65.7 Pg of carbonate C in grazing lands (Table 4.1).

Method #3 (STATSGO) is most accurate because the data come from soil survey information and, therefore, have received the most groundtruthing. Based on the STATSGO method, woodlands in the U.S. drylands contain about 10.6 Pg of carbonate C, shrublands contain about 18.4 Pg of carbonate C, and grasslands contain about 22.9 Pg carbonate C, for a total of 51.9 Pg of carbonate C (Table 4.1).

Schlesinger's estimates for the woodland category are smaller than STATSGO's estimates, probably because Schlesinger included forest with his woodland category. Schlesinger's (1982) shrubland and grassland categories are higher than STATSGO's, perhaps because shrublands and grasslands in Arizona contain more $CaCO_3$ because of greater aridity.

Annual Rates of Soil Inorganic Carbon Sequestration in U.S. Grazing Lands

To estimate the annual rate of C that soil carbonate sequesters, we assumed that no CO_2 is sequestered as carbonate C in calcareous bedrock terrain (Reaction 10). Therefore, we digitized areas of the western and midwestern U.S. dominated by limestone, dolostone, marls, and calcareous sandstones (Fig. 4.6). We based the location of these areas of calcareous bedrock on geologic maps by Oetking (1966, 1967), Feray (1968), Renfro and Feray (1972), Bennison (1973), Renfro (1973), Bennison (1984), and Raisz (1995). Next, we subtracted the areas of calcareous bedrock from the grazing land ecoregions (Fig. 4.4) to give the ecoregion areas with non-calcareous bedrock parent material (Table 4.2).

To obtain a range of sequestration rates, we used low-end and high-end ranges of $CaCO_3$ accumulation rates that the pedologic and geologic literature reported. For the low end, we used 1 g $CaCO_3/m^2/yr$ (Gile et al., 1981; Schlesinger,

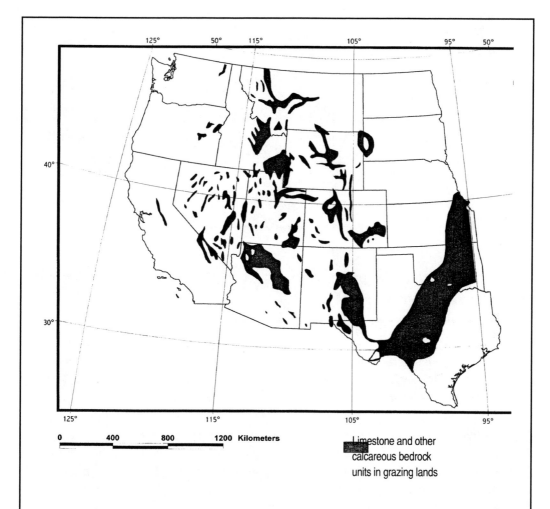

Figure 4.6. Map of limestone and other calcareous bedrock areas where carbonate-C sequestration is considered negligible for the calculations in Table 4.2.

1985). For the high end, we used 12 g $CaCO_3$/m^2/yr (Gile et al., 1981; Reheis et al., 1992, 1995). Furthermore, for the low end we assumed that 90% of the pedogenic carbonate that formed in grazing land soils was from limestone dust (Chadwick et al., 1994). Therefore, only 10%, or 0.1 g $CaCO_3$/m^2/yr is pedo-atmogenic carbonate. Of that amount, C sequestration would be 12% of the $CaCO_3$ accumulation rate (i.e., 12 molar grams of C per 100 molar grams of $CaCO_3$), or 0.012 g carbonate C/m^2/yr (Table 4.2).

For the high end, we assumed that only 10% of the pedogenic carbonate was from limestone dust. Thus, 90% of the 12 g $CaCO_3$/m^2/yr (i.e., 10.8 g $CaCO_3$/m^2/yr) is pedo-atmogenic carbonate. Of that amount, C sequestration would be 12% of the $CaCO_3$ accumulation rate, or 1.3 g carbonate C/m^2/yr (Table 4.2).

Table 4.2. Estimated annual rates of carbonate C sequestration by grazing land soils in the western and midwestern U.S.

Grazing Land Ecoregions	Ecoregion Areas	Area of Calcareous Bedrock within each Ecoregion	Ecoregion Area with Noncalcareous Parent Material	Low End 0.012 (g carb-C/m^2/yr)	High End 1.3 (g carb-C/m^2/yr)
	(Mha)	(Mha)	(Mha)	(Pg C/yr)	(Pg C/yr)
woodland	58.0	22.5	35.4	4.3×10^{-6}	4.6×10^{-4}
shrubland	89.0	10.7	78.4	9.4×10^{-6}	1.0×10^{-3}
grassland	158.0	9.4	148.6	1.8×10^{-5}	1.9×10^{-3}
	Total 305.0	42.6	262.4	3.2×10^{-5}	3.4×10^{-3}
				0.032 MMTC/yr	3.4 MMTC/yr
	Total% of U.S. 1442 MMTCE/yr [1] =			0.002%	0.2%
	Total% of global carbonate C sequestration rate[2] =			0.1%	15%

[1] *Annual carbon emission rate from Lal et al. (1998, p. 4).*
[2] *0.023 Pg carbonate-C/yr (Schlesinger, 1997, p. 365).*

After subtracting the areas of calcareous bedrock from the grazing land ecoregions, the low-end estimate is 3.2×10^{-5} Pg carbonate C/yr (or 0.032 MMTC/yr), which is only 0.002% of the annual C emissions in the U.S. (Table 4.2). The high-end estimate is 3.4×10^{-3} Pg SIC/yr (or 3.4 MMTC/yr), which is 0.2% of the annual C emissions in the U.S. (Table 4.2). In comparison with the estimated global sequestration rate of 0.023 Pg SIC/yr (Schlesinger, 1997, p. 365), the low-end estimate suggests that grazing lands in the U.S. sequester 0.1% of the global SIC amount, whereas the high-end estimate suggests that U.S. grazing lands sequester 15% of the annual global SIC (Table 4.2).

Sources of error in these estimates include:

1. We subtracted only calcareous bedrock areas from grazing land areas. Figure 4.6 does not include Quaternary calcareous deposits, such as calcareous loess, glacial till, and fluvial deposits, especially in the upper Midwest (e.g., Aandahl, 1982, p. 206). However, even if we had adequate maps of these unconsolidated deposits in the West and Midwest, we still would be uncertain whether the $CaCO_3$ would be primary carbonate or recycled pedogenic carbonate.

2. Similarly, we did not subtract alluvial fans skirting calcareous bedrock mountain peaks, especially those in the Basin and Range.

3. The assumption that calcareous bedrock soils do not contain pedo-atmogenic carbonate may be wrong. Pedo-atmogenic carbonate would form in such soils if Ca from igneous sources was added as dust, rain, or irrigation.

Follett, Kimble, and Lal, editors

4. We did not treat irrigated agricultural areas of the High Plains and elsewhere as special cases.

5. The accumulation rates reported in the literature may be over simplified. They are based on soil ages spanning a couple to tens of thousands of years. These accumulation rates assume a unidirectional buildup of carbonate and may overlook dynamic biogenic carbonate that occurs on roots and microorganisms.

Management to Sequester Atmospheric Carbon Dioxide as Soil Inorganic Carbon

Fostering SIC sequestration by altering the supply of calcium and bicarbonate

Under natural conditions, most carbonate forms at a rate of 1 to 12 g $CaCO_3$/ m^2/yr (Gile et al., 1981; Machette, 1985; Schlesinger, 1985). If management is to promote this process to sequester C as $CaCO_3$, several conditions are important.

First, this process is limited to the Pedocal zone of low rainfall and high pH (i.e., pedogenic carbonate C is not an option in humid regions). Second, a non-carbonate source of Ca is required, as shown by Reaction 10 versus Reaction 17 and Figure 4.2. This source could be from Ca silicates or some other non-carbonate source, possibly gypsum.

The question, then, is how could a non-carbonate source of Ca be obtained? One possibility is irrigation with groundwater residing in igneous terrane. For instance, large areas of the West meet this requirement, such as the Sierra Nevada groundwater recharge system, as well as many other mountain ranges in the Basin and Range in both the U.S. and Mexico. Regions dominated by limestone would have a lower potential for management to sequester C as $CaCO_3$ unless one could pump in a non-carbonate source of Ca.

Third, the HCO_3^- supply, generated by biotic respiration, is also important. Therefore, by increasing biotic respiration, possibly by adding manures and sludges, the HCO_3^- supply might be increased. However, based on isotope studies (e.g., Nordt et al., 1998), the amount of HCO_3^- derived from soil CO_2 greatly exceeds the amount of Ca needed to form $CaCO_3$. Therefore, Ca, not HCO_3^-, would be the limiting factor in sequestering C by $CaCO_3$ formation. Overall, an optimum amount of silicate Ca, HCO_3^- from respiration, water, and pH is required to sequester carbonate C.

Release of carbon dioxide from soil inorganic carbon

Grazing land soils are important not only as a potential sink for atmospheric CO_2, but also as a potential source of CO_2. Pedogenic $CaCO_3$ forms in the subsoil. Its exhumation and exposure to acidic rain and organic acids might release CO_2 as the result of congruent dissolution (Reaction 11). However, for this reaction to occur, enough acidity is necessary not only to dissolve $CaCO_3$ and produce HCO_3^-, but also to cause HCO_3^- to convert to CO_2 and H_2O, as Reaction 5 represents. In theory, congruent dissolution might occur not only where acidic rain contacts $CaCO_3$ but also where microbiotic crusts grow and excrete organic acids on exposed $CaCO_3$.

Moreover, erosion promotes the loss of OM (SOC) from soil because it disrupts plant growth that would otherwise bring C into the soil system via photosynthesis (Lal et al., 1998). The same process could apply to carbonate C, because erosion destroys root-microbial infrastructure. If roots and microbes precipitate a significant amount of carbonate (Lobova, 1967; Monger et al., 1991), the demise of this bio-pedologic system curtails $CaCO_3$ precipitation.

Conclusions

Processes of soil carbonate formation potentially leading to carbon sequestration

In contrast to C sequestration by OM, which involves a one-step process whereby C is converted from gaseous CO_2 to plant tissue C by photosynthesis, C in pedogenic $CaCO_3$ requires three additional steps. After C has been fixed by photosynthesis, it passes from OM back to gaseous CO_2 by respiration, then to aqueous HCO_3^- by dissolution and dissociation, and then to solid $CaCO_3$ by crystallization.

Depending on parent material, the formation of pedogenic carbonate can be grouped into two categories: (1) dissolution-reprecipitation of preexisting carbonate (Reaction 10) or (2) silicate Ca combining with HCO_3^- generated by respiration (Reaction 17). The important difference between these two reactions is that one mole of CO_2 is used to dissolve preexisting $CaCO_3$ in Reaction 10 and one mole is released during the reprecipitation of $CaCO_3$. Hence, no C is sequestered. In contrast, two moles of CO_2 are consumed to weather Ca silicate in Reaction 17, yet only one mole is released to the atmosphere with precipitation of $CaCO_3$. Therefore, one mole of C is sequestered by $CaCO_3$.

Whether or not pedogenic carbonate sequesters atmospheric CO_2 depends on the timescale. On the short-term timescale of decades and centuries, we can differentiate transfers and sinks based on whether the source of Ca is preexisting carbonates (transfers) or Ca silicates (sinks). On the long-term timescale of millions of years, even marine carbonate rock is a sink for atmospheric CO_2 (Berner, 1999).

Amount of soil inorganic carbon in U.S. grazing lands

Soil carbonate C, based on the STATSGO database and ecoregions of Bailey et al. (1994), totals, in woodlands in the western U.S., about 10.6 Pg C; in shrublands, about 18.4 Pg C; and in grasslands, about 22.9 Pg C; for a total of 51.9 Pg carbonate C (Table 4.1). If we extrapolate soil characterization data derived for Arizona ecosystems (Schlesinger, 1982) to larger regions of the West, then U.S. woodlands contain about 2.9 Pg C, shrublands contain about 27.6 Pg C, and grasslands contain about 35.2 Pg C, for a total of 65.7 Pg carbonate C. Furthermore, if we take a value of 930 Pg carbonate C as the global amount (Schlesinger, 1997; Eswaran et al., 1999), then U.S. grazing lands contain about 6 to 7% of the total carbonate C in world soils.

Annual rates of soil inorganic carbon sequestration in U.S. grazing lands

If we look at a range of carbonate C sequestration rates based on the low-end and high-end $CaCO_3$ accumulation rates that the literature reports, and subtract calcareous bedrock terrane from grazing lands based on the assumption that no carbonate C sequestration occurs there, we find sequestration rates ranging from 3.2×10^{-5} to 3.4×10^{-3} Pg carbonate C/yr. This range represents, on the low end, 0.002% of the annual U.S. C emission of 1442 MMTC (Table 4.2) to, on the high end, 0.2% of the annual U.S. C emissions.

Globally, if we assume a carbonate C sequestration rate of 0.023 Pg/yr (Schlesinger, 1985), then, on the low end, the U.S. sequesters 0.1% of the annual global amount and, on the high end, 15% of the annual global amount.

Soil inorganic carbon sequestration potential in U.S. grazing lands

In theory, soils with a high potential to sequester carbonate C are those in arid and semiarid lands that could be irrigated economically with water containing Ca from non-carbonate sources (Reaction 17). These soils are in areas with groundwater residing in sediments derived from igneous rocks, such as the Sierra Nevada system. Such igneous terranes occur in many areas of the western U.S. and Mexico.

Adding biosolids to certain grazing land soils also might increase respiration, and, consequently, bicarbonate content. If Ca from non-carbonate sources were available, the formation of pedogenic carbonate, and thus carbonate C, might be increased.

Needed research

An important point about the soil carbonate pool in terms of C sequestration is that it is not a homogeneous pool. Instead, it is composed of at least three sub-pools: primary carbonates, pedo-lithogenic carbonates, and pedo-atmogenic carbonates (Fig. 4.1). Before we can understand the potential of grazing land to sequester atmospheric CO_2, we need to measure the size and better understand the nature of the pedo-atmogenic carbonate pool.

The greatest error in estimating annual sequestration rates might be in the carbonate accumulation rates used (Table 4.1). Although these values are probably accurate on a timescale covering thousands of years, the rates may overlook more dynamic accumulations that occur on the decadal scale. This may be especially important when carbonates are formed biogenically in grazing land soils.

We need to understand if degradation of grazing lands contributes to the greenhouse effect as the result of erosion and the exposure of soil carbonates to acid conditions. If so, preventing grazing land degradation and restoring lands already degraded may be the best contribution that carbonates in grazing lands can make to mitigating the greenhouse effect. Research data also are needed on the impact of liming on the release of CO_2.

Acknowledgments

The authors thank Ronald Follett, Rebecca Kraimer, John W. Hawley, and Leland H. Gile for comments and improvements. We also appreciate the support we received from the New Mexico State University Agricultural Experiment Station.

References

Aandahl, A.R. 1982. *Soils of the Great Plains.* University of Nebraska Press. Lincoln, NE.

Adams, J.M. 1993. Caliche and the carbon cycle. *Nature* 361:213-214.

Anderson, J.P.E. 1982. Soil respiration. *In* A.L. Page (ed), *Methods of Soil Analysis, Part 2, Chemical and Microbiological Properties,* Am. Soc. Agronomy, Madison, WI, pp. 831-872.

Arkley, R.J. 1963. Calculation of carbonate and water movement in soil from climatic data. *Soil Sci.* 96:239-248.

Bailey, R.G., P.E. Avers, T. King, and W.H. McNabb. 1994. *Ecoregions and subregions of the U.S.* USDA-Forest Service. Washington, DC.

Bathurst, R.G.C. 1975. Carbonate sediments and their diagenesis. *Developments in Sedimentology 12.* Elsevier. Amsterdam.

Bennison, A.P. 1973. *Geological Highway Map of the Pacific Northwest Region.* American Association of Petroleum Geologists. Tulsa, OK.

Bennison, A.P. 1984. *Geological Highway Map of the Northern Great Plains.* American Association of Petroleum Geologists. Tulsa, OK.

Berner, R.A. 1992. Weathering, plants, and the long-term carbon cycle. *Geochimica et Cosmochimica Acta* 56:3225-3231.

Berner, R.A. 1993. Paleozoic atmospheric CO_2: importance of solar radiation and plant evolution. *Sci.* 261:68-70.

Berner, R.A. 1999. A new look at the long-term carbon cycle. *GSA Today* 9:1-6.

Berner, R.A., and A.C. Lasaga. 1989. Modeling the geochemical carbon cycle. *Sci. Am.* 260:74-81.

Bohn, H.L., B.L. McNeal, and G.A. O'Connor. *1985. Soil Chemistry.* 2nd ed. Wiley. New York.

Bretz, J.H., and L. Horberg. 1949. Caliche in southeastern New Mexico. *Journ. Geol.* 57:491-511.

Buyanovsky, G.A., and G.H. Wagner. 1983. Annual cycles of carbon dioxide level in soil air. *Soil Sci. Soc. Am. J.* 47:1139-1145.

Cerling, T.E., and J. Quade. 1993. Stable carbon and oxygen isotopes in soil carbonates. *In* P.K. Swart, K.C. Lohmann, J. McKenzie, and S. Savin (eds), *Climate Change in Continental Isotopic Records, AGU Geophys. Monogr.* 78:217-231.

Chadwick, O.A., E.F. Kelly, D.M. Merrits, and R.G. Amundson. 1994. Carbon dioxide consumption during soil development. *Biogeochemistry* 24:115-127.

Cropper, Jr., W.P., K.C. Ewel, and J.W. Raich. 1985. The measurement of soil CO_2 evolution *in situ. Pedobiologia* 28:35-40.

Dick-Peddie, W.A. 1993. *New Mexico Vegetation: Past, Present, and Future.* University of New Mexico Press. Albuquerque, NM.

Dokuchaev, V.V. 1883. *Russian Chernozem.* Translated by Israel Program of Scientific Translation, 1967. Jerusalem.

Doner, H.E., and W.C. Lynn. 1989. Carbonate, halide, sulfate, and sulfide minerals. *In* J.B. Dixon and S.B. Weed (eds), *Minerals in Soil Environments, Soil Sci. Soc. Am. Book Series,* No. 1, 2nd ed, Madison, WI, pp. 279-330.

Dregne, H.E. 1976. *Soils of Arid Regions.* Elsevier Scientific Publishing Company. Amsterdam.

Eswaran, H., P.F. Reich, J.M. Kimble, F.H. Beinroth, E. Padmanabhan, and P. Moncharoen. 2000. Global carbon sinks. *In* R. Lal, J.M. Kimble, H. Eswaran, and B.A. Stewart (eds), *Global Climate Change and Pedogenic Carbonates,* CRC Press, Boca Raton, FL, pp. 15-26.

Feray, D.E. 1968. *Geological Highway Map of the Pacific Southwest Region.* American Association of Petroleum Geologists. Tulsa, OK.

Gallegos, R.A. 1999. *Biogenic carbonate, desert vegetation, and stable carbon isotopes.* M.S. Thesis, New Mexico State University. Las Cruces, NM.

Gile, L.H., and R.B. Grossman. 1979. *The Desert Project soil monograph.* USDA-NRCS, Natl. Soil Survey Center. Lincoln, NE

Gile, L.H., F.F. Peterson, and R.B. Grossman. 1965. The K horizon — a master soil horizon of carbonate accumulation. *Soil Sci.* 99:74-82.

Gile, L.H., F.F. Peterson, and R.B. Grossman. 1966. Morphological and genetic sequences of carbonate accumulation in desert soils. *Soil Sci.* 101:347-360.

Gile, L.H., J.W. Hawley, and R.B. Grossman. 1981. *Soils and Geomorphology in the Basin and Range area of sSouthern New Mexico — Guidebook to the Desert Project.* New Mexico Bureau of Mines and Mineral Resources, Memoir 39. Socorro, NM.

Grossman, R.B., R.J. Ahrens, L.H. Gile, C.E. Montoya, and O.A. Chadwick. 1995. Areal evaluation of organic and carbonate carbon in a desert area of southern New Mexico. *In* R. Lal, J. Kimble, E. Livine, and B.A. Stewart (eds), *Soil and Global Change,* CRC Press, Boca Raton, FL, pp. 81-92.

Hawley, J.W., G.O. Bachman, and K. Manley. 1976. Quaternary stratigraphy in the Basin and Range and Great Plains provinces, New Mexico and western Texas. *In* W.C. Manhaney (ed), *Quaternary Stratigraphy of North America,* Dowden, Hutchinson, and Ross, Inc., Stroudsburg, PA, pp. 235-274.

Holechek, J.L., R.D. Pieper, and C.H. Herbel. 1989. *Range Management.* Regents/Prentice Hall. Englewood Cliffs, NJ.

Hunt, C.B. 1986. *Surficial Deposits of the United States.* Van Nostrand Reinhold Company. New York.

Jenny, H., and C.D. Leonard. 1934. Functional relationships between soil properties and rainfall. *Soil Sci.* 38:363-381.

Krauskopf, C. 1967. *Introduction to Geochemistry.* McGraw-Hill. New York.

Küchler, A.W. 1995. A physiognomic classification of vegetation. *In* E.B. Espenshade (ed), *Goode's World Atlas,* 19th edition, Rand McNally & Company, New York, NY, pp. 70-71.

Lal, R., J.M. Kimble, R.F. Follett, and C.V. Cole. 1998. *The Potential of U.S. Cropland to Sequester Carbon and Mitigate the Greenhouse Effect.* Sleeping Bear Press, Inc. Chelsea, MI.

Lal, R., et al. 1999. *Global Climate Change and Pedogenic Carbonates.* CRC Press. Boca Raton, FL.

Lobova, E. 1967. *Soils of the Desert Zone of the USSR.* Issued in translation by the Israel Program for Scientific Translation. Jerusalem.

Machette, M.N. 1985. Calcic soils of the southwestern United States. *In* D.L. Weide (ed), *Soils and Quaternary Geology of the Southwestern United States, Geological Soc. Am. Spec. Paper 203,* Boulder, CO, pp. 1-21.

MacMahon, J.A., 1985. *Deserts. National Audubon Society Nature Guides.* Alfred A. Knopf, Inc. New York.

Marbut, C.F. 1928. A scheme for soil classification. *Proc. First Int. Congr. Soil Sci.* 4:1-31.

Monger, H.C., L.A. Daugherty, W.C. Lindemann, and C.M. Liddell. 1991. Microbial precipitation of pedogenic calcite. *Geology* 19:997-1000.

Monger, H.C., and R.A. Gallegos. 1999. Biotic and abiotic processes and rates of pedogenic carbonate accumulation in the southwestern United States — relationship to atmospheric CO_2 sequestration. *In* R. Lal et al. (eds) *Global Climate Change and Pedogenic Carbonates*, CRC Press, Boca Raton, FL, pp. 273-289.

Monger, H.C., and E. Kelly. Soil silica. *In* J.B. Dixon and D. Schulze (eds), *Environmental Soil Mineralogy*, Soil Sci. Soc. Am. Book Series, Madison, WI (in press).

Morse, J.W., and F.T. Mackenzie. 1990. *Geochemistry of Sedimentary Carbonates. Developments in Sedimentology 48.* Elsevier. Amsterdam.

Nordt, L.C., C.T. Hallmark, L.P. Wilding, and T.W. Boutton. 1998. Quantifying pedogenic carbonate accumulations using stable carbon isotopes. *Geoderma* 82:115-136.

Oetking, P. 1966. *Geological Highway Map of the Mid-Continent Region.* American Association of Petroleum Geologists. Tulsa, OK.

Oetking, P. 1967. *Geological Highway Map of the Southern Rocky Mountain Region. Map no. 2.* American Association of Petroleum Geologists. Tulsa, OK.

Paul, E.A., R.F. Follett, S.W. Leavitt, A. Halvorson, G.A. Peterson, and D.J. Lyon. 1997. Radiocarbon dating for determination of soil organic matter pool sizes and dynamics. *Soil Sci. Soc. Am. J.* 61:1058-1067.

Rabenhorst, M.C., and L.P. Wilding. 1986. Pedogenesis on the Edwards Plateau, Texas: III. New model for the formation of petrocalcic horizons. *Soil Sci. Soc. Am. J.* 50:693-699.

Rabenhorst, M.C., L.P. Wilding, and L.T. West. 1984. Identification of pedogenic carbonates using stable carbon isotopes and microfabric analysis. *Soil Sci. Soc. Am. J.* 48:125-132.

Raisz, E. 1995. Physiography. *In* E.B. Espenshade (ed). *Goode's World Atlas,* 19th edition. Rand McNally & Company. New York, NY, pp. 66-67.

Reardon, E.J., G.B. Allison, and P. Fritz. 1979. Seasonal chemical and isotopic variations of soil CO_2 at Trout Creek, Ontario. *J. of Hydrol.* 43:355-371.

Reheis, M.C., J.M. Sowers, E.M. Taylor, L.D. McFadden, and J.W. Harden. 1992. Morphology and genesis of carbonate soils on the Kyle Canyon fan, Nevada, USA. *Geoderma* 52:303-342.

Reheis, M.C., J.C. Goodmacher, J.W. Harden, L.D. McFadden, T.K. Rockwell, R.R. Shroba, J.M. Sowers, and E.M. Taylor. 1995. Quaternary soils and dust deposition in southern Nevada and California. *Geol. Soc. Am. Bull.* 107:1003-1022.

Renfro. H.B. 1973. *Geological Highway Map of Texas.* American Association of Petroleum Geologists. Tulsa, OK.

Renfro, H.B., and D.E. Feray. 1972. *Geological Highway Map of the Northern Rocky Mountain Region.* American Association of Petroleum Geologists. Tulsa, OK.

Ruhe, R.V. 1967. *Geomorphic surfaces and surficial deposits in southern New Mexico.* New Mexico Bureau of Mines and Mineral Resources, Mem. 18. Socorro, NM.

Schlesinger, W.H. 1982. Carbon storage in caliche of arid soils: a case study from Arizona. *Soil Sci.* 133:247-255.

Schlesinger, W.H. 1985. The formation of caliche in soils of the Mojave Desert, California. *Geochimica et Cosmochimica Acta* 49:57-66.

Schlesinger, W.H. 1997. *Biogeochemistry: An Analysis of Global Change,* 2nd ed. Academic Press. New York.

Soil Survey Staff. 1996. *Soil Survey Laboratory Methods Manual. Soil Survey Invest. Rept. No. 42.* USDA-NRCS. Washington, DC.

Soil Survey Staff. 1998. *Keys to Soil Taxonomy.* 8th ed. USDA-NRCS. Washington, DC.

Terhune, C.L., and J.W. Harden. 1991. Seasonal variations of carbon dioxide concentrations in stony, coarse-textured desert soils of southern Nevada, USA. *Soil Sci.* 151:417-429.

Trumbore, S.E., O.A. Chadwick, and R. Amundson. 1996. Rapid exchange between soil carbon and atmospheric carbon dioxide driven by temperature change. *Science* 272:393-396.

Weast, R.C. 1986. *CRC Handbook of Chemistry and Physics.* CRC Press. Boca Raton, FL.

West, L.T., L.R. Wilding, and M.C. Rabenhorst. 1988. Differentialtion of pedogenic and lithogenic carbonate forms in Texas. *Geoderma* 43:271-287.

Soil Processes, Plant Processes, and Carbon Dynamics on U.S. Grazing Land

CHAPTER 5

Exploiting Heterogeneity of Soil Organic Matter in Rangelands: Benefits for Carbon Sequestration

S.B. Bird,[1] J.E. Herrick,[2] and M.M. Wander[3]

Introduction

Soil organic matter (SOM) has a large influence on the long-term sustainability of soil systems and the ecosystems and landscapes they support (Follett et al., 1987). SOM conservation and enhancement have positive effects on plant nutrient status, soil structure, and soil water holding capacity (DeJong et al., 1983; Soane, 1990; Kern and Johnson, 1993). The beneficial impacts of these factors on plant productivity create a positive feedback loop between net primary production (NPP) and C sequestration (Fig. 5.1).

Fundamental interactions between SOM, soil aggregation, soil quality, and C sequestration are well established (e.g., Doran and Parkin, 1994; Jastrow and Miller, 1997). SOM directly affects soil physical, chemical, and biological properties. Acting through these properties, SOM influences key soil functions, which in turn affect plant growth, movement of water, air, and nutrients, the storage of atmospheric C, impacts of pollutants on the soil system, and habitat quality of soil-dwelling organisms (Fig. 5.2).

While these relationships are widely accepted to be of great importance to atmospheric C storage in soils, the mechanisms and processes involved are not fully understood, particularly in relation to the relative importance of mechanisms in different ecosystems and landscapes. The horizontal and vertical distributions of SOM influence how the soil system responds to disturbances, and they affect the

[1] Soil Ecology Research Associate, ARS-USDA Jornada Experimental Range, Las Cruces, NM 88011.

[2] Soil Scientist, ARS-USDA Jornada Experimental Range, Las Cruces, NM 88011.

[3] Asst. Prof. of Soil Fertility/Ecology, Dept. of Natural Resources and Environmental Sciences, University of Illinois Champaign-Urbana, IL 61801.

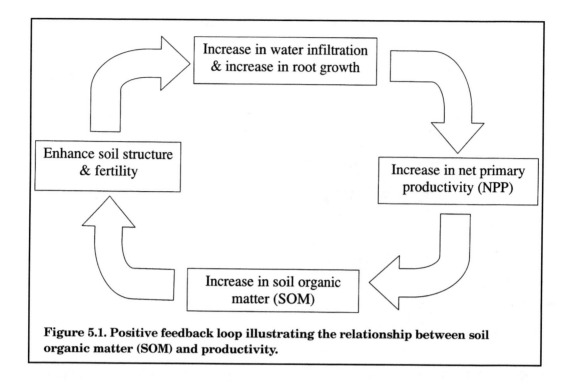

Figure 5.1. Positive feedback loop illustrating the relationship between soil organic matter (SOM) and productivity.

relationship between a disturbance and soil quality or its potential to sequester atmospheric C (Herrick et al., 1998) (Fig. 5.2).

SOM concentration and composition vary at differing spatial scales. At the aggregate level (below 2 cm), SOM distribution affects, and is affected by, soil aggregate formation and stabilization (Oades and Waters, 1991). At individual plant (2 mm to 20 m) and plant community (2 m to kilometers-wide) scales, temporal and spatial variability of fungal biomass distribution is particularly important for organic matter (OM) dynamics, and patterns of vegetative distribution interact with nutrient and water resource availability. At ecosystem and landscape scales (kilometers-wide areas), spatial patterns of SOM, vegetation, and water resources influence the flows of materials (such as organic compounds and soil particles), energy, and species between and among different landscape structural elements.

The high level of variability of soil resources at ecosystem and landscape scales often is viewed as a barrier to management. We argue that SOM heterogeneity actually provides management opportunities that can be exploited to increase C sequestration and/or enhance ecosystem function. In this chapter, after reviewing the distribution of and processes affecting C sequestration at aggregate, plant, and plant community scales and describing the similarities and differences of cropland and rangeland ecosystems, we propose extending to the landscape scale the aggregate hierarchy model Tisdall and Oades (1982) originally

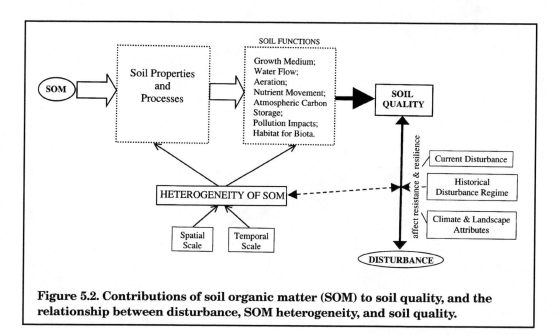

Figure 5.2. Contributions of soil organic matter (SOM) to soil quality, and the relationship between disturbance, SOM heterogeneity, and soil quality.

proposed and others advanced (e.g., Dexter, 1988; Elliott and Coleman, 1988; Kay, 1990; Waters and Oades, 1991).

This conceptual model describes aggregate formation and stabilization with an emphasis on the role SOM plays. This model can be extrapolated to soil structure considered across a range of scales. We show how applying this model can optimize C sequestration and/or rangeland function by targeting the application of management inputs to promote recovery and growth of species which increase C inputs and retention. This approach facilitates management for C sequestration by explicitly recognizing that the capacity of rangeland ecosystems to sequester C is a function of interactions among the spatial distribution of plant production, soil resources, and water resources.

Soil Organic Matter at the Aggregate Scale

SOM and soil aggregation link closely. The hierarchical model of aggregate formation and stabilization Tisdall and Oades proposed (1982) and others advanced recognizes differences in SOM composition and turnover rates at different scales of aggregate formation (Fig. 5.3). Researchers have noted the importance of soil biota, their activities, and their by-products to soil structure, and Jastrow and Miller (1991) suggested that the influence of biotic factors on soils can be interpreted at a variety of spatial scales.

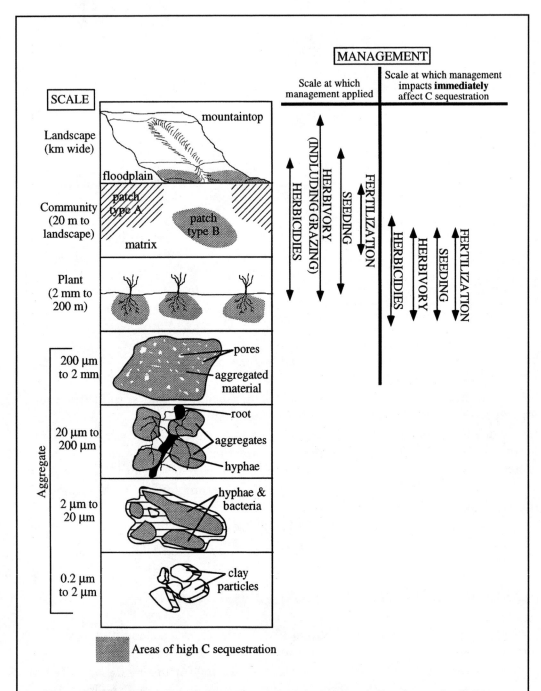

Figure 5.3. Extension of the hierarchical model of soil aggregation to include structural features recognizable at the plant, community, and landscape scales, showing areas of highest carbon sequestration and examples of scales of management applications and related impacts.

At the finest scale, sub-microaggregate units form between primary particles held together by organic and inorganic cements and electrostatic forces (Emerson et al., 1986). Organic bonds and bacteria and fungal debris bind together these clay microstructures with larger primary particles to form microaggregates of 2 to 20 μm in diameter (Waters and Oades, 1991). Microbial debris and fragments join these silt sized clusters to form aggregates 20 to 250 μm in diameter. In turn, these mesoaggregates are bound and held together in macroaggregates greater than 250 μm.

Tisdall and Oades (1982) defined *macroaggregates* (aggregates greater than 250 μm, obtained after wet-sieving) as those formed largely through enmeshment by roots and hyphae. Soil fauna and microorganisms can decompose the particulate organic matter (POM) at the core of a macroaggregate until a threshold is reached where this OM is no longer large enough to hold the aggregate together (Golchin et al., 1997). Thus, smaller aggregates are formed from a single large macroaggregate, the size distribution of which depends on the residual POM remaining.

Further fragmentation can occur as the decomposition process continues. With continued input of new POM into the soil system, this fragmentation is offset by the formation of new macroaggregates which may use the fragmented aggregates as structural units. There is therefore a link between the decomposition of OM and soil aggregation dynamics (Golchin et al., 1997). When cultivation disturbs soils, the OM associated with macroaggregates, which is more labile and more readily mineralized than SOM associated with microaggregates, is the primary source of OM released (Elliott, 1986; Oades et al., 1987; Waters and Oades, 1991).

The upper size limit for macroaggregates is normally considered 1000 to 2000 μm. However, critical information about naturally occurring structural features associated with larger scale processes may be lost if soil structural units larger than 2000 μm are overlooked. The characteristics, size, and shape of natural peds typically described by morphometric techniques are influenced by OM content (Dexter, 1988). Recently, Yang and Wander (1998) showed that dry-sieved aggregates, which ranged in size from 2 to 25 cm, were true aggregates in that SOM affected their formation and stabilization. Hence, large dry-sieved aggregates larger than 2000 μm should be considered in aggregate stabilization hierarchy theory.

Interactions between SOM, soil aggregation, climatic conditions, topography, and vegetative dynamics greatly affect an ecosystem's functioning and productivity. Functional relationships between SOM, soil structure, and water movement also are involved in these interactions. SOM fractions, aggregates, and aggregation stabilization agents (and associated environmental influences) interact to determine soil C storage capacity and stability and thus directly contribute to soil quality (see Follett, Ch. 3).

Soil Organic Matter at the Plant Scale

Spatial distribution of SOM in perennial rangelands, and especially savannas and shrublands, correlates highly with vegetative patterns and plant community dynamics (Smith et al., 1994; Schlesinger et al., 1990, 1996). Beneath plants, higher concentrations of SOM frequently are associated with more highly developed soil structure, including more stable aggregates and higher macropore densities. In essence, the soil associated with individual plants functions as a *superaggregate*. The plant serves as the link between individual aggregates (Tisdall and Oades, 1982) and landscape features (Fig. 5.3).

Growing plants affect aggregation in a number of ways. Plants influence aggregates *directly* through root-C inputs that promote aggregate formation by enmeshment and encapsulation (Tisdall and Oades, 1982; Dormaar and Foster, 1991). Plants influence aggregates *indirectly* by stimulating microbial activity (Angers and Mehuys, 1988; Miller and Jastrow, 1990; Perfect et al., 1990; Chantingny et al., 1997) and by exerting physical pressure that both consolidates and fragments soil (Matterechera et al., 1992). Direct organic inputs from plants occur via litter drop, root exudates, root mortality, and above- and belowground herbivory. Plant residues on the soil surface also can decrease soil aggregate breakdown, thus influencing SOM retention by reducing exposure to oxidation.

The nominal size and spatial distribution of plants and plant canopies affect not only the distribution of C inputs, but also the potential for these inputs to be redistributed within the community or lost completely from the system through wind and water erosion (for more detail, see Schuman et al., Ch. 11). Mycorrhizal and bacterial root symbioses also have direct effects on SOM composition and distribution and, in some arid ecosystems, mycorrhizal relationships can enhance plant productivity significantly.

Numerous interactions and feedbacks exist among SOM, soil aggregation, vegetation, and soil-dwelling organisms. Soil-dwelling species often respond to patchy resource distributions, exhibit clumping behavior, or remain in aggregated groups due to poor dispersal power, resulting in clumped distributions that can increase heterogeneity of SOM further. Mixing of SOM by soil biota also can homogenize SOM distribution (Kooistra and van Noordwijk, 1996).

Soil-dwelling microorganisms and invertebrate animals can influence SOM distribution, soil aggregate dynamics, and soil structure greatly. Through their roles in decomposition and macroorganic matter fragmentation, these organisms directly and indirectly affect SOM composition and spatial distribution. Soil texture and structure modify these relationships by affecting decomposer microenvironments, and the stability of their substrates and products, while surface litter depth and stability may determine the size of the microarthropod community (Cepeda, 1986).

Heterogeneous spatial distributions of microbial biomass and soil inverte-brates reflect the patchy distribution of soil resources in rangelands, particularly in arid systems (Santos et al., 1978; Freckman and Mankau, 1986; Smith et al., 1994). Creation of voids by motile invertebrates such as earthworms and milli-pedes alters pore size and soil aeration, and excreta deposition affects aggrega-tion and SOM storage. In addition, microinvertebrates (those less than 0.2 μm in diameter, such as protozoa and small mites) and mesoinvertebrates (those be-tween 0.2 and 10 μm in diameter, such as springtails) have many interrelation-ships with microorganisms in soil. These interactions affect SOM distribution, nutrient mineralization, and aggregation.

In warm arid and semiarid rangeland systems, termites are known to affect soil chemistry and soil structure significantly (Holt, 1987; Jones, 1990). Low levels of SOM content correlate with termite abundance (Nash and Whitford, 1995), and ter-mite gut flora metabolize OM more rapidly than free-living soil microbes (Lee and Wood, 1971; Butler and Butterfield, 1979). This evidence suggests that these organ-isms have a strong influence on SOM variation within their ecosystems.

Clearly, the interactions among vegetation, soil biota, SOM, and aggregation are complex at the plant scale. Kay (1997) argues that soil structure should be considered at scales appropriate to the processes under consideration, noting that the scales of soil structure relevant to these processes extend from microns to meters. Kay (1997) lists fine-scale processes associated with the impacts of soil structure on plant growth, the storage and provision of water, oxygen, and nutri-ents, and coarse-scale hydrologic processes controlling infiltration, drainage, storage, and evapotranspiration of water. From this argument follows a need to consider vegetative patches or complexes that exist at community or landscape scales when focusing on processes such as annual NPP and C sequestration.

Soil Organic Matter at the Plant Community and Landscape Scales

Complex gradients of soil water and SOM exist over landscapes due to the numerous interactions that occur among vegetation, soil texture and structure, seasonal weather conditions, topography, disturbance regimes, and water redis-tribution patterns. These relationships result in the characteristic distributions of biota over the landscape (Cornelius et al., 1991). In turn, weather and its effects on aggregation, biological processes, and human impacts govern dynamics of soil heterogeneity (particularly in the topsoil) (Kooistra and van Noordwijk, 1996).

The spatial distribution of water resources across semiarid and arid range-land landscapes is a particularly important factor affecting soil dynamics. Topo-graphic features and spatial distributions of vegetation greatly influence patterns of soil moisture heterogeneity, resulting in a high level of spatial variability of plant available water (Coronato and Bertillier, 1996). Topographic depressions and slopes sheltered from wind tend to support localized areas of higher nutrient levels, water content, and microbial biomass (Burke et al., 1989; Gallardo and Schlesinger, 1992). These areas have higher C fixation potential and therefore may be able to sequester C more rapidly than other locations within the landscape.

The Case for Rangelands

Rangelands versus croplands

The majority of research into relationships between SOM and soil structure has been conducted on cropland. Fine-scale processes associated with these interactions often have been assumed to be identical in rangeland systems, but this has not been demonstrated unequivocally.

Rangeland ecosystems differ in many ways from more intensively managed annual cropping systems. The environmental and/or soil conditions that exist in rangeland areas normally prevent crop production. Some rangeland soils, particularly in the humid tropics, suffer from fertility problems associated with low pH and high aluminum levels that lead to phosphorus deficiency. Most rangelands in the U.S. are limited by water (either too much or too little) or cold temperatures. Some areas of the intermountain west and California are limited by high salt levels (both natural and anthropogenic).

Rangelands also frequently are managed for more than one objective. For example, in the U.S. National Forests, grazing lands often are managed for livestock and timber production, wildlife conservation, and recreation. Rangeland vegetation tends to develop in close association with the soil and affects how the system responds to disturbance (Herrick et al., 1998). In contrast, croplands tend to be relatively fertile areas, where annual crops are cultivated on soils that have developed independently from the crops.

While there may be similarities in structure and function at the aggregate scale, at plant, plant community, ecosystem, and landscape levels, rangelands are clearly different from croplands. Patterns of vegetative distribution are noticeably more heterogeneous in rangeland systems (Cornelius et al., 1991). Perennial vegetation predominates, and individual plants often are clumped in patchy distributions across the landscape, especially in more arid regions where shrubs frequently dominate. Since vegetation has a large effect on SOM's distribution and aggregation, it follows that distribution of resources has a similarly clumped pattern.

Soil heterogeneity in rangelands

Recognizing and understanding soil heterogeneity is vital to understanding soil systems and the impacts of management practices (Kooistra and van Noordwijk, 1996). Despite this, our understanding of how soil heterogeneity contributes to ecological processes at different scales remains rudimentary.

Spatial distribution of SOM in rangelands is clearly one example of soil heterogeneity that deserves consideration when managing for system integrity and potential atmospheric C sequestration. In arid and semiarid rangeland landscapes, where sporadic and discontinuous rainfall has a large effect on plant growth (Gallardo and Schlesinger, 1992), a conspicuous pattern of resource-rich vegetative "islands" surrounded by resource-poor interspaces often occurs. This pattern reflects in the high variation of SOM and plant productivity recorded across rangeland landscapes.

In southwestern semiarid rangelands, the transition from grass-dominated to shrub-dominated systems has accentuated a patchy distribution of resources and vegetation. Over the past 100 years, shrubland has replaced large areas of grassland in the southwestern U.S., creating a shift toward a more heterogeneous soil resource distribution (Meir, 1985; Gallardo and Schlesinger, 1992) and an associated heterogeneous distribution of microbial biomass. Studies on the Jornada Experimental Range in New Mexico suggest that long-term grazing of semiarid grasslands has led to an increase in spatial and temporal heterogeneity of water, SOM, and other soil resources (Schlesinger et al., 1990).

Land use changes, primarily associated with higher cattle grazing intensity, road construction, and other landscape modifications, have redistributed water via overland flow, creating a highly patchy distribution of soil moisture (Schlesinger et al., 1989). Desert shrub species, with deeper and more extensive water-absorbing root systems than grasses, have intensified these patches, causing further localization of soil resources under shrub canopies, which act as foci for rainfall infiltration (Schlesinger, 1990). Erosion and gas emissions from shrub interspaces cause a loss of fertility in these areas, and nutrient cycling becomes further concentrated beneath shrub plants.

Over time, these shrub "islands" become the favored sites for shrub regeneration; fertility increases more; and a positive feedback loop of desertification begins. A change in scale of soil heterogeneity, from fine scale under grassland to coarse scale under shrubland, characterizes this transition process. This change in pattern of the distribution of soil resources can have a significant effect on the system's functioning (Schlesinger et al., 1996).

At the Jornada Experimental Range in New Mexico, Abrams and Jarrell (pers. comm.) found significant differences in soil chemical heterogeneity between shrub dunes and interdune spaces in the top 30 cm of the soil (Table 5.1). Heterogeneity increased with dune age, and the correlation between dune presence and

Table 5.1. Landscape and plant scale variability in total soil carbon in communities dominated by mesquite (*Prosopis glandulosa*) in the Jornada Del Muerto Basin, New Mexico, USA.

Depth:	Organic C		Inorganic C		Total C	
	0-10 cm MTC/ha	10-20 cm MTC/ha	0-10 cm MTC/ha	10-20 cm MTC/ha	0-10 cm MTC/ha	10-20 cm MTC/ha
LANDSCAPE SCALE						
Playa	24.7	15.2	4.9	4.8	29.7	20.1
Arroyo	6.2	7.6	0.7	1.0	6.9	8.6
Grassland	4.7	5.0	5.7	8.4	10.4	13.4
Dunes	3.7	2.7	0.03	0.07	3.8	2.8
PLANT SCALE	0-30 cm					
Mesquite	12.0					
Interspace	8.6					

Figures represent mean weight (MT) of carbon per hectare. Landscape scale data modified from Virginia et al. (1992). Plant scale data modified from A.D. Abrams and W.M. Jarrell (unpublished data).

soil nutrient content heightened, which did not occur on grassland sites. OC and total N were higher and bulk density was lower in dune soils than in interdune soils, regardless of dune age. N, P, and OC were redistributed in the interdunes, but not depleted. This phenomenon is associated with accumulation of resources under mesquite: by concentrating resources beneath individual plants, and depleting adjacent areas, mesquite gains a competitive advantage over other plant species.

Grassland degradation is associated with redistribution of water and nutrient resources in these dune communities (Virginia and Jarrell, 1983; Schlesinger et al., 1990). Once underway, this positive feedback mechanism is difficult to reverse. Hence, management strategies designed to enhance C sequestration in these soil systems need to be developed within the context of these changes in landscape structure and function.

Interactions and feedbacks between soil aggregation, climatic conditions, key biological processes, and human activity have a significant effect upon soil heterogeneity and SOM dynamics. Variation of SOM and soil moisture distribution in rangelands results in spatial heterogeneity at fine and coarse scales (Table 5.1).

Carbon Sequestration and Spatial Heterogeneity in Soil Structure

C sequestration, soil structure, and SOM dynamics closely link in both croplands and rangelands. However, resource variability in rangelands makes predicting changes in C storage in the soil difficult (Christensen, 1996). Nutrient availability and distribution, plus the size and activity of soil microbial biomass pools, play major roles in the processes associated with C sequestration. Understanding the spatial distribution of the soil components is therefore important for enhancing and sustaining those processes in rangelands.

Desertification is thought to lead to reductions in net primary productivity (NPP) and increases in atmospheric CO_2 release, while reducing the economic productivity of nonforested arid lands (Daily, 1995). The alterations in C cycling may decrease SOC storage, which is in equilibrium with C inputs. However, little is known about the effects of changes in plant community composition on C sequestration (Connin et al., 1995).

Connin et al. (1995) investigated the impact of community level shifts on C balance in desert communities associated with desertification in southern New Mexico. Published data were used to estimate shoot-to-root ratios for shrub and black grama grass (*Bouteloua eriopoda* Torr.) in order to convert aboveground biomass to total biomass. Relative to grasslands, mesquite communities were found to accumulate biomass C while losing SOC, whereas tarbush communities appeared to gain SOC while losing biomass C. At an ecosystem level, C storage appeared unchanged. The authors noted that soil texture and landscape location of these communities varied and may have contributed to the observed differences in SOC.

In a related study, Connin et al. (1997) isolated light (<2.0 g/cm³) and heavy (>2.0 g/cm³) density fractions to indirectly assess the contributions to SOM of recent and historic vegetation. This fractionation, followed by subsequent characterization of C isotopes, suggested that SOM origins had shifted from C_4 plants to C_3 plants and that C sequestration in shrubland roots was significant. Workers also excavated trenches in adjacent sites and reported notable differences in soil horizonation. Sampling was done to a 1-m depth in this study, an improvement on other studies not designed to investigate C sequestration but still not nearly deep enough to capture all of the biologically active organic and inorganic C.

Roots of some shrubs, which penetrate to extreme soil depths, have been observed to sequester C at a depth of 3 to 5 m (Gile et al., 1995, 1998). While this may contribute to C sequestration and the competitiveness of deep-rooted woody shrubs, it also may alter the shrubland's economic value and its resistance and resilience to disturbance. The majority of similar studies focus sampling at the surface and shallow rooting depths (0 to 30 cm) because the vast majority of biotic and abiotic processes are regulated and highly concentrated at the surface of desert soils.

More studies which include deep sampling and control of confounding soil and landscape factors clearly are needed to assess different communities' potential to sequester C in arid land systems accurately.

Trigger Sites:
Capitalizing on Spatial Variability
to Increase Carbon Sequestration

Chapter 11 (Schuman et al.) discusses how the most dramatic changes in C sequestration are likely to result from modifying plant communities' composition. Range managers have been trying to alter plant community composition for millennia, but only in the past several decades has their tool box included much more than fire. Historically, range improvement practices have been applied primarily to increase forage availability for domestic livestock and, in some areas, for wildlife. A more recent objective has been to limit soil erosion. Attempts have been made to increase primary productivity by introducing or reintroducing one or more desirable plant species or by removing an undesirable species.

Historic attempts to modify plant community composition, increase productivity, and limit erosion have met with mixed success and frequently proved uneconomical. Differences in soil texture and depth, slope and aspect, and redistribution of water and nutrients, as well as large spatial and temporal differences in seed banks and pressure by native and domesticated herbivores, have defeated many attempts to apply agronomic practices such as plowing and seeding in much of the arid Southwest (Roundy and Biedenbender, 1995). Seedings in these areas often fail, and the soil surface disturbance associated with seedbed preparation often leads to increased erosion.

One review of 6 years of seeding trials in the Chihuahuan Desert showed that virtually no reseeding attempts produced a positive net economic return and many ended in complete failure (Ethridge et al., 1997). These limitations are confounded by the fact that, even when successful, it is not economically feasible to apply these capital- and labor-intensive practices across millions of hectares of degraded rangeland.

However, reseeding can be extremely successful in more mesic regions of the U.S. (Johnson and Bradshaw, 1979). Changes in species composition facilitated by reseeding in these areas may lead to significant changes in C sequestration potential (Schuman et al., Ch. 11). Reseeding of devegetated land, such as mine land, can increase C sequestration in these areas.

In most cases, range improvement practices have been applied at the pasture or section level (tens to thousands of hectares in area). Thus, the highly variable nature of resource availability generally has been treated as a barrier to developing effective range improvement strategies. This variability may be turned to an advantage by exploiting, rather than trying to overcome, spatial differences in resource availability. The discussion above, together with a review of range improvement projects in the Southwest (Roundy and Biedenbender, 1995), reveals that susceptibility to change varies widely across the landscape. By targeting limited resources to those areas which have the highest resource availability, the probability of success can increase significantly (Herrick et al., 1997).

These high potential areas can be thought of as *trigger sites*, both because a change can be triggered relatively easily and because a change in these sites frequently has the potential to trigger changes in the surrounding area. Trigger sites may be characterized by increased resource availability, more efficient use of existing resources, or both.

Trigger sites may range from a square meter to tens of hectares. For example, a trigger site may be as small as the area behind a temporary litter dam. Behind and within this dam, infiltration increases, due to higher residence time as runoff is slowed. A higher infiltration capacity can develop due to increased macroporosity generated by litter decomposers, or due to reduced physical crusting as a result of aggregate stabilization by decomposition by-products and protection from rain. The dam also traps seeds, and the cooler soil surface temperatures and higher availabile soil moisture enhances the microenvironment for establishing seedlings (Fowler, 1986). At coarser spatial scales, a run-in zone caused by micro-topographic variability may share many of the same characteristics of the litter dam but also have more silt and clay, resulting in higher soil moisture retention capacity.

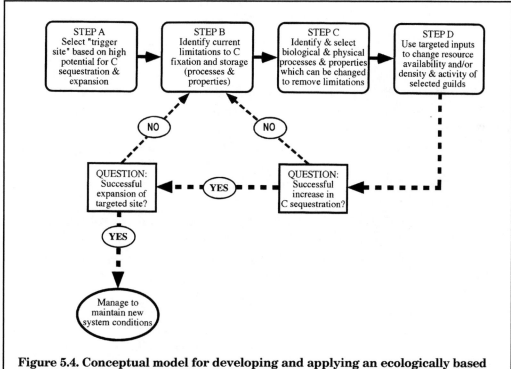

Figure 5.4. Conceptual model for developing and applying an ecologically based landscape approach to increasing carbon sequestration on rangelands (adapted from Herrick et al., 1997).

Herrick et al. (1997) described the trigger site concept for application to relatively small patches within landscapes. However, as Figure 5.3 suggests, the approach Figure 5.4 outlines could easily extend to the watershed or regional scale, by applying similar selection criteria to specific parts of the landscape and then selecting trigger sites within these areas. With a land systems approach (e.g., Dent and Young, 1981) together with remote sensing and GIS, this approach easily could apply across relatively large areas.

Carbon Sequestration and Other Ecosystem Functions

Ultimately, management decisions based on the potential for increased C sequestration require congruence between C accumulation and the fulfillment of other management objectives, such as retention of soil and water resources and conservation of biodiversity. In cropland systems, C sequestration and soil quality nearly always correlate positively. But it is not clear whether this same relationship extends to rangelands, where shrub-dominated systems may sometimes sequester more SOC but also frequently are associated with higher rates of soil erosion (Gibbens et al., 1983) and lower biodiversity (Huenneke, 1995).

Conclusions

The potential to increase storage of atmospheric CO_2-C by affecting soil structure through management practices in croplands has been documented well. (Karlen and Cambardella, 1996; Lal, 1997). Rangelands pose challenges due to their high degree of spatial variability associated with SOM, soil aggregation, and other resources.

This spatial variability also presents unique opportunities for management. Managers can exploit it by targeting inputs to those parts of the landscape which already have a high potential for enhanced productivity and C storage, due to their greater available resources.

Managers can implement this approach immediately. However, a more holistic understanding of rangeland systems and the relationships between soil structure, SOM, and nutrient and water distribution, at a variety of spatial scales, could enhance its success significantly. Finally, the relationships, both within and between plant communities, between C sequestration and other ecosystem functions, such as retention of soil and water resources and conservation of biodiversity, need clarifying.

Acknowledgments

This work was supported in part by a grant from the International Aridlands Commission. We thank Joel Brown, Greg Forbes, and Justin Van Zee for helpful suggestions and feedback while reviewing an earlier draft of this chapter.

References

Angers, D.A., and G.R. Mehuys. 1988. Effects of cropping on carbohydrate content and water-stable aggregation of a clay soil. *Can. J. Soil Sci.* 69:373-380.

Burke, I.C., C.M. Yonker, W.J. Parton, C.V. Cole, K. Flash, and D.S. Schimel. 1989. Texture, climate and cultivation effects on soil organic matter content in U.S. grassland soils. *Soil Sci. Soc. Am. J.* 53:800-805.

Butler, J.H.A., and J.C. Butterfield. 1979. Digestion of lignin by termites. *Soil Biol. Biochem.* 11:507-513.

Cepeda, P.J.G. 1986. *Spatial and temporal patterns of decomposition and microarthropod assemblages in decomposing surface leaf-litter on a Chihuahuan Desert watershed.* Ph.D. Dissertation, New Mexico State University. Las Cruces, New Mexico.

Chantigny, M.H., D.A. Angers, D. Prévost, L-P. Vézina, and F-P. Chalifour. 1997. Soil aggregation and fungal and bacterial biomass under annual and perennial cropping systems. *Soil Sci. Soc. Am. J.* 61:262-267.

Christensen, B.T. 1996. Carbon in primary and secondary organomineral complexes. *In* M.R. Carter and B.A. Stewart (eds), *Structure and Organic Matter Storage in Agricultural Soils*, CRC-Lewis, Boca Raton, FL, pp. 97-165.

Connin, S.L, R.A. Virginia, C.P. Chamberlain, and L.P. Huenneke. 1995. Dynamics of carbon storage in degraded arid land environments: a case study from the Jornada Experimental Range, New Mexico, USA. *In* V.R. Squires, E.P. Glenn, and A.T. Ayoub (eds), *Drylands and Global Change in the Twenty-First Century*, University of Arizona, AZ.

Connin, S.L., R.A. Virginia, and C.P. Chamberlain. 1997. Carbon isotopes reveal soil organic matter dynamics following arid land shrub expansion. *Oecologia* 110:374-386.

Cornelius, J.M., P.R. Kemp, J.A. Ludwig, and G.L. Cunningham. 1991. The distribution of vascular plant species and guilds in space and time along a desert gradient. *J. Veg. Sci.* 2:59-72.

Coronato, F.R., and M.B. Bertillier. 1996. Precipitation and landscape related effects on soil moisture in semi-arid rangelands of Patagonia. *J. Arid Environ.* 34:1-9.

Daily, G. 1995. Restoring value to the world's degraded lands. *Science* 269: 350-354.

DeJong, R., C.A. Campbell, and W. Nicholaichuk. 1983. Water retention and its relationship to soil organic matter and particle size distributions for disturbed samples. *Can. J. Soil Sci.* 63:291-302.

Dent, D., and A. Young. 1981. *Soil Survey and Land Evaluation.* George Allen and Unwin. London,.

Dexter, A.R. 1988. Advances in the characterization of soil structure. *Soil Tillage Res.* 11:199-238.

Doran, J.W., and T. B. Parkin. 1994. Defining and assessing soil quality. *In* J.W. Doran, D.C. Coleman, D.F. Bezdicek, and B.A. Stewart (eds), *Defining Soil Quality for a Sustainable Environment, SSSA Special Publication 35*, Soil Science Society of America, Madison, WI, pp. 3-21.

Dormaar, J.F., and R.C. Foster. 1991. Nascent aggregates in the rhizosphere of perennial ryegrass (*Lolium perenne L.*). *Can. J. Soil Sci.* 71: 465-474.

Elliott, E.T. 1986. Aggregate structure and carbon, nitrogen, and phosphorus in native and cultivated soils. *Soil Sci. Soc. Am. J.* 50:627-633.

Elliott, E.T., and D.C. Coleman. 1988. Let the soil work for us. *Ecol. Bull.* 39:23-32.

Emerson, W.W., R.C. Forster, and J.M. Oades. 1986. Organo-mineral complexes in relation to soil aggregation and structure. *In* P.M. Huamf and M. Schnitzer (eds), *Interactions of Soil Minerals with Natural Organics and Microbes*, Soil Science Society of America, Madison, WI, pp. 521-548.

Ethridge, D.E., R.D. Sherwood, R.E. Sosebee, and C.H. Herbel. 1997. Economic feasibility of rangeland seeding in the arid southwest. *J. Range Manag.* 50:185-190.

Follett, R.F., S.C. Gupta, and P.C. Hunt. 1987. Conservation practices: relation to management of plant nutrients for crop production. *In* R.F. Follett (ed), *Soil Fertility and Organic Matter as Critical Components of Production Systems, SSSA Special Publication 19*, Soil Science Society of America, Madison, WI, pp. 19-51.

Fowler, N.L. 1986. Microsite requirements for germination and establishment of three grass species. *Am. Mid. Nat.* 115(1):131-145.

Freckman, D.W., and R. Mankau. 1986. Abundance, distribution, biomass and energetics of soil nematodes in a northern Mojave Desert ecosystem. *Pedobiologia* 29:129-142.

Gallardo, A., and W.H. Schlesinger. 1992. Carbon and nitrogen limitations of soil microbial biomass in desert ecosystems. *Biogeochem.* 18:1-17.

Gibbens, R.P., J.M. Tromble, J.T. Hennessy, and M. Cardenas. 1983. Soil movement in mesquite dunelands and former grasslands of southern New Mexico from 1933 to 1980. *J. Range Manage.* 36:145-148.

Gile, L.H., R.P. Gibbens, and J.M. Lenz. 1995. Soils and sediments associated with remarkable, deeply-penetrating roots of crucifixion thorn (*Koeberlinia spinosa* Zucc.). *J. Arid Environ.* 31:137-151.

Gile, L.H., R.P. Gibbens, and J.M. Lenz. 1998. Soil-induced variability in root systems of creosotebush (*Larrea tridentata*) and tarbush (*Flourensia cernua*). *J. Arid Environ.* 39:57-78.

Golchin, A., J.A. Baldock, and J.M. Oades. 1997. A model linking organic matter decomposition, chemistry, and aggregate dynamics. *In* R. Lal, J.M. Kimble, R.F. Follett, and B.A. Stewart (eds), *Soil Processes and the Carbon Cycle,* CRC Press, Boca Raton, FL, pp. 245-266.

Herrick, J.E., K.M. Havstad, and D.P. Coffin. 1997. Rethinking remediation technologies for desertified landscapes. *J. Soil Water Conserv.* 52(4):220-225.

Herrick, J.E., M.A. Weltz, J.D. Reeder, G.E. Schuman, and J.R. Simaton. 1998. Range-land soil erosion and soil quality: role of soil resistance, resilience, and disturbance regime. *In* R. Lal (ed), *Soil Quality and Soil Erosion*, Soil and Water Conservation Society, Ankeny, IA, pp. 209-233.

Holt, J.A. 1987. Carbon mineralization in semi-arid northeastern Australia: the role of termites. *J. Trop. Ecol.* 3(3):255-263.

Huenneke, L.F. 1995. Shrublands and grasslands of the Jornada Long-Term Ecological Research Site: desertification and plant community structure in the northern Chihuahuan Desert. *In* J.R. Barrow, E.D. McArthur, R.E. Sosebee, and R.J. Tausch (compilers), *Proceedings: Shrubland Ecosystem Dynamics in a Changing Environment*, U.S. Forest Service Intermountain Research Station, Ogden, UT, pp.48-50.

Jastrow, J.D., and R.M. Miller. 1991. Methods for assessing the effects of biota on soil structure. *Agric. Ecosyst. Environ.* 34:279-303.

Jastrow, J.D., and R.M. Miller. 1997. Soil aggregate stabilization and carbon sequestration: feedbacks through organomineral associations. *In* R. Lal, J.M. Kimble, R.A. Follett, and B.A. Stewart (eds), *Soil Processes and the Carbon Cycle*, CRC Press, Boca Raton, FL, pp. 207-223.

Johnson, M.S., and A.D. Bradshaw. 1979. Ecological principles for the restoration of disturbed and degraded land. *Appl. Biol.* 4:141-200.

Jones, J.A. 1990. Termites, soil fertility and carbon cycling in dry tropical Africa: a hypothesis. *J. Trop. Ecol.* 6:291-305.

Karlen, D.L., and C.A. Cambardella. 1996. Conservation strategies for improving soil quality and organic matter storage. *In* M.R. Carter and B.A. Stewart (eds), *Structure and Organic Matter Storage in Agricultural Soils*, CRC-Lewis, Boca Raton, FL, pp. 395-420.

Kay, B.D. 1990. Rates of change of soil structure under different cropping systems. *Adv. Soil Sci.* 12:1-52.

Kay, B.D. 1997. Soil structure and organic carbon: a review. In R. Lal, J.M. Kimble, R.F. Follett, and B.A. Stewart (eds), *Soil Processes and the Carbon Cycle*, CRC Press, Boca Raton, FL, pp. 169-197.

Kern, J.S., and M.G. Johnson. 1993. Conservation tillage impacts on national soil and atmospheric carbon levels. *Soil Sci. Soc. Am. J.* 57:200-210.

Kimble, J.M., R.F. Follett, and B.A. Stewart (eds). *Soil Processes and the Carbon Cycle*. CRC Press, Boca Raton, FL.

Kooistra, M.J., and M. van Noordwijk. 1996. Soil architecture and distribution of organic matter. *In* M.R. Carter and B.A. Stewart (eds), *Structure and Organic Matter Storage in Agricultural Soils*, CRC-Lewis, Boca Raton, FL, pp. 15-56.

Lal, R. 1997. Residue management, conservation tillage, and soil restoration for mitigating greenhouse effect by CO_2 enrichment. *Soil Tillage Res.* 43:81-107.

Lee, K.E., and T.G. Wood. 1971. Physical and chemical effects on soils of some Australian termites, and their pedological significance. *Pedobiologia* 11:376-409.

Matterechera, S.A., A.R. Dexter, and A.M. Alston. 1992. Formation of aggregates by plant roots in homogenized soils. *Plant Soil* 69:69-79.

Miller, R.M., and J.D. Jastrow. 1990. Hierarchy of root and mycorrhizal fungal interactions with soil aggregation. *Soil Biol. Biochem.* 22:579-584.

Nash, M.H., and W.G. Whitford. 1995. Subterranean termites: regulators of soil organic matter in the Chihuahuan Desert. *Biol. Fert. Soils* 19:15-18.

Oades, J.M., and A.M.A. Waters, 1991. Aggregate hierarchy in soils. *Aust. J. Soil Res.* 29:815-828.

Oades, J.M., A.M. Vassallo, A.M.A. Waters, and M.A. Wilson. 1987. Characterization of organic matter in particle size and density fractions from a red-brown earth by solid state 13C NMR. *Aust. J. Soil Res.* 25:71-82.

Perfect, E., B.D. Kay, W.K.P. van Loon, W. Sheard, and T. Pojasok. 1990. Factors influencing soil structural stability within a growing season. *Soil Sci. Soc. Am. J.* 54:173-179.

Roundy, B.A., and S.H. Biedenbender. 1995. Revegetation in the desert grassland. *In* M.P. McClaran and T.R. Van Devender (eds), *The Desert Grassland*, The University of Arizona Press, Tucson, AZ, pp. 265-303.

Santos, P.F., E. DePree, and W.G. Whitford. 1978. Spatial distribution of litter and microarthropods in a Chihuahuan desert ecosystem. *J. Arid Environ.* 1:41-48.

Schlesinger, W.H. 1990. Evidence from chronosequence studies for a low carbon-storage potential of soils. *Nature* 348:232-234.

Schlesinger, W.H., P.J. Fonteyn, and W.A. Reiners. 1989. Effects of overland flow on plant water relations, erosion, and soil water percolation on a Mojave desert landscape. *Soil Sci. Soc. Am. J.* 53:1567-1572.

Schlesinger, W.H., J.F. Reynolds, G.L. Cunningham, L.F. Huenneke, W.M. Jarrell, R.A. Virginia, and W.G. Whitford. 1990. Biological feedbacks in global desertification. *Science* 247:1043-1048.

Schlesinger, W.H., J.A. Raikes, A.E. Hartley, and A.F. Cross. 1996. On the spatial pattern of soil nutrients in desert ecosystems. *Ecology* 77:364-374.

Smith, J.L., J.L. Halvorson, and H. Bolton, Jr. 1994. Spatial relationships of soil microbial biomass and C and N mineralization in a semi-arid shrub-steppe ecosystem. *Soil Biol. Biochem.* 26:1151-1159.

Soane, B.D. 1990. The role of organic matter in soil compactability: a review of some practical aspects. *Soil Tillage Res.* 16:179-201.

Tisdall, J.M., and Oades, J.M. 1982. Organic matter and water-stable aggregates in soils. *J. Soil Sci.* 33:141-163.

Virginia, R.A., and W.M. Jarrell. 1983. Soil properties in a mesquite-dominated Sonoran Desert ecosystem. *Soil Sci. Soc. Am. J.* 47:138-144.

Waters, A.G., and J.M. Oades. 1991. Organic matter in water stable aggregates. *In* W.S. Wilson (ed), *Advances in Soil Organic Matter Research: The Impact on Agriculture and the Environment*, The Royal Society of Chemistry, Melksham, England, pp. 163-174.

Yang, X.M., and M.M. Wander. 1998. Temporal changes in dry aggregate size and stability: tillage and crop effects on a silty loam Mollisol in Illinois. *Soil Tillage Res.* 49:173-183.

CHAPTER **6**

Root Biomass and Microbial Processes

J.D. Reeder,[1] C.D. Franks,[2]
and D.G. Milchunas[3]

Introduction

Increases in atmospheric CO_2 measured over the last few decades are likely to continue and lead to significant global climate changes during the coming decades. To predict with any confidence these climatic and ecological consequences, we must identify globally significant sources and sinks of carbon (C) and define the relationships between them (Tans et al., 1990; Canadell et al., 1996a).

Simulation model analyses and measurements of the $^{13}C/^{12}C$ ratio of atmospheric CO_2 have indicated the existence of a large terrestrial C sink at temperate latitudes in the northern hemisphere, but the nature of this sink remains unknown (Tans et al., 1990; Ciais et al., 1995). Grazing lands in temperate latitudes may represent a significant portion of this C sink. U.S. publicly and privately owned grazing lands (rangelands and pastures) constitute about 336 Mha (Sobecki et al., Ch. 2). Grasses usually dominate the plant communities of these grazed ecosystems, but the potential for grassland ecosystems to sequester C is not well understood.

Fundamental to evaluating whether grasslands are a source or sink for C is understanding the magnitude and dynamics of root biomass and the soil microbial processes that mediate C cycling in the soil. This is because plant roots provide the primary input of organic C (OC) to grassland soils (Dormaar, 1992), and microbial oxidation of soil organic C (SOC) occupies a key position in the global C cycle by providing the principal means by which terrestrial OC is returned to the atmosphere as CO_2 (Zibilske, 1994).

[1]USDA-ARS, 1701 Centre Ave., Fort Collins, CO 80526, U.S.A.

[2]USDA-NRCS, National Soil Survey Center, 100 Centennial Mall N., Lincoln, NE 68508, U.S.A.

[3]Dept. Rangeland Ecosystem Science and Natural Resource Ecology Laboratory, Colorado State University, Fort Collins, CO 80523, U.S.A.

The dense, fibrous root systems of grasses produce soils high in organic matter (OM), of which about 58% is C. Grasslands may contain 40,000 MMT of C in the top 2 m of the soil profile, or about 12.6 ± 4.3 kg/m^2 (Follett, Ch. 3). This large reservoir of C stored in the soil organic matter (SOM) represents about 90% of the total C in most temperate grassland systems (Burke et al., 1997).

SOC is significant for more than just overall pool size, however. Recalcitrant C fractions of the SOM have the longest residence times relative to plant biomass (i.e., hundreds to thousands of years), so changes in these fractions have the largest effects on the capacity of grassland ecosystems to sequester C long-term (Canadell et al., 1996a).

To understand the role of grazing lands in C sequestration, it is important to understand how ecosystems dominated by grasses differ from forest and agricultural ecosystems, in the proportion and magnitude of total plant biomass that is below ground and in the type, quantity, and temporal distribution of plant residues entering the soil (Tate, 1987). This chapter discusses the biomass distribution and turnover of roots and root C in grassland systems; the role of soil microorganisms in C cycling; and the effects of management practices on root biomass, microbial activity, and C sequestration in grassland ecosystems.

Biomass, Distribution, and Turnover of Roots and Root Carbon

Root morphology and function

Plants' roots serve several important roles. The root system provides water and nutrients to the plant from the surrounding soil and soil solution; provides a concentration gradient enabling translocation of nutrients throughout the plant; serves as a carbohydrate storage organ for plants; supports and anchors the plant; and reduces erosion around the plant.

Root morphology varies with type of plant. Grasses form a dense fibrous root system consisting of many individual roots of nearly equal length and diameter (Salsbury and Ross, 1992) (Fig. 6.1). New root growth occurs from the tips of roots, and older roots attain a maximum diameter when cells fully expand (Russell, 1977). In comparison, shrubs, half shrubs, and forbs in grazing lands form a taproot system consisting of a single large main root and small lateral roots (Salsbury and Ross, 1992) (Fig. 6.1). Older roots can progressively increase in diameter, and the mechanical strength of these thick roots provides support to large, woody shoot systems (Russell, 1977).

Functional roots can be divided into three principal zones: *the root tip*, a region of rapid growth, exudation, and cell sloughage; *the root hair zone*, a region of nutrient uptake and mycorrhizal symbiosis; and *the zone of suberization and lat-*

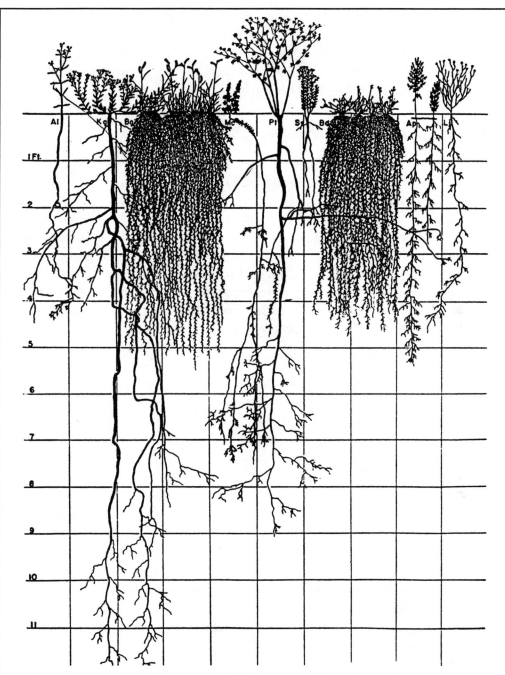

Figure 6.1. Roots of different grassland plants. Grasses: Bg, blue grama (*Bouteloua gracilis*); Bd, buffalo grass (*Buchloe dactyloides*). Forbs: Al, narrow-leafed 4-o'clock (*Allionia linearis*); Kg, prairie false boneset (*Kuhnia gultinosa*); Mc, globemallow (*Malvastrum coccineum*); Pt, a legume (*Psoralea tenuiflora*); Ss, *Sideranthus spinulosis*; Ap, western ragweed (*Ambrosia psilostachya*); Li, skeleton weed (*Lygodesmia juncea*) (Thorp, 1948).

eral growth, where nutrient uptake decreases (Moore et al., 1991). Although representing a small percentage of total root biomass, root hairs and the tips of young growing roots are the primary locations for absorption of water and dissolved nutrients (Baron, 1967), and they absorb as much as 100 times more water than suberized roots (Ares, 1976; Coleman, 1976).

Root biomass and distribution within the soil profile

Understanding and predicting grassland ecosystem C flux requires an accurate assessment of belowground plant biomass, distribution, and turnover. However, problems of definition and methodology hamper all evaluations of root biomass and turnover. The amount of root biomass present at a particular site depends on the life span of the roots and the decomposition rate of dead roots (Sims and Coupland, 1979), but the term "root biomass" is difficult to define clearly with respect to whether a root is live or dead and to when a dead root becomes SOM.

The different procedures used to collect root biomass variably capture detrital roots, as well as the finest roots, root hairs, and rhizosphere microorganisms, so the selected root-soil separation procedures determine when root C is defined as SOM. Various methods have been developed to establish the relative proportions of live and dead root biomass (Dormaar, 1992; Redmann, 1992); all of these techniques have limitations, and few have actual users (Dormaar, 1992). Because determining the ratio of live:dead root material is so difficult, it frequently is not even attempted in studies evaluating root biomass. In this discussion, therefore, "root biomass" refers to (live + dead) root material, with various degrees of fine root loss and/or SOM inclusion.

Root:shoot ratios

Grasslands differ from forest and agricultural ecosystems in both the proportion and the magnitude of total plant biomass that is below ground. In a synthesis of existing databases, Jackson et al. (1996) reported that the root:shoot ratios of temperate grasslands (avg. 3.7) were more than an order of magnitude higher than those of temperate forests (0.18 to 0.23) or annual croplands (avg. 0.10). However, the average root biomass of temperate grasslands (1.4 kg/m^2) was three times lower than the average root biomass of temperate deciduous or coniferous forests (4.2 to 4.4 kg/m^2), though it was higher by an order of magnitude than the average root biomass of annual cropping systems (0.15 kg/m^2).

Although the average root:shoot ratio of temperate grasslands is around 3.7 (Jackson et al., 1996), the range is quite broad across different grassland ecosystems. Belowground net primary production in temperate grasslands ranges from 40 to 85% of total net primary production (Milchunas and Laurenroth, 1992), resulting in root:shoot ratios ranging from about 1:1 in cool-mesic systems to 8:1 in warm-arid habitats (Coleman, 1976).

Rooting depths

Compared to some other ecosystems, grasslands are considered shallowly rooted. Temperate grasslands, as well as tundra and boreal forests, generally display shallower rooting profiles (avg. 2.6, 0.5, and 2.0 m, respectively) than do deserts (avg. 9.5 m) or temperate coniferous forests (avg. 3.9 m) (Jackson et al., 1996; Canadell et al., 1996b). Within grassland habitats, the plant community includes three rooting types, based on depth: widely spreading, *superficially rooted* (0 to 10cm) species such as cacti; *shallowly rooted* species such as grasses, which have the majority of their dense fibrous root systems in the uppermost 40 cm of the soil, although some roots usually penetrate much deeper; and *deeply rooted* species, which include shrubs, half-shrubs, and forbs with primary taproot systems often penetrating to depths >1 m but with lateral roots in the upper soil layers (Laurenroth and Milchunas, 1992) (Fig. 6.1).

In ecosystems dominated by grasses, about 75 to 80% of total root biomass is in the top 30 cm of soil, and about 44 to 57% is in the top 10 cm (Sims et al., 1978; Lorenz, 1977; Jackson et al., 1996). Although the root biomass of all grasses generally decreases rapidly with depth, individual species differ in rooting pattern (Weaver, 1958). For example, blue grama, a warm season (C_4) grass, has a dense root system with about 85% of its root mass in the top 15 cm of the soil, whereas the majority of the root biomass of cool season (C_3) grasses, such as needle-and-thread and western wheatgrass, distributes more uniformly in the top 30 cm of the soil profile (Coupland and Johnson, 1965).

Biomass variation with habitat

Table 6.1 presents root biomasses and distributions in the soil profiles of nine North American grasslands. The data indicate that the largest root biomasses occurred in the mixed grass prairie habitats (range 1100 to 2200 g/m²), while the desert grassland had the smallest amount of root material (avg. 159 g/m²).

As Sims and Coupland (1979) discuss, the data illustrate several important points. Average biomass of total underground parts generally decreases with increasing average temperature, as indicated in the mixed grass sites by the decrease with latitude in average total underground biomass — from 1845 g/m² at Matador (Sask.), to 1458 g/m² at Dickinson (ND), 1345 at Cottonwood (SD), and 1121 g/m² at Hays (KS). Belowground biomass within the short grass prairie sites similarly declined from the Pawnee (CO) site (avg. 1120 g/m²) to the Pantex (TX) site (avg. 712 g/m²).

Root biomass generally increases with increasing average precipitation, as the three southernmost sites suggest, where mean total underground biomass de-

Table 6.1. Biomass of underground plant parts in g/m² in nine natural temperate grasslands.[1]

	Tall Grass Site			Mixed Grass Sites							
	Osage, OK[3]			Matador, Sask			Dickenson, ND	Cottonwood, SD			Hays, KS
	1970	1971	1972	1968	1969	1970	1970	1970	1971	1972	1970
MAT[2]	15.2	15.2	14.9	4.1	2.5	2.8	4.1	8.0	7.8	5.7	11.6
Roots, cm											
0-10	516	624	435	842			998	483	782	689	793
10-20	214	176	104	312			160	185	207	263	328
20-30	320	242	140	216			108	133	156	189	bedrock
30-40				184			77	103	115	157	
40-50				167			58	86	98	123	
50-60				124			57	80	75	111	
Total Roots	1050	1042	679	1845	1898[4]	2220[4]	1458	1070	1433	1532	1121

	Mountain Site	Short Grass Sites						Desert Site		
	Bridger, MT	Pawnee, CO			Pantex, TX			Joranada, NM		
	1972	1970	1971	1972	1970	1971	1972	1970	1971	1972
MAT[2]	2.9	7.0	8.0	8.0	14.3	13.6	13.2	14.3	14.7	13.4
Roots, cm										
0-10	889	1039	573	449	556	223	373	96	89	73
10-20	312	168	123	97	112	50	128	57	41	28
20-30	203		92	79	108	36	82	40	33	19
30-40	144	212	68	61	71		64	--- caleche layer --		
40-50	91		66	58	78		60			
50-60	---	152	48	73	142		55			
Total Roots	1639	1571	970	817	1067	309	762	193	163	120

[1]Data from Sims and Coupland (1979) and Sims and Singh (1978).
[2]MAT = mean annual temperature, °C.
[3]Root biomass data for 30-50 cm not available.
[4]Profile distribution data not available.

creased with increasing aridity — from tall grass prairie (Osage, OK, 926 g/m²), to short grass steppe (Pantex, TX, 712 g/m²), to desert grassland (Jornada, NM, 159 g/m²). SOC content also generally increases with precipitation and decreases with temperature (Burke et al., 1989), which reflects the importance of the contribution of roots to the formation of SOM.

Annual variation

The data in Table 6.1 also illustrate the wide yearly variations in root biomass common in grassland systems (Sims et al., 1978; Coupland, 1992). These annual variations result from several factors. One is yearly variation in climatic factors (precipitation, temperature, evapotranspiration, and solar radiation), which affect net primary production (Sims et al., 1978) and plant species composition (Milchunas et al., 1989).

Another factor is inadequate sampling to address field heterogeneity in root mass and profile distribution. Plant patchiness causes the wide variation in root mass and distribution that occur in grassland ecosystems (Charley, 1977; Milchunas and Laurenroth, 1989), as do differences in plant species composition associated with topography and soil type (Laurenroth and Milchunas, 1992). Finally, since root biomass may vary intraseasonally, part of the yearly variation may result from variations in the time of the year of sampling. Fine roots frequently contribute the majority of belowground production, and their life expectancy can be as short as a few weeks (Jackson et al., 1996), resulting in measurable variation in root biomass during the growing season.

Wide intraseasonal fluctuations in root biomass have been reported for tall grass (Kucera, 1992), mixed grass (Laurenroth and Whitman, 1977; Coupland, 1992), and short grass (Ares, 1976) habitats. In habitats containing a large cool season grass component, maximum root biomass usually occurs in late spring or early summer (Coleman, 1976; Coupland, 1992), whereas in habitats dominated by warm season grasses, maximum root biomass usually occurs toward the end of the growing season (Ares, 1976; Parton et al., 1978). However, fluctuations in root biomass relate to temperature and precipitation (Laurenroth and Whitman, 1977), so erratic temperature and precipitation patterns can suppress or accelerate plant production and alter the time at which maximum root biomass occurs.

A long-term record of root biomass at a short grass site (Fig. 6.2) demonstrates the wide intraseasonal and annual fluctuations that one can encounter, even given a large number of root samples (Milchunas and Laurenroth, 2000). No consistent seasonal patterns in root biomass or in overwinter loss patterns appeared over a period of 13 years, although longer trends in increases and decreases of root biomass were apparent. When conservative statistical constraints are placed on seasonal peaks and troughs, root biomass often does not differ significantly throughout the year, due to conditions simultaneously favorable for both growth and decomposition.

Sampling depth

Most studies of SOM's dynamics in grazing lands have limited sampling to the top 30 to 40 cm of the soil, because that portion of the soil profile contains the highest concentrations of SOM, as well as approximately 80% of the root biomass.

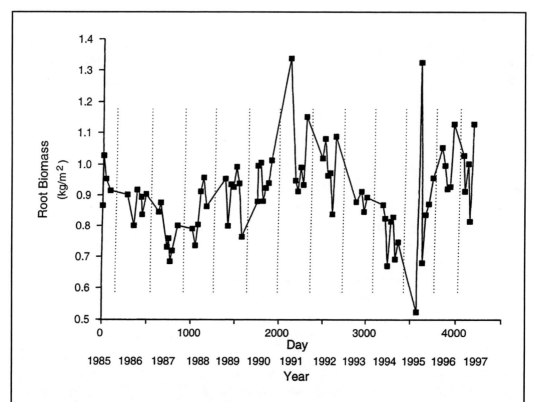

Figure 6.2. Seasonal root biomass dynamics over 13 years at a short grass steppe site in northern Colorado. Each point represents 40 soil cores (6.6 cm diam. x 20 cm) (Milchunas and Laurenroth, 2000).

However, maximum rooting depth for temperate grasslands has been estimated at 2.6 ± 0.2 m, and rooting depths as high as 6 m have been observed (Canadel et al., 1996b).

Few studies of grasslands have evaluated deep SOM and root dynamics, despite the fact that approximately 40% of soil C can be stored below 30 cm (Weaver et al., 1935). Deep roots provide C to the soil, driving the active C cycle deep into the subsoil (Trumbore et al., 1995; Nepstad et al., 1994; Richter and Markewitz, 1995). Deep soil C accumulations in grassland soils are disproportionately higher than root biomass distribution would suggest (Weaver et al., 1935; Gill, 1998), because the cooler and more oxygen-limited environment of the subsoil lowers decomposition rates.

To be accurate, assessments of the potential C sequestration of grazing lands must include quantification of the mass and turnover of subsoil roots and SOM. Recent studies (e.g., Gill, 1998; Gill et al., 1999) to evaluate the mechanisms controlling decomposition rates and distribution of surface and subsoil C in grasslands should improve our understanding of their C storage capacity.

Turnover of roots and root carbon

Because the root system includes functional, living material, as well as dead biomass at various stages of decomposition, the ability to establish the ratio of live to dead root material is critical in determining rates of transfer of root C to SOM. However, with root growth, death, and decomposition occurring simultaneously, accurate measurements of these various sources of C loss to the soil are difficult to obtain, and results vary depending on the procedures used.

Dormaar (1992) suggested that procedures Singh and Coleman (1973) proposed, using ^{14}C, have more potential than other methods for ascertaining functional root biomass at different times during the growing season. However, these procedures are elaborate and require expensive equipment. The scarcity of databases on soil:root relationships results directly from the elaborate, labor-intensive procedures which direct examinations of the underground components require (Dormaar, 1992).

Exudates and sloughing

Plant C is released to the soil either by the death and decomposition of roots or by exudation and sloughing from living roots. Organic exudates and sloughed root materials can represent significant losses from the roots, but their roles in C cycling are not well understood (Redmann, 1992; Paul and Clark, 1996). Mucigel, a polysaccharide material produced at the tip of a growing root, is secreted from the root cap as a lubricant as the root moves through the soil, and it provides an excellent energy source for soil microorganisms (Salisbury and Ross, 1992). Root exudates are low molecular weight compounds, such as sugars, amino acids, organic acids, and enzymes, that leak from plant roots and control the metabolic activity of the microbial community (Tate, 1987).

Root exudation of C has been reported to represent 6 to 17% of the net C fixed by photosynthesis during one growing season in grass species such as western wheatgrass and blue grama (Biondini et al., 1988; Milchunas et al., 1985; Milchunas and Laurenroth, 1992). Others have reported that more than one third of the photosynthate reaching healthy roots can be lost to the soil as sloughed cap cells, mucilages, soluble exudates and lysates, and decaying root hairs and outer cortical cells (Foster, 1988; Bedunah and Sosebee, 1995).

Decomposition

The majority of plant C that flows to the soil does so through root death and decomposition. Coleman (1976) suggested that annual total root turnover is generally lower in more mesic habitats, such as tall grass prairie (about 25% per year), than in more xeric systems, such as short grass steppe (>50% per year). However, the rate of root turnover depends on a number of factors. Rates vary widely from year to year, depending on climatic conditions during the growing season.

Winter mortality also can affect annual rate of root turnover. Dormaar et al. (1981) reported winter losses of over 50% of blue grama root mass at a Canadian mixed grass site. In comparison, Milchunas and Laurenroth (1992) did not observe any consistent winter declines in root biomass at a short grass steppe site in Colorado dominated by the same species, and they estimated root turnover to be 8 years by a ^{14}C-decay method.

Another factor in the overall turnover rate of living roots is the species composition of the plant community. Plant species differ in the relative resistance of their roots to decomposition (Weaver, 1958; Caldwell, 1979) because the proportions and size of their structural roots, fine roots, and root hairs differ. These functional zones of roots vary in decomposition rate (Paul and Van Veen, 1978). Much of root tip and root hair regions are short-lived, with 30% to 60% dying within a few weeks after cessation of elongation (Ares, 1976). Dead fine roots and root hairs decompose rapidly (Dormaar, 1992). In comparison, decomposition of the bulk of roots, the suberized zone, is slower because of higher lignin and cellulose contents (Moore et al., 1991).

The roots of different plant species in grasslands vary in distribution within the soil, in chemical composition, and in their uptake of nutrients from different regions in the soil at different times and in different intensities (Caldwell, 1979; Dormaar, 1992). These differences result in significant differences in the quantity, distribution, and quality of OM in the soil profile (Dormaar, 1975; Lutwick and Dormaar, 1976); thus they cause the C sequestration potential of different grazing land habitats to vary.

Measurement problems

Methods of estimating root production and turnover are the primary limitation to our understanding of belowground C dynamics. Traditional root harvest procedures and calculations of productivity by positive increments in biomass have serious biases and errors. Thus the conclusions of earlier studies, which estimated root production and turnover based on positive increments in growth, have come into question.

Overestimations of productivity of up to 700% can occur due to a statistical artifact (Singh et al., 1984; Laurenroth et al., 1986; Sala et al., 1988a, b). In other cases, annual estimates of zero production can be obtained, due either to simultaneous rates of growth and decomposition or to the statistical constraint used to define an increase in biomass (Hansson and Andren, 1986; Kurz and Kimmins, 1987; Fogel, 1990; Milchunas and Laurenroth, 1992). Problems associated with new methods, such as the minirhizotron or C-isotope decay method, currently are being explored (e.g., Milchunas and Laurenroth, 1992; Pages and Bengough, 1997).

Given the current status of methodology, we readily can address questions about C sequestration in the roots of grazing lands through estimates of root biomass, but we cannot with confidence now determine what factors control root turnover and the flow of plant C to the soil.

Microbial Processes and Carbon Cycling

The biota (living component) of soils consists of microflora (bacteria, fungi, and algae) and microfauna (protozoa), the meso- and macrofauna (e.g., nematodes, mites, insects, and earthworms), and the belowground portion of higher plants, all in intimate association with each other as well as with the abiotic (non-living) constituents of the soil (Mitchell, 1986; Tate, 1987). The microflora and microfauna constitute the living component of SOM and represent about 0.3 to 5% of the total SOC (Anderson and Domsch, 1980) and up to 5% of the total soil N (Howarth and Paul, 1994). Microorganisms perform chemical reactions that provide plant nutrients and regulate the flux of C between the atmosphere and terrestrial ecosystems (Tate, 1987).

Metabolic processes

Metabolic processes are the chemical reactions within a living organism by which it produces and maintains its cells and makes energy available. As with all living organisms, soil microorganisms require both a source of energy and substrates (raw materials) for biosynthesis of new microbial cells.

Organisms are either of two types, depending on the sources from which they derive C and energy. Green plants and a few soil bacteria are *autotrophs,* or producers, because they synthesize OM (all their carbohydrates, proteins, vitamins, and enzymes) entirely from inorganic materials (Jenny, 1980). Autotrophs cycle atmospheric CO_2 into organic forms by using CO_2 as a sole C source. Autotrophs gain energy from either photosynthesis or the oxidation of inorganic substrates.

Most soil microorganisms and all fauna, however, are *heterotrophs*, organisms that require organic substrates both as a source of C and as a source of energy (Tortora et al., 1982; Alexander, 1977). Heterotrophs are the decomposers, transforming plant, animal, and microbial residues into inorganic substances, and thus they are responsible for recycling terrestrial OC to atmospheric CO_2. The products of microbial decomposition of organic substrates are energy, new living microbial cells, organic metabolite waste products, water, CO_2, cations, and anions (Jenny, 1980). Microorganisms further metabolize organic waste products to produce soil humus (SOM), the large pool of highly polymerized and chemically complex organic compounds that resist decomposition.

Factors affecting microbial activity rate

Physical and chemical limitations

Physical or chemical limitations to the soil microbial community result in variations in the decomposition rate of OM and, hence, variations in the rate of C cycling. Researchers usually identify the primary constraints on microbial activity as soil temperature, soil moisture, and availability of substrate C (Follett, 1998), although other factors also affect microbial activity — e.g., soil texture, soil aggregation, and significant interactions between temperature, water potential, and nutrient supply (Tate, 1987; Mitchell, 1986).

Temperature is one of the most important environmental conditions determining how rapidly OM decomposes. Maximum rates are considered to occur at about 30 to 35°C (86 to 95°F); above or below this optimal range, the decomposition rate generally declines. Soil C content and turnover time negatively correlate with mean annual temperature (Schimel et al., 1994). Thus, the average C content of grassland soils of the Central Great Plains decreases along temperature transects from Canada to southern Texas, reflecting increased decomposition rates associated with increasing mean annual temperature (Amelung et al., 1998; Burke et al., 1989).

Soil water content affects not only the moisture available to organisms but also soil aeration status, the concentration of soluble materials, and the pH of the soil solution (Paul and Clark, 1996). The optimal soil water condition for microbial activity is at approximately field capacity soil water content, which corresponds to approximately 60% of soil pore space filled with water (Linn and Doran, 1984). Microbial activity decreases as the soil becomes either waterlogged or droughty, but different groups of microorganisms display varying degrees of tolerance to waterlogged or droughty conditions. Bacterial activity declines sharply as soil dries, because movement to new locations of nutrients becomes restricted (Sommers et al., 1980; Griffin, 1980). Fungi are generally more tolerant of water stress than bacteria, because they can extend hyphae across dry, air-filled pores and actively explore for nutrients and moisture (Orchard and Cook, 1983).

Soil texture influences the amount, distribution, and turnover rate of SOM. The amount of OC in soil generally increases as clay content increases (Burke et al., 1989; Schimel et al., 1994). Decomposition rates of organic substrates tend to be lower in clay soils, in part because adsorption onto clay surfaces physically protects OM from microbial decomposition (Hassink et al., 1993). Significant interactions between soil texture, temperature, and water potential affect the amount and decomposition rate of SOM.

Soil aggregation determines the size and distribution of pores, as well as the spatial arrangement by type and amount of OM within the soil fabric, and as such is of prime importance in controlling microbial activity and OC turnover (Paul and Clark, 1996). OC incorporated and physically stabilized within aggregates is less exposed to microbial decomposition than free OM found between aggregates (Six et al., 1998). Soil pore space, as determined by aggregate structure, provides a great variety of habitat niches for the soil biota, as well as sites for OM decomposition (Elliott, 1980; Killham et al., 1993).

Because pore and pore neck size determine soil organisms' entry into and occupancy of pores, different populations are associated with different size pores. Nematode biomass correlates with pores 30 to 90 µm in diameter, while fungi and protozoa generally require pore neck sizes >6 µm for entry (Paul and Clark, 1996), and amoebae can occupy pores as small as 8 µm (Bamforth, 1988). Most soil bacteria inhabit pores of <6 µm neck diameter, and pores 0.3 to 3 µm in diameter are most favorable for bacterial colonization (Killham et al., 1993; Paul and Clark, 1996). Bacteria thus frequently reside within the interiors of microaggreagates, which contain pores mainly in the range of 0.2 to 6 µm diameter (Paul and Clark, 1996).

Pore neck size and diameter also affect water-retaining capacity and consequently affect the growth and survival of the inhabiting microbial population. In a drying soil, major pore systems lying between aggregates drain readily, while aggregates themselves may remain saturated. Aeration thus initially improves for microorganisms lying on the edge of a major pore but not necessarily for microorganisms within aggregates (Griffin, 1980).

Better aeration in the larger pores of wet soils corresponds to a higher potential for microbial activity and a faster decomposition rate of C substrate in larger pores than in smaller pores (Killham et al., 1993). As a wet soil dries, however, aeration improves within the aggregates, and the interiors of aggregates may then become protected sites where microorganisms survive temporarily adverse drought conditions (Foster, 1988). The location of an organic substrate within the soil matrix is thus a major factor in determining the rate at which organic residues decompose and at which C cycles between organic and inorganic forms.

Chemical composition of organic substrates

Chemical composition of the organic substrate is of prime importance in controlling microbial activity and C turnover. The largest fraction of OC entering the soil is that contributed by plant residues, which contain variable concentrations of proteins, hemicelluloses, cellulose, and lignin, as well as soluble substances such as sugars, amino sugars, organic acids, and amino acids (Paul and Clark, 1996). Nitrogen (N) is a key nutrient for microbial growth, and in mineral soils the level of available N is usually a limiting factor preventing maximum rates of microbial activity and OM decomposition (Alexander, 1977).

The C:N ratio of an organic substrate is a general indicator of rate of decomposition. Decomposition generally is more rapid for organic substrates with a low C:N ratio (high N content, such as proteins or amino acids) and slower for substrates with a high C:N ratio (low N content, such as cellulose, hemicellulose, or lignin). Soil biota rapidly metabolize root exudates, which include amino acids, carbohydrates, organic acids, enzymes, and vitamins (Tate, 1987), but suberized roots decompose more slowly because of higher lignin and cellulose contents. Supplemental N stimulates decomposition of these N-poor substrates.

Although the C:N ratio of the organic substrate generally indicates the rate of substrate decomposition, the ratio of lignin:N content of plant residues better predicts decomposition rate (Scott and Binkley, 1997; Aber et al., 1990; Melillo et al., 1982). Lignin does not easily decompose, and it protects other more readily decomposable components, so microorganisms less readily use substrates rich in lignin than they do lignin-poor substrates.

Microflora and fauna interactions

The dynamic interchange between soil microflora and fauna is important in regulating both the flux and availability of C (Mitchell, 1986). The microflora, primarily bacteria and fungi, are responsible for most of the annual turnover of C fixed by photosynthesis (Clark and Paul, 1970). Through their activity, most organic residues are decomposed and OC released as CO_2 (Vargas and Hattori, 1986).

Fauna (protozoa and nematodes) contribute less to the direct decomposition of organic residues, but they modify the physiochemical environment and influence microbial activity through grazing (Heal and Dighton, 1986). Grazing on microflora by fauna may stimulate microbial activity and result in increased mineralization of both C and N (Ingham et al., 1985; Hassink et al., 1993; Rutherford and Juma, 1992; Elliott and Coleman, 1977; Hunt et al., 1984, 1987). Pore size and distribution limit this predatory activity, since fauna generally are restricted to pores >30 μm, whereas a large proportion of bacteria may occupy pores <3 μm in diameter (Hassink et al., 1993). This physical separation explains why bacteria are able to endure in soil despite the presence of predators in the same habitat (Vargas and Hattori, 1986).

The interrelationship of pore size and soil texture also may influence the rate of faunal predation on microorganisms. The relative amount of large pores usually is higher in sandy soils than in loamy and clay soils, so predation may be more intense in sandy soils and thus strongly affect the decomposition rate of SOM and the mineralization rates of C and N (Hassink et al., 1993). Hassink et al. (1993) reported that nematodes grazing on bacteria increased N mineralization rates, which in the long run could result in an increase in the SOM C:N ratio. They speculated that a higher grazing intensity in sandy soils than in more finely textured soils might contribute to observed large differences in the C:N ratios of OM in sandy soils (C:N ~18) compared to loamy or clay soils (C:N ~10).

Microarthropods, such as mites, interact both directly and indirectly with soil microorganisms to regulate the rate of decomposition of SOM (Elkins and Whitford, 1982). Indirect influences include their fragmenting litter into small pieces, which increases surface area and thereby increases the potential for microbial activity, and consuming plant residues, which enriches residues via passage through the gut, thereby enhancing residue quality as a microbial substrate and accelerating its decomposition (Hendrix et al., 1986). Microarthropods affect soil microorganisms directly by grazing on bacterial and fungal biomass, which enhances nutrient availability for further microbial growth and thus stimulates microbial activity (Elkins and Whitford, 1982; Hendrix et al., 1986). As they move through the litter, microarthropods also disperse fungi and bacteria (Elkins and Whitford, 1982).

The activities of macroarthropods and large herbivores are also important in C cycling in grasslands. Macroarthropods such as earthworms, aphids, ants, and grasshoppers fragment and/or consume large quantities of plant material and thus channel energy, C, and nutrient flow within the ecosystem (Charley 1977; Laurenroth and Milchunas, 1992). Burrowing and excavation also transport plant residues from the soil surface and mix them within the soil matrix (Berry, 1994).

Aboveground consumption by the larger herbivores (domestic livestock and wildlife) may be important with respect to redistribution of C and nutrients within the ecosystem (Charley, 1977), for example, by stimulating the cycling of C and N from aboveground plant components to the soil (Schuman et al., 1999). (See Ch. 9 through 13 for reviews of the effects of livestock grazing on C cycling.)

Microbial biomass, composition, and distribution in grassland soils

Biomass and species composition

Considerable variation exists in reported microbial biomass of grassland soils because of differences in climate, soil physical and chemical properties, the type and amount of plant cover, the season of the year for sampling, and the method used to measure microbial populations (Ulehlova, 1992). In spite of this variation, however, the magnitude of the microbial biomass of temperate grasslands exceeds that of all other heterotrophs (the large and small mammals, birds, reptiles, insects, and all other invertebrates) combined and is second only to the biomass of the primary producers (i.e., the higher plant community) (Paul et al., 1979).

Estimated mean values of grassland soil microfloral biomass in the top 30 cm of soil range from 635 g/m^2 in a tall grass prairie (Ulehlova, 1992), to 223 g/m^2 in a

mixed grass prairie (Coupland, 1992), to 65 g/m^2 in a short grass steppe (Laurenroth and Milchunas, 1992). These data demonstrate that microbial biomass decreases with decreased plant productivity, reflecting the decreased rate of addition of plant residues to the soil (Voroney et al., 1981).

In the aboveground food web, fauna and fungi usually are the principal agents
of decay of cured aboveground standing dead vegetation. As this standing litter
becomes incorporated into surface litter and the moisture content increases, the
number of bacteria associated with the litter increases (Clark and Paul, 1970).
Fungi become less important as litter breaks down and bacteria gradually replace
them in the substrate succession (Kucera, 1992).

In the belowground food web, fungi usually constitute the major portion of soil
microbial biomass (Paul et al., 1979); detrital decomposition in many terrestrial
systems, such as forests or no-till agricultural ecosystems, is fungal-oriented
(Hendrix et al., 1986). The belowground food web of grasslands can be either fungal or bacterial, depending in part on climate and type of vegetation. Cooler,
moister locales appear to be a more suitable habitat for fungi and fungal litter
decomposition.

Kucera (1992) reported that fungal biomass predominated over bacterial biomass
in tall grass prairie soils, but northern locales appeared to provide a more suitable
habitat for fungi than southern regions. Fungi also dominate the microbial biomass
of mixed grass prairies; mean values of fungi and bacteria + actinomycetes (30 cm
depth) in northern mixed grass prairie are estimated at 177 and 46 g/m^2, respectively
(Coupland, 1992). In drier, warmer regions where less litter accumulates, fungi play
a less important role in litter decomposition (Kucera, 1992). In the short grass steppe,
bacteria and actinomycetes dominate detrital decomposition (Ingham et al., 1986;
Hunt et al., 1987).

Microbial distribution in the soil profile

Soil microbial biomass distributes unevenly in a soil profile (Ulehlova, 1992).
Since the availability of C substrate limits heterotrophic microbial activity, microbial biomass strongly correlates with SOM (Fromm et al., 1993) and with root biomass (Foster, 1988). Thus, microbial biomass usually concentrates in the surface
soil (A horizon) of the grassland soil and tends to decrease with increasing depth.

For example, Laurenroth and Milchunas (1992) reported that 53% of fungal
biomass and 25% of bacterial biomass occurred in the top 3 cm of soil of a short
grass steppe. In a northern mixed grass prairie, 38% of fungal biomass and 35% of
bacterial biomass occurred in the top 10 cm of the soil (Clark and Paul, 1970). In
native rangelands across a precipitation gradient from Colorado to Iowa, 75 to
90% of total soil microbial biomass was observed in the A horizons (0 to 10 cm), 59
to 79% of which was found in the top 5 cm of the soil profiles (Pruessner et al.
1995; Follett, 1998).

Although microbial biomass generally concentrates in the top few cm of the soil, microbial activity at the immediate soil surface tends to be pulsed, periodically limited by extremes in temperature and moisture. The temperature at the soil surface may reach as high as 70°C (158°F) at midday and show a diurnal fluctuation of 50°C (122°F) (Paul and Clark, 1996). High temperatures combined with the desiccating effects of wind and evaporation can result in air-dry surface soil (Donahue et al., 1983).

In contrast to soil microorganisms, soil microarthropods have shown a bimodal vertical distribution in the short grass steppe (Leetham and Milchunas, 1985). A maximum microarthropod biomass was observed at the soil surface (0 to 5 cm), coinciding with a large concentration of plant roots, while a lesser peak occurred at a depth of 25 to 40 cm, a zone with relatively high, and relatively stable, mean soil water content.

Root/microbial interactions

The rhizosphere is the soil volume adjacent to plant roots and is a zone of heightened microbial activity. Root surfaces are the main locations for soil organisms of all types (Foster, 1988). The plant root system supplies metabolic energy for microorganisms via exudates and root turnover (rhizodeposition), while the microfloral population at the root/soil interface enhances root function by increasing absorption of water and nutrients (Smucker and Safir, 1986). Within the rhizosphere, therefore, a mutually beneficial interaction exists in which plants and the soil biota use each other's metabolic wastes to their own advantage (Moore et al., 1991).

Soil microorganisms' net mineralization of OM makes N available for plant use, and predatory grazing of the microflora by protozoa, nematodes, and microarthropods enhances the exchange of N and C between plants and microbes (Moore et al., 1991). Many rhizobacteria also secrete organic materials which chemically weather soil minerals, releasing elements that promote plant growth (Foster, 1988).

The symbiotic interaction between plants and vesicular arbuscular mycorrhizal (VAM) fungi is an example of the mutually beneficial interaction between plants and the soil biota within the rhizosphere. These fungi assist the root by assuming many of the functions that root hairs otherwise would accomplish (Salisbury and Ross, 1992). Upon infecting host plant roots, VAM fungal hyphae grow out into the soil, effectively extending the zone of ion absorption and transportation of P and N to the plant. In return, the fungi receive C from the plant (Moore et al., 1991). Mycorrhizal hyphae also bind soil particles, preventing the development of gaps around roots during drying and shrinkage of soil, thus helping maintain liquid continuity and surfaces for water films (Coleman et al., 1983).

Grasslands differ from other ecosystems in the temporal distribution of OM inputs and in the proportion of rhizosphere effect in the soil. These differences contribute to the characteristically high degree of OM accumulation in grassland soils (Tate, 1987). OM input into grassland soils, and consequent stimulation of microbial activity, occurs throughout the growing season because roots and shoots are produced continuously. Because of the high rooting density, the majority of surface soil of most temperate humid grasslands may be loosely defined as *rhizosphere soil* (Foster, 1988; Tate, 1987). In semiarid or arid grasslands, however, water limitation leads to roots' uneven occupation of the soil and the formation of "resource islands" where soil materials accumulate beneath individual plants (Burke et al., 1998).

Potential Strategies to Improve C Sequestration by Managing Root and Microbial Biomass

Grazing lands total about 336 Mha within the U.S. and have been estimated to contain 40,000 MMT of C in the top 2 m of the soil profile (Follett, Ch. 3). Although grazing lands are gaining recognition as an important factor in mitigating global climate change resulting from increasing atmospheric CO_2, the potential for these ecosystems to sequester or release C is not well understood.

Studies needed

About 90% of the total C in most temperate grassland systems resides below ground, primarily in the SOM and secondarily in the root biomass. However, because of the complexity of grazing land systems and the difficulties associated with studying belowground processes, relatively few studies have evaluated either the basic ecology of the belowground components of grazing land ecosystems or the impacts of management practices on them. Moreover, the few studies available have reported inconsistent belowground responses to management practices.

Basic research is needed to improve our understanding of how management practices affect the partitioning of plant C to the roots; the timing, quantity, and form of root C transfer to the soil; and the composition and activity of the microbial community that controls the return of SOC to atmospheric CO_2. Basic research also is needed to improve our understanding of the potential for deep profile grassland roots and SOM to sequester C. Finally, research is needed to develop standard methods to quantify C transformations in the soil, particularly if assessments of C credits are to be made for increasing C storage in grazing lands. Standardized methods are needed because management practices may result in only small changes in soil C concentration that are near detection limits and must be extrapolated over large soil volumes (Nepstad et al., 1994).

General strategies

In spite of our limited understanding of the belowground components of grazing lands, we can discuss, in general terms, approaches to managing these ecosystems with the objective of enhancing belowground C sequestration. Since the quantity of OC in the soil depends on the balance between primary production and the rate of decomposition (Paul and Clark, 1996), increasing C storage in grassland soils requires management practices that increase root production and/or decrease belowground microbial decomposition processes.

Management practices to decrease microbial decomposition processes include practices that enhance plant production (i.e., increase the fixation of atmospheric C) without adding supplemental N. The resulting lower quality (higher C:N ratios) of plant litter, at least in the short term, would increase microbial immobilization of N and decrease the decomposition rate of plant residues and soil humus.

Management practices such as grazing or burning also can alter the structure and activity of the microbial community because of changes in the patterns of litter accumulation and in the quality of litter returned to the soil (Dormaar, 1975; Kucera, 1992; Schuman et al., 1999). However, neither the duration of these changes to the microbial community nor their long-term impacts on C sequestration are understood well. Development of strategies to manage belowground decomposition will require a better understanding of the short- and long-term effects of management practices on belowground microbial processes.

Strategies to increase root production include reclaiming degraded grazing lands and establishing perennial grasses on degraded croplands. Reestablishing perennial grasslands where invaded annual grasslands now exist, or establishing perennial grasslands on croplands historically planted to annual crops, creates extensive permanent root systems that turn over during a period of years, rather than the shallow rooting systems of annual grasses which turn over annually. Reestablishing the perennial grass component of mixed shrub/grass rangelands where the perennial grass component has been lost would be similarly beneficial.

Management practices to establish desired perennial grasses, such as irrigating, adding fertilizer, or seeding improved plant varieties, usually are economically feasible options for managed pastures (see Ch. 12). However, these practices often are not economically feasible on native rangelands, particularly in arid or semiarid locations where low and erratic precipitation limits plant response to intensive management practices.

In addition to concerns about the economic feasibility of intensive management practices on native grasslands, the positive and negative impacts on C sequestration also have not been evaluated sufficiently. Tillage practices have been used to increase plant production by furrowing rangelands for improved water storage, or by developing supplemental pastures of seeded "improved" grass species. Some success in increasing plant production has occurred from pitting rangelands to improve water storage (Wight and White, 1974), and nutrients released by tillage-induced increases in the mineralization of SOM have increased the establishment and production of grasses on furrowed rangelands (Wight and Siddoway, 1972).

However, tillage practices expose SOC to loss by increasing the potential for wind and water erosion and by enhancing the rate of microbial decomposition. OM decomposition increases with tillage through mixing, aggregate degradation, and consequent exposure of interaggregate OM to decomposing microorganisms and changes in temperature, moisture, and aeration (Burke et al., 1995; Reeder et al., 1998).

Some of these potential losses may be offset by the fact that tillage mixes the surface and subsurface soil, which vertically redistributes SOC and decreases the amount of OC exposed at the soil surface (Reeder et al., 1998). Additionally, increased plant production as a result of seeding "improved" grass species may offset losses of soil C due to tillage (Reeder et al., 1998). Finally, changing the species composition of the plant community will alter the distribution, mass, and chemical composition of the root system and thus will alter the quantity, distribution, and quality of OM in the soil profile.

Consequences of periodic drought also must be considered. Drought and grazing of tolerant native species may provide greater stability than improved species during periods of stress, even though native species may be less productive during favorable periods. Management practices involving tillage and a change in plant community composition therefore must be evaluated for both short- and long-term changes in C storage potential.

Global vs. local policies

Because the monetary return of reseeding, fertilizing, or irrigating native rangelands frequently does not offset the costs of these management practices, efforts to enhance C sequestration in native rangelands in some cases may be limited primarily to management of the timing and intensity of grazing (see Ch. 11). Reported effects of grazing on root biomass and soil C are inconsistent across locations around the world, probably because of differences in grazing strategy and intensity, and in climatic and edaphic conditions, as well as differences among plant communities in the evolutionary history of grazing before the introduction of domestic livestock (Milchunas and Laurenroth, 1993).

At the global scale, and within grazing levels not considered abusive, the attributes of an ecosystem (e.g., plant community productivity and evolutionary adaptations of plants to grazing) have been found to be more important than the intensity and/or duration of livestock grazing in explaining impacts of grazing on plant communities' composition and primary productivity. At the pasture/ranch scale, however, traditional range management and the fine-tuning of grazing systems (combinations of intensities, durations, rest/deferments, etc.) become important management decisions.

The global nature of the climate change problem lends itself to a consideration of policy decisions at the larger, national/global scale. This is a scale at which management of grazing lands has received little attention, yet potentially the greatest influence on C sequestration could result from broad policy decisions being applied uniformly across different grazing land ecosystems. No general relationships exist between grazing and changes in composition of plant species, root mass, or soil C content (Milchunis and Laurenroth, 1993). Rather than applying uniform standards across all grazing lands, national policy concerning grazing by domestic livestock (e.g., fees or incentives through conservation programs) should reflect differences in the potential impacts of the grazing on different grazing land ecosystems (Milchunas et al., 1998).

At the local scale, the key to maintaining or even increasing soil C stocks in grazing lands is to maintain grass productivity and root inputs into the soil by fine-tuning and adaptively modifying traditional range management techniques (Trumbore et al., 1995). Managing for more deeply rooted species or for plants with a higher root:shoot ratio may be too narrow a focus to be practical, and emphasis should be on management practices that sustain a plant community appropriate to the local climate and soil and that maintain that system's productive capacity.

References

Aber, J.D., J.M. Melillo, and C.A. McClaugherty. 1990. Predicting long-term patterns of mass loss, nitrogen dynamics, and soil organic matter formation from initial fine litter chemistry in temperate forest ecosystems. *Can. J. Bot.* 68:2201-2208.

Alexander, Martin. 1977. *Introduction to Soil Microbiology,* 2nd ed. John Wiley and Sons. New York.

Amelung, W., W. Zech, X. Zhng, R.F. Follett, H. Tiessen, E. Knox, and K.W. Flach. 1998. Carbon, nitrogen and sulfur pools in particle-size fractions as influenced by climate. *Soil Sci. Soc. Am. J.* 62:172-181.

Anderson, J.P.E., and K.H. Domsch. 1980. Quantities of plant nutrients in the microbial biomass of selected soils. *Soil Sci.* 130:211-216.

Ares, J. 1976. Dynamics of the root system of blue grama. *J. Range Manage.* 29:208-213.

Bamforth, S.S. 1988. Interactions between protozoa and other organisms. *Agric. Ecosys. and Environ.* 24:229-234.

Baron, W.M.M. 1967. *Organization in Plants.* Edward Arnold Publishers. London.

Bedunah, D.J., and R.E. Sosebee (eds). 1995. *Wildland Plants: Physiological Ecology and Developmental Morphology.* Society for Range Mangement. 1839 York Street, Denver, CO 80206.

Berry, E.C. 1994. Earthworms and other fauna in the soil. *In* J.L. Hatfield and B.A. Stewart (eds), *Soil Biology. Effects on Soil Quality. Advances in Soil Science*, Lewis Publishers, Boca Raton, FL, pp. 61-90.

Biondini, M., D.A. Klein, and E.F. Redente. 1988. Carbon and nitrogen losses through root exudation by *Agropyron Cristatum*, *A. Smithii* and *Bouteloua Gracilis*. *Soil Biol. Biochem.* 20:477-482.

Burke, I.C., W.K. Laurenroth, and D.P. Coffin. 1995. Soil organic matter recovery in semiarid grasslands: implications for the conservation reserve program. *Ecol. Apps.* 53:793-801.

Burke, I.C., W.K. Laurenroth, and D.G. Milchunas. 1997. Biogeochemistry of managed grasslands in central North America. *In* E.A. Paul, K. Paustian, E.T. Elliott, and C.V. Cole (eds), *Soil Organic Matter in Temperate Agroecosystems: Long-Term Experiments in North America*, CRC Press, Boca Raton, FL, pp. 85-102.

Burke, I.C., W.K. Laurenroth, M.A. Vinton, P.B. Hook, R.H. Kelly, H.E. Epstein, M.R. Aguial, M.D.Robles, M.O. Aguilera, K.L. Murphy, and R.A.Gill. 1998. Plant-soil interactions in temperate grasslands. *Biogeochem.* 42:121-143.

Burke, I.C., C.M. Yonker, W.J. Parton, C.V. Cole, K. Flach, and D.S. Schimel. 1989. Texture, climate and cultivation effects on soil organic matter content in U.S. grassland soils. *Soil Sci. Soc. Am. J.* 53:800-805.

Caldwell, M.M. 1979. Root structure: the considerable cost of belowground function. *In* O.T. Solbrig et al. (eds), *Topics in Plant Population Biology*, Columbia University Press, New York.

Canadell, J.G., L.F. Pitelka, and J.S. Ingram. 1996a. The effects of elevated [CO_2] on plant-soil carbon belowground: A summary and synthesis. *Plant and Soil* 187:391-400.

Canadell, J., R.B. Jackson, J.R. Ehleringer, H.A. Mooney, O.E. Sala, and E.D Schulze. 1996b. Maximum rooting depth of vegetation types at the global scale. *Oecologia* 108:583-595.

Charley, J.L. 1977. Mineral cycling in rangeland ecosystems. *In* R.E. Sosebee (ed), *Rangeland Plant Physiology, Range Science Series No. 4,* Soc. Range Management, Denver, CO, pp. 215-255.

Ciais, P., P.P. Tans, M. Trolier, J.W.C. White, and R.J. Francey. 1995. A large northern hemisphere terrestrial CO_2 sink indicated the $^{13}C/^{12}C$ ratio of atmospheric CO_2. *Science* 269:1098-1102.

Clark, F.E., and E.A. Paul. 1970. The microflora of grassland. *In* N.C. Brady et al. (eds), *Advances in Agronomy 22*, Academic Press, London, pp. 375-435.

Coleman, D.C. 1976. A review of root production processes and their influence on soil biota in terrestrial ecosystems. *In* J.M. Anderson and A. Macfaydyen (eds), *The Roll of Terrestrial and Aquatic Organisms in Decomposition Processes*, Blackwell Scientific Publishers, Oxford, pp. 417-434.

Coleman, D.C., C.P.P. Reid, and C.V. Cole. 1983. Biological strategies of nutrient cycling in soil systems. *Adv. Ecol. Res.* 13:1-55.

Coupland, R.T. 1992. Mixed prairie. *In* Robert T. Coupland (ed), *Natural Grasslands, Introduction and Western Hemisphere, Ecosystems of the World 8a*, Elsevier, Amsterdam, pp. 151-182.

Coupland, R.T., and R.E. Johnson. 1965. Rooting characteristics of native grassland species in Saskatchewan. *J. Ecol.* 53:475-507.

Donahue, R.L., R.W. Miller, and J.C. Shickluna. 1983. *Soils: An introduction to soils and plant growth.* Prentice-Hall Inc. Englewood Cliffs, NJ.

Dormaar, J.F. 1975. Susceptibility of organic matter of Chernozemic Ah horizons to biological decomposition. *Can. J. Soil Sci.* 55:473-480.

Dormaar, J.F. 1992. Decomposition as a process in natural grasslands. *In* R.T. Coupland (ed), *Natural Grasslands, Introduction and Western Hemisphere, Ecosystems of the World 8a*, Elsevier, Amsterdam, pp. 121-136.

Dormaar, J.F., S. Smoliak, and A. Johnston. 1981. Seasonal fluctuations of blue grama roots and chemical characteristics. *J. Range Manage.* 34:62-64.

Elkins, N.Z., and W.G. Whitford. 1982. The role of microarthropods and nematodes in decomposition in a semi-arid ecosystem. *Oecologia* 55:303-310.

Elliott, E.T. 1980. Habitable pore space and microbial trophic interactions. *Oikos* 35:327-335.

Elliott, E.T., and D.C. Coleman. 1977. Soil protozoan dynamics in a short grass prairie. *Soil Biol. Biochem* 9:113-118.

Fogel, R. 1985. Roots as primary producers in belowground ecosystems. *In* A.H. Fitter, D. Atkinson, D.J. Read, and M.B. Usher (eds), *Ecological Interactions in Soil, Br. Ecol. Soc. Sp. Pub. No. 4*, Blackwell Scientific Publications, Oxford, England, pp. 23-35.

Follett, R.F. 1998. CRP and microbial biomass dynamics in temperate climates. *In* R. Lal et al. (eds), *Management of Carbon Sequestration in Soil*, CRC Press, Boca Raton, FL, pp. 305-322.

Foster, R.C. 1988. Microenvironments of soil microorganisms. *Biol. Fertil. Soils* 6:189-201.

Fromm, H., K. Winter, J. Filser, R. Hantschel, and F. Beese. 1993. The influence of soil type and cultivation system on the spatial distributions of the soil fauna and microorganisms and their interactions. *Geoderma* 60:109-118.

Gill, R.A. 1998. *Biogic controls over the depth distribution of soil organic matter.* Doctoral Dissertation in Ecology, Colorado State University. Fort Collins, CO.

Gill, R.A., I.C. Burke, D.G. Milchunas, and W.K. Laurenroth. 1999. Relationship between root biomass and soil organic matter pools in the short grass steppe of eastern Colorado. *Ecosystems* 2:226-236.

Griffin, D.M. 1980. Water potential as a selective factor in the microbial ecology of soils. *In* J.F. Parr et al. (eds), *Water Potential Relations in Soil Microbiology*, Soil Science Society of America, Madison, WI, pp. 141-151.

Hansson, A.L., and O. Andrén. 1986. Belowground plant production in a perennial grass ley (*Festuca pratensis* Huds.) assessed with different methods. *J. Appl. Ecol.* 23:657-666.

Hassink, J., L.A. Bouwman, K.B. Zwart, and L. Brussaard. 1993. Relationships between habitable pore space, soil biota and mineralization rates in grassland soils. *Soil Biol. Biochem.* 25:47-55.

Heal, O.W., and J. Dighton. 1986. Nutrient cycling and decomposition in natural terrestrial ecosystems. *In* M.J. Mitchell and J.P. Nakas (eds), *Microfloral and Faunal Interactions in Natural and Agro-Ecosystems,* Martinus Nijhoff/Dr. W. Junk Publishers, Hingham, MA, pp. 14-73.

Hendrix, P.F., R.W. Parmelee, D.A. Crossley, Jr., D.C. Coleman, E.P. Odum, and P.M. Groffman. 1986. Detritus food webs in conventional and no-tillage agroecosystems. *BioScience* 36(6):374-380.

Howarth, W.R., and E.A. Paul. 1994. Microbial biomass. *In* R.W. Weaver et al. (eds), *Methods of Soil Analysis, Part 2 — Microbiological and biochemical properties, SSSA Book Series No. 5,* Soil Science Society of America, Inc., Madison, WI, pp. 753-773.

Hunt, H.W., D.C. Coleman, C.V. Cole, R.E. Ingham, E.T. Elliott, and L.E. Woods. 1984. Simulation model of a food web with bacteria, amoebae, and nematodes in soil. *In* M.J. Klug and C.A. Reddy (eds), *Current Perspectives in Microbial Ecology*, Amer. Society for Microbiology, Washington, DC, pp. 346-352.

Hunt, H.W., D.C. Coleman, E.R. Ingham, R.E. Ingham, E.T. Elliott, J.C. Moore, S.L. Rose, C.P.P. Reid, and C.R. Morley. 1987. The detrital food web in a short grass prairie. *Biol. Fertil. Soils* 3:57-68.

Ingham, E.R., J.A. Trofymow, R.N. Ames, H.W. Hunt, C.R. Morley, J.C. Moore, and D.C. Coleman. 1986. Trophic interactions and nitrogen cycling in a semi-arid grassland soil. I. Seasonal dynamics of the natural populations, their interactions and effects on nitrogen cycling. *J. Appl. Ecol.* 23:597-614.

Ingham, R.E., J.A. Trofymow, E.R. Ingham, and D.C. Coleman. 1985. Interactions of bacteria, fungi, and their nematode grazers: Effects on nutrient cycling and plant growth. *Ecol. Monogr.* 55:119-140.

Jackson, R.B., J. Canadell, J.R. Ehleringer, H.A. Mooney, O.E. Sala, and E.D. Schulze. 1996. A global analysis of root distributions for terrestrial biomes. *Oecologia* 108:389-411.

Jenny, H. 1980. *The Soil Resource: Origins and Behavior. Ecological Studies 37.* Springer-Verlag. New York.

Killham, K., M. Amato, and J.N. Ladd. 1993. Effect of substrate location in soil and soil pore-water regime on carbon turnover. *Soil Biol. Biochem.* 25:57-62.

Kucera, C.L. 1992. Tall grass prairie. *In* R.T. Coupland (ed), *Natural Grasslands, Introduction and Western Hemisphere, Ecosystems of the World 8a,* Elsevier, Amsterdam, pp. 227-268.

Kurz, W.A., and J.P. Kimmins. 1987. Analysis of some sources of error in methods used to determine fine root production in forest ecosystems: a simulation approach. *Can. J. For. Res.* 17:909-912.

Lauenroth, W.K., H.W. Hunt, D.M. Swift, and J.S. Singh. 1986. Estimating aboveground net primary production in grasslands: a simulation approach. *Ecol. Model.* 33:297-314.

Laurenroth, W.K., and D.G. Milchunas. 1992. Short grass steppe. *In* R.T. Coupland (ed), *Natural Grasslands, Introduction and Western Hemisphere, Ecosystems of the World 8a*, Elsevier, Amsterdam, pp. 183-226.

Laurenroth, W.K., and W.C. Whitman. 1977. Dynamics of dry matter production in a mixed grass prairie in western North Dakota. *Oecologia* 27:339-351.

Leethan, J.W., and D.G. Milchunas. 1985. The composition and distribution of soil microarthropods in the short grass steppe in relation to soil water, root biomass, and grazing by cattle. *Pedobiologia* 28:311-325.

Linn, D.M., and J.W. Doran. 1984. Effect of water-filled pore space on carbon dioxide and nitrous oxide production in tilled and nontilled soils. *Soil Sci. Soc. Am. J.* 48:1257-1272

Lorenz, R.J. 1977. Changes in root weight and distribution in response to fertilization and harvest treatment of mixed prairie. *In* J.K. Marshall (ed), *The belowground ecosystem: a synthesis of plant-associated processes, Range Science Dept Science Series No. 26*, Colorado State University, Fort Collins, CO, pp. 63-72.

Lutwick, L.E., and J.F. Dormaar. 1976. Relationships between the nature of soil organic matter and root lignins of grasses in a zonal sequence of Chernozemic soils. *Can. J. Soil Sci.* 56:363-371.

Melillo, J.M., J.D. Aber, and J.F. Muratore. 1982. Nitrogen and lignin control of hardwood leaf litter decomposition dynamics. *Ecology* 63:621-626.

Milchunas, D.G., and W.K. Laurenroth. 1989. Three-dimensional distribution of plant biomass in relation to grazing and topography in the short grass steppe. *Oikos* 55:82-86.

Milchunas, D.G., and W.K. Laurenroth. 1992.Carbon dynamics and estimates of primary production by harvest, C^{14} dilution and C^{14} turnover. *Ecology* 73:593-607.

Milchunas, D.G., and W.K. Laurenroth. 1993. Quantitative effects of grazing on vegetation and soils over a global range of environments. *Ecol. Monogr.* 63(4):327-366.

Milchunas, D.G., and W.K. Laurenroth. 2000. Belowground primary production by carbon isotope decay and long-term root biomass dynamics: an update. *Ecosystems* (submitted).

Milchunas, D.G., W.K. Laurenroth ,and I.C. Burke. 1998. Livestock grazing: animal and plant biodiversity of short grass steppe and the relationship to ecosystem function. *OKOS* 83:65-74.

Milchunas, D.G., W.K. Laurenroth, P.L. Chapman, and M.K. Kazempour. 1989. Effects of grazing, topography, and precipitation on the structure of a semiarid grassland. *Vegetatio* 80:11-23.

Milchunas, D.G., W.K. Laurenroth, J.S. Singh, C.V. Cole, and H.W. Hunt. 1985. Root turnover and production by ^{14}C dilution: implications of carbon partitioning in plants. *Plant and Soil* 88:353-365.

Mitchell, M.J. 1986. Introduction. *In* M.J. Mitchell and J.P. Nakas (eds), *Microfloral and Faunal Interactions in Natural and Agro-Ecosystems*, Martinus Nijhoff/Dr. W. Junk Publishers, Hingham, MA, pp. 1-14.

Moore, J.C., H.W. Hunt, and E.T. Elliott. 1991. Ecosystem perspectives, soil organisms and herbivores. *In* Pedro Barbosa et al. (eds), *Microbial Mediation of Plant-Herbivore Interactions*, John Wiley & Sons, New York, pp. 105-140.

Nepstad, D.C., C.R. De Carvalho, E.A. Davidso, P.H. Jopp, P.A. Lefebvre, G.H. Negrelros, E.D. da Silva, T.A. Stone, S.E. Trumbore, and S. Vlelra. 1994. The role of deep roots in the hydrological and carbon cycles of amazonian forests and pastures. *Nature* 372:666-669.

Orchard, V.A., and F.J. Cook. 1983. Relationship between soil respiration and soil moisture. *Soil Biol. Biochem.* 15(4):447-453.

Pages, L., and A. G. Bengough. 1997. Modeling minirhizotron observations to test experimental procedures. *Plant and Soil* 189:81-89.

Parton, W.J., J.S. Singh, and D.C. Coleman. 1978. A model of production and turnover of roots in short grass prairie. *J. Appl. Ecol.* 47:515-542.

Paul, E.A., and F.E. Clark. 1996. *Soil Microbiology and Biochemistry*. Academic Press, Inc. San Diego CA.

Paul, E.A., F.E. Clark, and V.O. Biederbeck. 1979. Microorganisms. *In* R.T. Coupland (ed), *Grassland ecosystems of the world: Analysis of grasslands and their uses. International Biological Programme 18*, Cambridge University Press, Cambridge, pp. 87-97.

Paul, E.A., and J.A. Van Veen. 1978. The use of tracers to determine the dynamic nature of organic matter. *Trans. 11ᵗʰ Int. Congr. Soil Science*. Edmonton, Alberta, 3:61-102.

Pruessner, E.G., R.F. Follett, J.M. Kimble, and S.E. Samson. 1995. Soil Properties and soil carbon storage within soil profiles of the historical grasslands of the US — Part 2. *Agron. Abstr.* 87:285.

Redmann, R.E. 1992. Primary productivity. *In* R.T. Coupland (ed), *Natural Grasslands, Introduction and Western Hemisphere, Ecosystems of the World 8a*, Elsevier, Amsterdam, pp. 75-93.

Reeder, J.D., G.E. Schuman, and R.A. Bowman. 1998. Soil C and N changes on conservation reserve program lands in the Central Great Plains. *Soil & Tillage Rsch.* 47:339-349.

Richter, D.D., and D. Makewitz. 1995. How deep is soil? *BioScience* 45:600-609.

Russell, R.S. 1977. *Plant Root Systems: Their Function and Interaction with the Soil*. McGraw-Hill (UK) Limited. London.

Rutherford, P.M., and N.G. Juma. 1992. Influence of texture on habitable pore space and bacterial-protozoal populations in soil. *Biol. and Fertil. of Soils* 12:221-227.

Sala, O., W.J. Parton, L. Joyce, and W.K. Laurenroth. 1988. Primary production of the central grassland region of the United States. *Ecology* 69:45-49.

Sala, O.E., M.E. Biondini, and W.K. Lauenroth. 1988. Bias in estimates of primary production: an analytical solution. *Ecol. Model.* 44:43-55.

Salsbury, F.B., and C.W. Ross. 1992. *Plant Physiology*, 4th Ed. Wadsworth Publishing Co, Inc. Belmont, CA.

Schimel, D.S., B.H. Braswell, E.A. Holland, R. McKeown, D.S. Ojima, T.H. Painter, W.J. Parton, and A.R. Townsend. 1994. Climatic, edaphic, and biotic controls over storage and turnover of carbon in soils. *Global Biogeochem. Cycles* 8(3):279-293.

Schuman, G.E., J.D. Reeder, J.T. Manley, R.H. Hart, and W.A. Manley. 1999. Impact of grazing management on the carbon and nitrogen balance of a mixed grass rangeland. *Ecol. Apps.* 9:65-71.

Scott, N.A., and D. Binkley. 1997. Foliage litter quality and annual net N mineralization: comparison across North American forest sites. *Oecologia* 111:151-159.

Sims, P.L., and R.T. Coupland. 1979. Producers. *In* R.T. Coupland (ed), *Grassland ecosystems of the world: Analysis of grasslands and their uses, Internat. biol. Programme 18*, Cambridge University Press, Cambridge, pp. 49-72.

Sims, P.L., and J.S. Singh. 1978. The structure and function of ten western North American grasslands. II. Intra-seasonal dynamics in primary producer compartments. *J. Ecol.* 66:547-572.

Sims, P.L., J.S. Singh, and W.K. Laurenroth. 1978. The structure and function of ten western North American Grasslands. I. Abiotic and vegetational characteristics. *J. Ecol.* 66:251-285.

Singh, J.S., and D.C. Coleman. 1973. A technique for evaluating functional root biomass in grassland ecoysystems. *Can. J. Bot.* 51:1867-1870.

Singh, J.S., W.K. Lauenroth, H.W. Hunt, and D.M. Swift. 1984. Bias and random errors in estimators of net root production: a simulation approach. *Ecol.* 65:1760-1764.

Six, J., ET. Elliott, K. Paustian, and J.W. Doran. 1998. Aggregation and soil organic matter accumulation in cultivated and native grassland soils. *Soil Sci. Soc. Am. J.* 62:1367-1377.

Smucker, A.J.M., and G.R. Safir. 1986. Root and soil microbial interactions which influence the availability of photoassimilate carbon to the rhizosphere. *In* M.J. Mitchell and J.P. Nakas (eds), *Microfloral and Faunal Interactions in Natural and Agro-Ecosystems*, Martinus Nijhoff/Dr. W. Junk Publishers, Hingham, MA, pp. 203-244.

Sommers, L.E., C.M. Gilmour, R.E. Wildung, and S.M. Beck. 1980. The effect of water potential on decomposition processes in soils. *In* J.F. Parr et al. (eds), *Water Potential Relations in Soil Microbiology*, Soil Science Society of America, Madison, WI, pp. 97-117.

Tans, P.P., I.Y. Fung, and T. Takahashi. 1990. Observational constraints on the global atmospheric CO_2 budget. *Sci.* 247:1431-1438.

Tate, R.L. III. 1987. *Soil Organic Matter — Biological and Ecological Effectors.* Wiley (Interscience). New York.

Thorp, J. 1948. How soils develop under grass. *In* A. Stefferud (ed), *Grass: The Yearbook of Agriculture*, USDA, Washington, DC, pp. 55-66.

Tortora, G.J., B.R. Funke, and C.L. Case. 1982. *Microbiology, An Introduction.* Benjamin/Cummings Publishing, New York.

Trumbore, S.E., E.A. Davidson, P.B. de Camagro, D.C. Nepstad, and L.A. Martielli. 1995. Belowground cycling of carbon in forests and pastures of Eastern Amazonia. *Global Biogeochem. Cycles* 9:515-528.

Ulehlova, B. 1992. Micro-organisms. *In* R.T. Coupland (ed), *Natural Grasslands: Introduction and Western Hemisphere, Ecosystems of the World 8A,* Elsevier, Amsterdam, pp. 95-119.

Vargas, R., and T. Hattori. 1986. Protozoan predation of bacterial cells in soil aggregates. *FEMS Microbiol. Ecol.* 38:233-242.

Voroney, R.P., J.A. Van Veen, and E.A. Paul. 1981. Organic C dynamics in grassland soils. 2. Model validation and simulation of the long-term effects of cultivation and rainfall erosion. *Can. J. Soil Sci.* 61:211-224.

Weaver, J.E. 1958. Summary and interpretation of underground development in natural grassland communities. *Ecol. Monogr.* 28:55-78.

Weaver, J.E., V.H. Hougen, and M.D. Weldon. 1935. Relation of root distribution to organic matter in prairie soil. *Bot. Gaz.* 96: 389-420.

Wight, J.R., and L.M. White. 1974. Interseeding and pitting on a sandy range site in eastern Montana. *J. Range Manage.* 27:206-210.

Wight, J.R., and F.H. Siddoway. 1972. Improving precipitation-use efficiency on rangeland by surface modification. *J. Soil Water Conserv.* 27:170-174.

Zibilske, L.M. 1994. Carbon mineralization. *In* R.W. Weaver et al. (eds), *Methods of Soil Analysis. Part 2 — Microbiological and Biochemical Properties, SSSA Book Series No. 5*, Soil Science Society of America, Inc., Madison, WI, pp. 835-863.

CHAPTER 7

Carbon Dioxide Fluxes over Three Great Plains Grasslands

A.B. Frank,[1] P.L. Sims,[2] J.A. Bradford,[3]
P.C. Mielnick,[4] W.A. Dugas,[5] and H.S. Mayeux[6]

Introduction

Atmospheric CO_2 concentration is increasing as a result of changes in land use and burning of fossil fuels (Watson et al., 1996). Numerous attempts have been made to identify the sizes of C sources and sinks (Rastetter et al., 1992; Schimel, 1995; Fan et al., 1998). Terrestrial grassland ecosystems, which comprise 32% of the earth's natural vegetation (Adams et al., 1990), play a significant role in uptake of atmospheric CO_2 and the global C cycle (Batjes, 1998; Sundquist, 1993). Dixon et al. (1994) estimated that tropical forests contain about 40% of the C stored in terrestrial biomass. Others concluded that temperate regions' ecosystems also may function as an important C sink that could contribute to balancing the global C budget (Fan et al., 1998; Keeling et al., 1996; Schimel, 1995; Rastetter et al., 1992; Gifford, 1994).

Grasslands may be an important terrestrial ecosystem in the global C cycle because they comprise a large part of the world's land area (World Resource Institute, 1986) and remain, along with other types of rangeland, a dominant landscape in the U.S. (Sims and Risser, 1999). Several short-term studies suggest that grasslands function as a sink for atmospheric CO_2 during the growing season, but data for annual periods are limited.

[1] Plant Physiologist, USDA, Agricultural Research Service, Mandan, ND 58554.

[2] Research Leader/Rangeland Scientist, USDA-Agricultural Research Service, Woodward, OK 73801.

[3] Plant Physiologist, USDA-Agricultural Research Service, Woodward, OK 73801.

[4] Micrometeorologist, Texas Agric. Exp. Sta., Temple, TX 76502.

[5] Professor and Resident Director, Texas Agric. Exp. Sta., Temple, TX 76502.

[6] Laboratory Director/Rangeland Scientist, USDA-Agricultural Research Service, El Reno, OK 73036.

Kim et al. (1992) reported that a prairie site dominated by warm season tall grasses fixed 4.1 g $CO_2/m^2/d$ from May through October. During the period when plants were senescent, the ecosystem CO_2 budget was in balance with the atmosphere, but during droughts and after plant senescence, the ecosystem released about 3 g $CO_2/m^2/d$ into the atmosphere. Dugas et al. (1999) reported that, on an annual basis, in 1993 and 1994, a tall grass prairie site fixed 0.7 g $CO_2/m^2/d$. Both of these studies showed that maximum CO_2 fixation rates generally coincided with maximum leaf area and biomass accumulation during the growing season.

The magnitude of this potential C sink is still in question (Houghton et al., 1999), but estimates of that magnitude might be based best on long-term measurements of net ecosystem CO_2 fluxes across a wide range of grasslands typical of the varied environments in which they occur. In this chapter, we provide a case study which demonstrates that CO_2 fluxes can be measured over rangeland vegetation in remote areas for extended periods. We also attempt to extrapolate flux measurements made over much of the year to annual values.

We determined CO_2 fluxes for three diverse grassland ecosystems that transect a wide climatic gradient on the Great Plains of the U.S. from about 31°N to 46°N. The three sites include a northern Great Plains mid-grass prairie; a southern Great Plains mixed grass prairie consisting of tall, mid, and short grasses; and a southern Great Plains tall grass prairie. Measurements were made from 1995 through 1997.

Methods

Site description

The three native grassland sites included in this study were at Mandan, ND; Woodward, OK; and Temple, TX. These locations are part of a network of 10 grassland sites included in the Rangeland Carbon Dioxide Flux Project of the USDA, Agricultural Research Service (Svejcar et al., 1997). The three sites lie along a north-south gradient in the North American Great Plains (Fig. 7.1). Mandan is about 1900 km north of Temple and 1200 km north of Woodward. The Mandan site is dominated by C_3 grasses, whereas the Woodward and Temple sites are dominated by C_4 grasses.

The Mandan site is at the Northern Great Plains Research Laboratory, Mandan, ND (latitude 46°46'N, longitude 100°55'W, elevation 518 m), on slopes of 2 to 10%. The vegetation as determined by point frame procedures (25 frames, 50 hits/frame) in 1995 and 1997 is typical of a northern Great Plains mid-prairie ecosystem dominated by blue grama (*Bouteloua gracilis* [H.B.K.] Lag. ex Griffiths), needle-and-thread (*Stipa comata* Trin. and Rupr.), Carex (*Carex* spp.), little bluestem (*Schizachyrium scoparium* [Michx.] Nash), side-oats grama (*Bouteloua curtipen-*

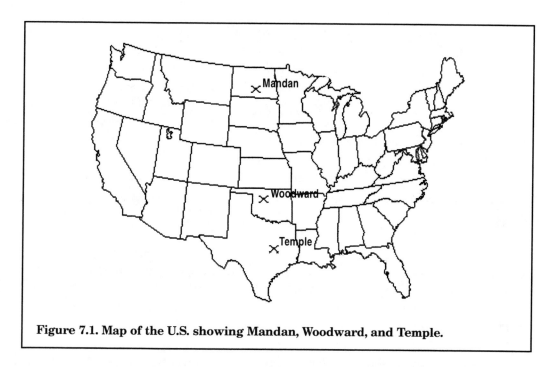

Figure 7.1. Map of the U.S. showing Mandan, Woodward, and Temple.

dula [Michx.] Torr.), and Kentucky bluegrass (*Poa pratensis* L.). The grassland is on a Werner-Sen-Chama soil complex (loam, silt loam, and silty clay loam Entic and Typic Haploborolls). This site has never had fertilizer or herbicides applied and was last grazed in 1992.

The centrally located site is at the Southern Plains Range Research Station, Woodward, OK (latitude 36°36'N, longitude 99°35'W, elevation 630 m), on about 2% slopes. This site is a mixed grass prairie comprised of tall grasses (sand bluestem [*Andropogon gerardii* var. *paucipilus* (Nash) Fern], little bluestem, Indian grass [*Sorghastrum nutans* (L.) Nash], and switchgrass [*Panicum virgatum* L]); an array of mid-grasses (sand dropseed [*Sporobolus cryptandrus* (Torr.) Gray], sand paspalum [*Paspalum setaceum* Michx.], sand lovegrass [*Eragrostis trichodes* (Nutt.) Wood], and fall witchgrass [*Digitaria cognata* (J. A. Schultes) Pilger]); and an understory of short grasses (blue grama and buffalograss [*Buchloe dactyloides* (Nutt)]) (Berg, 1994).

Texas bluegrass (*Poa arachnifera* Torr.) is the primary cool season perennial grass in the Woodward site's ecosystem, although Carex spp. and Canadian wild-rye (*Elymus canadensis* L.) are also present. Pratt soils (sandy, mixed mesic Lamellic Haplustalfs) occurred on the more level areas, and Tivoli soils (mixed, thermic Typic Ustipsamments) occurred on upper slopes. This site has never been fertilized, but herbicides, generally 2, 4-D, were used to control sand sagebrush (*Artemisia filifolia* Torr.). This site was grazed annually at the recommended stocking rates through 1994. During the years of flux measurements, the site was grazed during the winter for short periods to remove standing dead material.

The southernmost site is at the Blackland Research Center, Temple, TX (latitude 31°06'N, longitude 97°20'W, elevation 219 m), on about 2% slope. This site is dominated by big bluestem (*Andropogon gerardii* Vitman), little bluestem, and Indian grass. Johnsongrass (*Sorghum halopense* [L.] Pers.) and annual herbs also were present on the site. The site is on a Houston Black soil (fine smectitic, thermic Udic Haplusterts). This grassland had not been sprayed with herbicides, grazed, fertilized, or burned for 50 years until being burned on 22 February 1995.

CO_2 flux

The CO_2 flux over each site was measured using Bowen ratio/energy balance (BREB) instrumentation (Model 023/CO_2 Bowen ratio system, Campbell Scientific, Inc., Logan, UT). Fluxes were calculated at 20-minute intervals, using methods described by Dugas (1993) and Dugas et al. (1999). Bowen ratios were calculated from temperature and humidity gradients, measured every 2 seconds at 1 and 2 m above the canopy. Sensible heat flux was calculated from the Bowen ratio, average net radiation (Model Q*7 net radiometer, REBS, Seattle, WA), and soil heat flux that was calculated from two soil heat flux plates (Model HFT, REBS) and soil temperature measured above the plates.

Net radiometers were calibrated against a laboratory standard (Model 7.2, REBS) over grass each year before use. Sensor sensitivities were constant. The turbulent diffusivity, assumed equal for heat, water vapor, and CO_2, was calculated using the 20-minute sensible heat flux and temperature gradient measurements. Twenty-minute averages of CO_2 flux, corrected for vapor density differences at the two heights (Webb et al., 1980), were calculated as a product of turbulent diffusivity and the 20-minute average CO_2 gradient, measured every 2 seconds at 1 and 2 m above the vegetation.

When the BREB method for calculating turbulent diffusivity was not valid, because of disparities between the sign of the flux and the gradient, it was calculated using wind speed, atmospheric stability, and canopy height, as Dugas et al. (1999) described. This alternate method of calculation of diffusivity was used almost exclusively at night and then ranged from approximately 10% of the time at Mandan to 15% at Temple.

CO_2 and water vapor concentration gradients between the two heights were measured with an infrared gas analyzer Model 6262, (Li-Cor, Inc., Lincoln, NE, USA). The gas analyzers were calibrated at least biweekly, using a gas mixture of near ambient CO_2 concentration. Fluxes were not corrected for temperature differences because, in a separate test, fine wire thermocouple measurements indicated that the temperature of the air streams from the two arms did not differ when entering the gas analyzer (unpublished observations).

All data generated from the BREB system at each location were captured with a model 21X data logger (Campbell Scientific, Inc.). The periods of data collection appear in Table 7.1. Occasionally, when equipment malfunctioned or maintenance was required, the 20-minute CO_2 flux data were not available. Those periods of missing data were not used to calculate daily fluxes. Seasonal fluxes were calculated from flux measurements made during the periods Table 7.1 shows, whereas annual fluxes included measured seasonal fluxes plus an estimate of net CO_2 losses during the winter.

Since CO_2 losses from the grasslands at Mandan and Woodward during the dormant or winter period were not continuously available, estimates of net CO_2 loss were based on the last 7 and 14 days of flux measurements in years 1995, 1996, and 1997 and averaged –0.5 and –0.7 g $CO_2/m^2/d$, for Mandan and Woodward, respectively. Unpublished data obtained during the winter season since 1997 at both Mandan and Woodward were similar to these estimates.

Estimates for Temple were based on actual winter flux measurements made in 1997 and averaged –2.1 g $CO_2/m^2/d$. Monthly CO_2 flux averages for each year were calculated as the product of the average daily flux in a month and the number of days in that month. Except for months at the beginning and ending of measurement periods, the number of days included in the monthly averages was greater than 26, except only 20 days in July 1997 were available for Temple. The 3-year monthly averages were calculated from each year's monthly fluxes.

The Woodward and Mandan sites had at least 200 m of fetch in all directions. Instrumentation at Temple was located in the northwest corner of the grassland. When winds were from the southeast to southwest, the predominant wind direction, fetch was greater than 120 m. Dugas et al. (1999) showed that fluxes at Temple, normalized for net radiation, were not significantly different when wind direction resulted in reduced fetch. Thus, all data at Temple were used, regardless of wind direction.

Biomass and leaf area

Plant biomass and green leaf area were measured by clipping 0.25 m² quadrats (four quadrats were clipped at Mandan and Woodward and six at Temple) on selected dates (Table 7.2). The clipping sites at Mandan were about 40 m in each principal direction from the tower; at Woodward, each clip site was from 25 to 75 m in each direction from the tower; and at Temple, each clipping site was about 35 to 70 m to the east, south, and west from the tower.

At Mandan and Woodward, leaves were separated from stems, and leaf area was measured with a belt-driven photoelectric area meter. Leaves and stems were

Table 7.1. Data collection period, with day of year and number of days when Bowen ratio/energy balance data were available to calculate seasonal CO$_2$ flux at Mandan, Woodward, and Temple.

Location	Year	Data Collection Period	CO$_2$ Flux ----d----
Mandan	1995	27 May (147) - 30 Oct (303)	157
	1996	22 Apr (113) - 2 Nov (306)	194
	1997	22 Apr (112) - 30 Oct (303)	192
Woodward	1995	19 Feb (50) - 20 Dec (354)	277
	1996	4 May (125) - 26 Dec (361)	195
	1997	1 Apr (91) - 17 Dec (351)	261
Temple	1995	4 Feb (35) - 26 Nov (330)	264
	1996	7 Mar (67) - 15 Dec (350)	257
	1997	12 Feb (43) - 4 Oct (277)	210

Table 7.2. Dates when aboveground biomass, leaf area, and root biomass were measured or calculated.

Location	Year	Aboveground Biomass and Leaf Area	Root Biomass
Mandan	1995	5 May; 2, 12, 29 Jun; 17 Jul; 3, 30 Aug; 10 Oct	30 Jun
	1996	24 Apr; 17 May; 6, 21 Jun; 9, 22 Jul; 12 Aug; 9, 30 Sep; 16 Oct	15 Jul
	1997	23 Apr; 12, 28 May; 10, 23 Jun; 8, 28 Jul; 19 Aug; 8, 29 Sep; 20 Oct	23 Jul
Woodward	1995[1]	1 Jun; 5 Jul; 7 Aug	6 Jul
	1996	13 Jun; 16 Jul; 14 Aug	17 Jul
	1997	3 Jun; 2 Jul; 1 Aug	3 Jul
Temple[2]	1995	7 Feb; 10 Mar; 27 Apr; 19 Jun; 4 Aug; 12 Oct; 30 Nov	4 Aug
	1996	6 Mar; 30 Apr; 25 Jun 6 Aug; 4 Nov	6 Aug
	1997	16 Apr; 2, 16 Jun; 1, 14, 28 Jul; 12, 26 Aug; 8 Sep; 13 Oct; 14 Nov	28 Jul

[1] *Leaf area was not measured at Woodward in 1995.*
[2] *Leaf area at Temple was calculated from biomass and specific leaf area relationships.*

oven-dried and weighed to obtain total aboveground biomass. Leaf area for Temple was determined from biomass and specific leaf area determinations (Dugas et al., 1999). Only green leaf tissue is included in all LAI data.

Plant height was measured at least 3 times each growing season and was used in calculation of turbulent diffusivity (Dugas et al., 1999). Root biomass was measured (Table 7.2) by taking two to six soil cores (6.6 cm diameter, except 4.3 cm diameter at Temple) to a 50-cm depth at Woodward and Temple, and to a 60-cm depth at Mandan. Roots were washed, oven-dried, and weighed. All roots (live and dead) were included in the biomass sample.

Results

Precipitation and temperature

Precipitation varied among locations and years, and it often deviated from long-term monthly averages at each site (Fig. 7.2). Greatest monthly precipitation occurred in the summer at Mandan and Woodward and tended to occur in the spring at Temple. Annual precipitation was always greatest at Temple.

Daily mean air temperatures were near average during the study period, with typically higher temperatures at Temple than at Woodward or Mandan (data not shown). For example, the 3-year average July temperature was 18, 26, and 29°C at Mandan, Woodward, and Temple, respectively.

CO_2 flux

Daily CO_2 fluxes at each location (Fig. 7.3) were near zero or slightly negative (CO_2 efflux) at the beginning of the measurement period in late winter and early spring. They became positive (i.e., CO_2 uptake) during the spring and summer as temperatures and LAI increased. They returned to near zero during the late autumn and early winter because of reduced LAI (plant senescence), soil water, radiation, and temperatures.

Variability in daily fluxes within a season was high and probably resulted from changes in radiation. Fluxes were particularly sensitive to precipitation. For example, short-term droughts reduced CO_2 fluxes at Mandan to near zero during the fall of 1995 and 1996 and the spring of 1997, at Woodward during the fall of 1995 through June of 1996, and at Temple during July and August 1996.

The diurnal CO_2 flux pattern, averaged over 20-day periods, was similar for all sites, differing only in magnitude, during periods of active uptake (May to July) and reduced uptake (September to October) (Fig. 7.4). Differences in CO_2 uptake during May to July followed the aboveground biomass production, which was greatest at Temple and least for Mandan. Nighttime respiratory CO_2 efflux

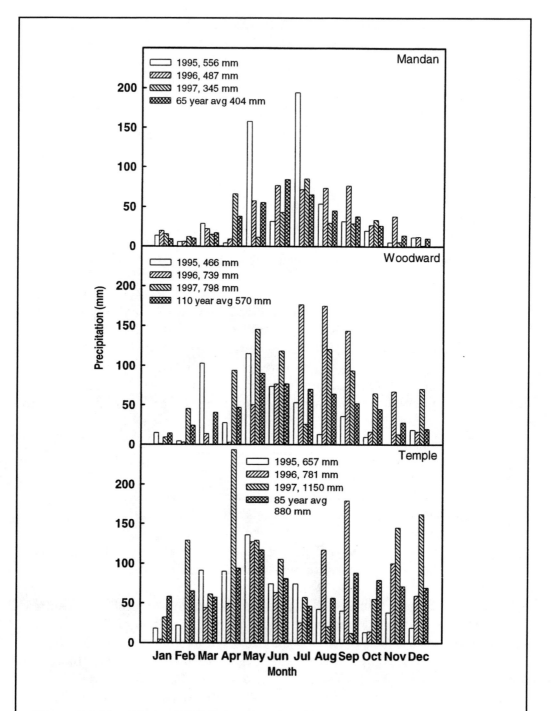

Figure 7.2. Monthly precipitation at Mandan, Woodward, and Temple in 1995, 1996, and 1997. Total annual and long-term average annual precipitation are shown with the bar legends.

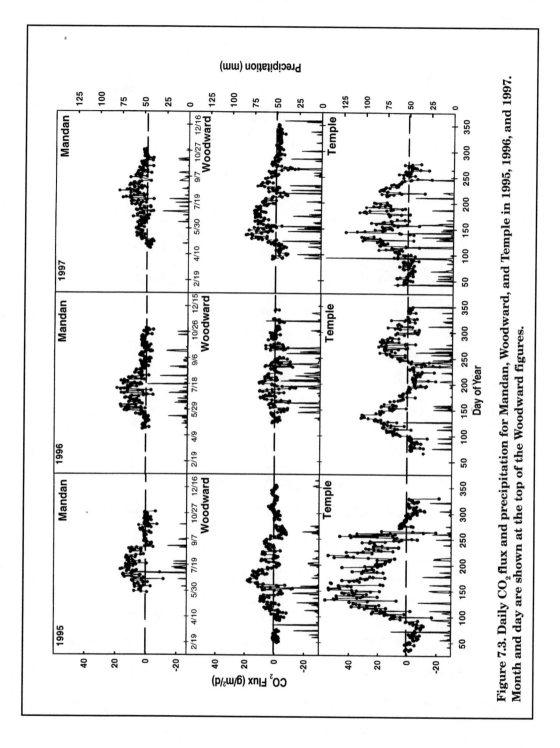

Figure 7.3. Daily CO₂ flux and precipitation for Mandan, Woodward, and Temple in 1995, 1996, and 1997. Month and day are shown at the top of the Woodward figures.

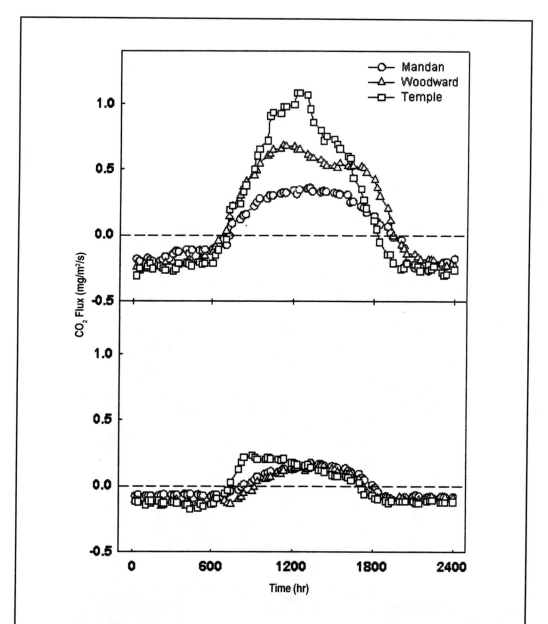

Figure 7.4. Diurnal CO$_2$ fluxes averaged for 20-day periods in 1997 during mid-season growth (top) and late season growth (bottom). Period of data for mid-season growth: July 16 to Aug. 5 for Mandan, June 29 to July 19 for Woodward, May 22 to June 11 for Temple. Period of data for late-season growth: Sept. 14 to Oct. 4 for Mandan and Temple, Sept. 17 to Oct. 7 for Woodward. Time in Central Standard Time.

patterns were also similar at each site, with slightly greater total efflux at Temple and Woodward than at Mandan during both May to July and September to October. The differences probably resulted from lower temperatures and less precipitation at the most northern site of Mandan.

Maximum monthly CO_2 fluxes averaged across 3 years occurred during July at Mandan, June at Woodward, and May at Temple (Fig. 7.5). Fluxes peaked earlier at Temple and Woodward than at Mandan, which coincided with earlier warming temperatures, earlier peak LAI (Fig. 7.6), and greater precipitation than at Mandan. The timing of maximum CO_2 fluxes at Mandan corresponded more closely to the time of peak green biomass and LAI (Fig. 7.6) than at Woodward and Temple, where drought periods often reduced CO_2 uptake after initial spring growth.

Total seasonal CO_2 uptake during the measurement periods averaged across the 3 years was 407 (181 days), 404 (244 days), and 1947 g/m^2 (244 days) at Mandan, Woodward, and Temple, respectively (Table 7.3). The number of days included in the seasonal flux measurements (Table 7.1) varied with location and years, so we also calculated average daily fluxes (Table 7.3) to normalize the flux data. Average seasonal daily fluxes were greatest for Temple (8.7 g/m^2), followed by Mandan (2.3 g/m^2) and Woodward (1.6 g/m^2). The range in seasonal CO_2 fluxes from 1995 to 1997 was nearly 2-fold at Mandan, 15-fold at Woodward, and 3-fold at Temple.

Precipitation total and distribution (Fig. 7.2) affected biomass accumulation and subsequently lead to variability in annual fluxes at all locations. Seasonal CO_2 fluxes at Mandan decreased from 1995 to 1997 as annual precipitation decreased from 556 mm in 1995 to 345 mm in 1997. Distribution of precipitation strongly influenced the large variability between annual fluxes at Woodward, especially in 1996. For example, the annual precipitation in 1996 (739 mm) was greater than the long-term average (570 mm), but the amount received during the critical biomass accumulation period or until July 1 was only 50% of the long-term average.

The variability in fluxes at Temple were due to the effects of the burn in 1995 and distribution of precipitation. The greatest seasonal CO_2 uptake at Temple was in 1995, following the burn in February, and the lowest seasonal fluxes were in 1996, the year of a severe summer drought. The large yearly variations in fluxes at all sites probably were affected by factors other than precipitation, such as radiation and temperature, which underscores the need for extended measurements over several years to accurately assess ecosystem CO_2 fluxes.

The BREB method for measuring CO_2 fluxes is less robust at night, especially during stable atmospheric conditions. We included all nighttime BREB measurements in our calculations of the seasonal fluxes Table 7.3 presents, because the relationship between CO_2 fluxes measured during the night by the BREB method

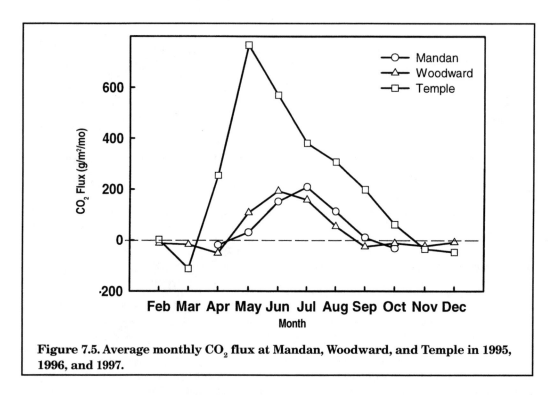

Figure 7.5. Average monthly CO$_2$ flux at Mandan, Woodward, and Temple in 1995, 1996, and 1997.

and sum of the night plant and soil respiratory losses suggest that the BREB method only slightly underestimated CO$_2$ losses at night (Fig. 7.7).

Biomass and leaf area

Peak, green, aboveground biomass typically occurred in June, July, or August, depending on site, precipitation, and year (Fig. 7.6). Biomass was generally greatest at Temple, followed by Woodward and Mandan. Peak aboveground biomass averaged, across 3 years, was 1028, 3234, and 3834 kg/ha at Mandan, Woodward, and Temple, respectively.

LAI tracked green biomass production at all locations and at peak biomass was 0.5, 1.2, and 2.7 at Mandan, Woodward, and Temple, respectively. The variation in green biomass and LAI across locations and years was high, which is typical in the Great Plains where climatic conditions vary widely across years.

This was especially evident at Woodward in 1996, when spring precipitation was considerably below average through June, and at Temple in 1996, when below average June and July precipitation (Fig. 7.2) markedly reduced biomass (Fig. 7.6). This drought period at Temple in 1996 was followed by ample August and September precipitation (Fig. 7.2) which, along with the 1995 burn effects, resulted in maximum green biomass in October.

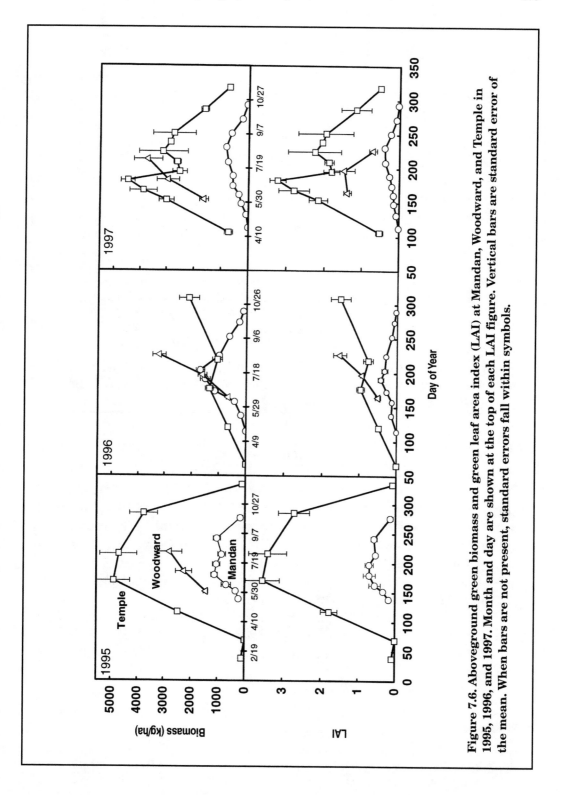

Figure 7.6. Aboveground green biomass and green leaf area index (LAI) at Mandan, Woodward, and Temple in 1995, 1996, and 1997. Month and day are shown at the top of each LAI figure. Vertical bars are standard error of the mean. When bars are not present, standard errors fall within symbols.

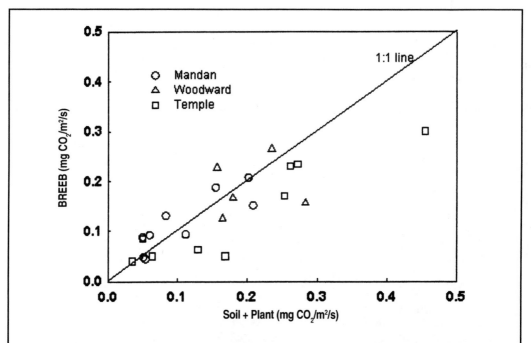

Figure 7.7. Relationship between nighttime CO_2 fluxes measured by the BREB method versus respiratory losses estimated from soil CO_2 efflux measurements and calculated plant respiratory losses.

Biomass and LAI differences among sites reflected precipitation and perhaps the natural production potential of each site, which is greater for Temple, followed by Woodward and Mandan. Root biomass, with standard error, on dates of peak aboveground biomass averaged 15,630 (1,295), 17,169 (2139), and 19,352 (1,472) kg/ha at Mandan, Woodward, and Temple, respectively. Total peak green aboveground biomass plus root biomass averaged 16,658, 19,697, and 23,186 kg/ha at Mandan, Woodward, and Temple, respectively.

These root biomass data compare favorably with earlier studies of root biomass in the central and northern Great Plains mixed grass and tall grass prairies reported by Sims and Singh (1978a), who showed root biomass to be about 30,000 kg/ha in an ungrazed North Dakota mixed grass prairie, about 16,000 kg/ha in a Kansas mixed grass prairie, and about 8000 kg/ha in a central Oklahoma tall grass prairie.

Discussion

The greater seasonal CO_2 flux at Temple (Table 7.3) reflected the greater biomass and LAI (Fig. 7.6) of the tall grass prairie and a longer period of active pho-

Table 7.3. Seasonal, annual, and daily CO_2 fluxes at Mandan, Woodward, and Temple in 1995, 1996, and 1997. Seasonal flux is the measured CO_2 flux for the data collection periods Table 7.2 specifies.

Seasonal Year	Seasonal Flux	Annual Flux	Seasonal Daily Flux	Annual Daily Flux
		-----g/m^2-----		
		Mandan		
1995	538	434	3.4	1.2
1996	395	310	2.0	0.8
1997	287	201	1.5	0.6
Average	407	315	2.3	0.9
		Woodward		
1995	432	370	1.6	1.1
1996	49	−70	0.3	−0.1
1997	731	658	2.8	1.8
Average	404	319	1.6	0.9
		Temple		
1995	3046	2832	13.4	7.8
1996	1048	819	4.1	2.2
1997	1746	1417	8.5	3.9
Average	1947	1689	8.7	4.6

tosynthesis for CO_2 uptake than was present at Mandan and Woodward. For the 3 years considered here, average seasonal fluxes for Mandan (407 g/m^2) and Woodward (404 g/m^2) were similar. Based on the greater precipitation, LAI, aboveground green, and total biomass at Woodward than Mandan, one would expect the CO_2 uptake to be greater at Woodward. However, an early autumn drought in 1995 and a spring-summer drought in 1996 at Woodward significantly reduced green biomass and seasonal CO_2 uptake in those years.

Available soil water strongly influences retention of green, CO_2-fixing leaf area in grasslands. Rogler and Haas (1947) showed that antecedent soil water, along with precipitation amount and distribution, were all important factors controlling grassland biomass production. Also, Sims and Singh (1978b) showed that aboveground net primary production and belowground net primary production across 10 western North American grasslands correlated most highly with long-term mean annual temperature, growing season precipitation, and actual evapotranspiration.

Accurate estimates of annual net CO_2 fluxes must account for the release of CO_2 to the atmosphere during the period when plants are dormant. For the 3 years of this study, the average length of the dormancy period, when net CO_2 uptake was less than zero, was about 220, 190, and 150 days at Mandan, Woodward, and Temple, respectively (Fig. 7.3). Estimates of CO_2 loss for Mandan and Woodward, based on the last 7 days (Mandan) and 14 days (Woodward), was –0.5 and –0.7 g $CO_2/m^2/d$ for Mandan and Woodward, respectively.

The estimate for Temple, based on actual winter flux measurements, was –2.1 g $CO_2/m^2/d$. The lower soil temperatures at Woodward and especially at Mandan vs. Temple easily could account for the lower dormant period's CO_2 loss to the atmosphere at Mandan and Woodward. Mielnick and Dugas (2000) showed that soil respiration strongly relates to soil temperature and that soil respiration is near zero when soil temperature approaches $0^{\circ}C$.

The estimated daily flux rates for the dormant season (–0.5, –0.7, and –2.1 g $CO_2/m^2/d$ for Mandan, Woodward, and Temple, respectively) were summed for the nonmeasurement period and added to the seasonal flux calculations in Table 7.3 to calculate annual fluxes. Based on these estimates for the dormant season, annual net CO_2 uptake averaged 315, 319, and 1689 g/m²/yr at Mandan, Woodward, and Temple, respectively (Table 7.3). These annual fluxes are equivalent to daily CO_2 uptake of 0.9, 0.9, and 4.6 g $CO_2/m^2/d$ or total annual C fixed of 856, 867, and 4594 kg C/ha/yr at Mandan, Woodward, and Temple, respectively.

Sims and Singh (1978b) suggested that the C budget of native grasslands should be near equilibrium. They also recognized that the dynamics of the above- and belowground biomass across the grasslands may affect the C budget (Sims and Singh, 1978a). Dugas et al. (1999) interpreted estimated annual fluxes of 500 and 800 kg C/ha/yr at the Temple site in 1993 and 1994 as "in approximate equilibrium," at least in relation to estimated annual fluxes that were 10 times the amounts in an adjacent perennial grass pasture newly established on cropland.

Results reported here suggest that grasslands may represent a small sink for atmospheric C. However, we realize that sampling errors are present when measuring fluxes and in estimating dormant period fluxes. The estimates of dormant period fluxes we used were based on flux measurements made during the early phase of the plant dormancy period at Mandan and Woodward, and during the mid-dormancy period at Temple. Fluxes from respiratory losses during the winter season at Mandan and Woodward would probably be less because of lower expected temperatures during the winter.

The relationship between respiratory losses which Figure 7.7 presents suggest that the BREB method slightly underestimated nighttime CO_2 losses. A simple sensitivity analysis, based on doubling the CO_2 flux during the dormant period, to allow for the underestimation by the BREB method, would reduce the net annual CO_2 flux by about 29%, 26%, and 15% at Mandan, Woodward, and Temple, respectively. Even at these reduced annual flux rates, the grasslands included in this study were net sequesters of atmospheric CO_2.

Limited annual CO_2 flux data for grasslands exists in the literature to compare with our results. Kim et al. (1992) reported that a tall grass prairie site in the central Great Plains fixed 750 g CO_2/m^2 from May through October, but only 250 g CO_2/m^2 from late June to August 2 years later; however, they did not estimate annual net fluxes. The Mandan and Woodward annual net CO_2 uptake reported here, of 315 and 319 g $CO_2/m^2/yr$, compares favorably with Kim et al. (1992) on an annual basis and with the preburn CO_2 uptake of 250 g $CO_2/m^2/yr$ which Dugas et al. reported (1999) for a tall grass prairie.

The daily CO_2 fluxes measured at Temple in 1995, 1996, and 1997 (Fig. 7.3) differ from those measured at the same site in 1993 and 1994 (Dugas et al., 1999). In our study, net uptake of CO_2 at Temple occurred from April through November, whereas Dugas et al. (1999) reported that uptake only occurred from April through July in 1993 and 1994. Calculated annual uptake rates also were considerably less prior to the 1995 burn than reported here. Before the burn, the annual net CO_2 uptake was about 250 g/m²/yr. The buildup of surface litter and detritus at this site, which was substantial (10,000 kg/ha) because of the long interval between burns, limited productivity through reduced energy flow because of shading and a lower rate of nutrient cycling.

Removal of this dead surface material may have increased CO_2 uptake from the atmosphere, especially in 1995, because of:

1. substantially increased canopy light penetration (Knapp, 1984; Knapp et al., 1993) that increased CO_2 uptake throughout the canopy (Schimel et al., 1991) and increased canopy photosynthesis rates;
2. slight increases in leaf photosynthesis rates (Svejcar and Browning, 1988) and stomatal conductance (Knapp and Seastedt, 1986); and
3. less available CO_2 from decomposition of surface litter (Bremer et al. 1998).

Aboveground biomass production in a tall grass prairie increased by more than 75% after a burn (Knapp and Seastedt, 1986). At Temple, CO_2 fluxes were lower in 1996 and 1997 than in 1995 but were still greater than those for the same site in 1993 and 1994 (Dugas et al., 1999). This suggests that the effects of the fire in 1995 on CO_2 uptake may have continued for at least 3 years at this site.

The relationship between biomass accumulation calculated from total C uptake (1.5 g CO_2 equals 1 g dry matter) and biomass accumulation was determined between the first clipping and a clipping at peak biomass production (root biomass based on 3:1 root/top ratio) to evaluate the utility of the BREB method for determining net primary productivity of grasslands (Fig. 7.8). Total CO_2 uptake was calculated from fluxes measured between the biomass clipping dates.

Soil microbial respiration of 4 g $CO_2/m^2/d$ (Dugas et al.,1999) was added to the CO_2 flux for the period to obtain total site CO_2 capture. The relationship between biomass obtained by sampling and biomass estimated from CO_2 flux was scattered below the 1:1 line (Fig. 7.8). Overall biomass measurements from clipping trended higher than biomass calculated from flux data, especially for Mandan and Woodward. The underestimation of biomass accumulation from flux data may have been partly due to the inclusion of all root material in the total root biomass.

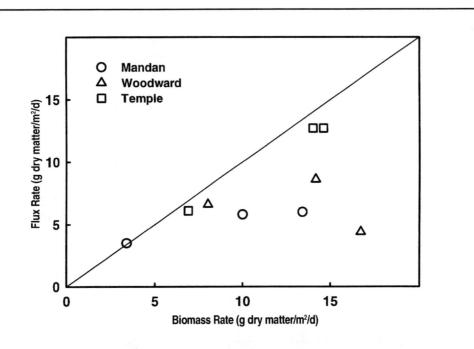

Figure 7.8. Dry matter accumulation rate calculated from daily CO$_2$ flux versus that calculated from hand clipped samples prior to peak biomass production at Mandan, Woodward, and Temple in 1995, 1996, and 1997. The 1:1 line is shown.

Conclusions

The transfer of plant C to stable soil C was not measured in this study. However, if 10% of the C fixed was stored in soils as humus, as suggested by Lal et al. (1998), then 86, 87, and 459 kg C/ha/yr would be stored annually in the soils at the Mandan, Woodward, and Temple sites, respectively. There are 51 Mha of grassland in the U.S. Great Plains states of North Dakota, South Dakota, Nebraska, Kansas, Oklahoma, and Texas (USDA Forest Service, 1980). The Mandan site is representative of about 14.4 Mha, the Woodward site of 16.4 Mha, and the Temple site of 20 Mha of these grasslands.

Based on the data from 1995 through 1997, about 1.2, 1.5, and 9.2 x10^6 Mg C/yr would be stored annually in soils of ecosystems represented by the Mandan, Woodward, and Temple sites, respectively. However, since the Temple results reported here may not be truly representative because of the burn effects, Temple pre-burn flux for 1993 and 1994 of 250 g CO$_2$/m^2/yr (Dugas et al., 1999) plus the results from this study of 1689 g CO$_2$/m^2/yr would result in a storage rate of 303 kg C/ha/yr. After adjustments, this would result in storage of 6.0 x 10^6 Mg C/yr in the grasslands representative of the Temple site.

Follett, Kimble, and Lal, editors

The grassland ecosystems represented in this study contain vegetation that is typical of native grasslands in each region, but they differ in that these sites were not grazed during the periods of flux measurements. Grazing would reduce plant aboveground biomass and net C uptake. Although the CO_2 flux data presented are for only a 3-year period that included highly variable precipitation amounts, it was anticipated that net fluxes would be near zero, as Sims and Singh (1978b) suggested for other grasslands. This would be consistent with the suggestion of Cole (1996) that grassland soils have little capacity to store additional C, mainly because of the inherently low soil fertility and low precipitation to support greater biomass production.

In contrast, the CO_2 fluxes reported here document a net positive CO_2 flux for the 3 sites evaluated. Batjes (1998), Gifford (1994), and Schimel (1995) suggest that C storage in these systems may be the result of increasing atmospheric CO_2 producing a CO_2 fertilization effect, which would increase photosynthesis and improve plants' efficiency of water use, and both processes would increase biomass C and possibly soil C. Morgan et al. (1994) showed that both C_3 and C_4 native prairie grasses exhibited increased photosynthetic capacity with increasing CO_2 concentration which, along with the extensive fibrous root system characteristic of grasses, provides further evidence for C storage in grassland ecosystems.

The CO_2 fluxes measured in this study suggest these 3 grassland sites function as small net sinks for atmospheric CO_2. The site at Temple appears to have the greater sink potential than those at Mandan and Woodward. All 3 grasslands sites in our study are located near the western edge of the region in North America that Fan et al. (1998) identified as a large terrestrial sink for atmospheric CO_2. They estimated 70 to 100% of the terrestrial sink in North America was located in the broadleaf plant (forests) regions south of $51°$ N latitude. Our CO_2 flux data suggest that the Great Plains grasslands may be contributing significantly to this North American sink.

Further clarification will be linked to our understanding of how the terrestrial grassland sink responds to land use management and climate. More definitive information on belowground biomass production and root turnover is also needed.

References

Adams, J.M., H. Faire, L. Faire-Richard, J.M. McGlade, and F.L. Woodward. 1990. Increases in terrestrial carbon storage from the last glacial maximum to the present. *Nature* 348:711-714.

Batjes, N.H. 1998. Mitigation of atmospheric CO_2 concentrations by increased carbon sequestration in the soil. *Biol. Fertil. Soils* 27:230-235.

Berg, W.A. 1994. Sand sagebrush-mixed prairie (SRM 722). *In* T.N. Shiflet (ed), *Rangeland Cover Types of the United States*, Society for Range Manage, Denver, CO, pp 99.

Bremer, D.J., J.M. Ham, C.E. Owensby, and A.K. Knapp. 1998. Response of soil respiration to clipping and grazing in a tallgrass prairie. *J. Environ. Qual.* 27:1539-1548.

Cole, V. 1996. Agricultural options for mitigation of greenhouse gas emissions. *In* R.T. Watson, M.C. Zinyowerea, and R.H. Moss (eds), *Climate Change 1995. Impacts, Adaptations and Mitigation of Climate Change: Scientific-Technical Analyses*, Cambridge University Press, Cambridge, pp. 747-771.

Dixon, R.K., S. Brown, R.A. Houghton, A.M. Solomon, M.C. Trexler, and J. Wisniewski. 1994. Carbon pools and flux of global forest ecosystems. *Sci.* 263:185-190.

Dugas, W.A. 1993. Micrometeorological and chamber measurements of CO_2 flux from bare soil. *Agric. For. Meteorol.* 67:115-128.

Dugas, W.A., M.L. Heuer, and H.S. Mayeux. 1999. Carbon dioxide fluxes over bermudagrass, native prairie, and sorghum. *Agric. For. Meteorol.* 93:121-139.

Fan, S., M. Gloor, J. Mahlman, S. Pacala, J. Sarmiento, T. Takahashi, and P. Tans. 1998. A large terrestrial carbon sink in North America implied by atmospheric and oceanic carbon dioxide data and models. *Sci.* 282:442-446.

Gifford, R.M. 1994. The global carbon cycle: A viewpoint on the missing sink. *Aust. J. Plant Physiol.* 21:1-15.

Houghton, R.A., J.L. Hacker, and K.T. Lawrence. 1999. The U.S. carbon budget: Contributions from land-use change. *Sci.* 285:574-578.

Keeling, C.D., J.F.S. Chin, and T.P. Whorf. 1996. Increased activity of northern vegetation inferred from atmospheric CO_2 measurements. *Nature* 382:146-149.

Kim, J.S., B. Verma, and R.J. Clement. 1992. Carbon dioxide budget in a temperate grassland ecosystem. *J. Geophys. Res.* 97:6057-6063.

Knapp, A.K. 1984. Post-burn differences in solar radiation, leaf temperature and water stress influencing production in a lowland prairie. *Am. J. Bot.* 71:220-227.

Knapp, A.K., J.T. Fahnestock, S.P. Hamburg, L.B. Statland, T. R. Seastedt, and D.S. Schimel. 1993. Landscape patterns in soil-plant water relations and primary production in tallgrass prairie. *Ecology* 74:549-560.

Knapp, A.K., and T.R. Seastedt. 1986. Detritus accumulation limits productivity of tallgrass prairie. *Bioscience* 36:662-668.

Lal, R., J. Kimble, E. Levine, and C. Whitman. 1998. Towards improving the global database on soil carbon. *In* R. Lal, J. Kimble, E. Levine, and B.A. Stewart (eds), *Soils and Global Change*, Lewis, Boca Raton, FL, pp. 343-436.

Mielnick, P.C., and W. A. Dugas. 2000. Soil CO_2 flux in a tall grass prairie. *Soil Biol. and Biochem.* 32:221-228.

Morgan, J.A., H.W. Hunt, C. A. Monz, and D.R. Lecain. 1994. Consequence of growth at two carbon dioxide concentrations and two temperatures for leaf gas exchange in *Pascopyrum smithii* (C_3) and *Bouteloua gracilis* (C_4). *Plant, Cell and Environ.* 17:1023-1033.

Rastetter, E.B., R.B. McKane, G.R. Shaver, and J.M. Melillo. 1992. Changes in C storage by terrestrial ecosystems: How C-N interactions restrict responses to CO_2 and temperature. *Water, Air, and Soil Poll.* 64:327-344.

Rogler, G.A., and H.J. Haas. 1947. Range production as related to soil moisture and precipitation on the Northern Great Plains. *Agron. J.* 39:378-389.

Schimel, D.S. 1995. Terrestrial ecosystems and the carbon cycle. *Global Change Biol.* 1:77-91.

Schimel, D.S., T.G.F. Kittel, A.K. Knapp, T.R. Seastedt, W.J. Parton, and V.B. Brown. 1991. Physiological interactions along resource gradients in a tallgrass prairie. *Ecology* 72:672-684.

Sims, P.L. and J.S. Singh. 1971. Herbage dynamics and net primary production in certain ungrazed and grazed grasslands in North America. *In* N. R. French (ed), *Preliminary Analysis of Structure and Function in Grasslands, Range Science Dept Science Series No. 10,* Colorado State Univ, Ft Collins, CO, pp. 59-124.

Sims, P.L. and Singh, J.S. 1978a The structure and function of ten western North American grasslands. II. Intraseasonal dynamics in primary producer compartments. *J. Ecol.* 66:547-572.

Sims, P.L. and J.S. Singh. 1978b. The structure and function of ten western North American Grasslands. IV. Compartmental transfers and energy flow within the ecosystem. *J. Ecol.* 66:983-1009.

Sims, P.L. and P.G. Risser. 1999. Chapter 9, Grasslands. *In* M.G. Barbour and D.W. Billings (eds), *North American Terrestrial Vegetation*, 2nd ed., Cambridge University Press, New York, pp. 321-354.

Sundquist, E.T. 1993. The global carbon dioxide budget. *Sci.* 259:934-941.

Svejcar, T.J., and J.A. Browning. 1988. Growth and gas exchange of *Andropogon gerardii* as influenced by burning. *J. Range Manage.* 41:239-244.

Svejcar, T., H. Mayeux, and R. Angell. 1997. The rangeland carbon dioxide flux project. *Rangelands* 19:16-18.

USDA Forest Service. 1980. *An assessment of the forest and range land situation in the United States. FS-345.* U.S. Government Printing Office. Washington, DC.

Watson, R.T., M.C. Zinyowerea, and R.H. Moss (eds). 1996. *Climate change 1995. Impacts, adaptations and mitigation of climate change: scientific-technical analyses. Contribution of working group II to the second assessment report of the Intergovernmental-Panel on Climate Change.* Cambridge University Press. Cambridge.

Webb, E.K., G.I. Pearman, and R. Leuning. 1980. Correction of flux measurements for density effects due to heat and water vapor transfer. *Q.J.R. Meteorol. Soc.* 106:85-100.

World Resources Institute. 1986. *World Resources, 1986.* Basic Books, Inc. New York.

CHAPTER **8**

Carbon Sequestration in Arctic and Alpine Tundra and Mountain Meadow Ecosystems

K.L. Povirk,[1] J.M. Welker,[2] and G.F. Vance[3]

Introduction

Evaluating the potential of cold-region environments to sequester C involves understanding many biogeochemical processes and interactions. This should include feedback relationships involving indigenous and anthropogenic biogeochemical activities. Human needs, desires, and opinions, as well as actions, can and do influence public policy about federally owned lands, including those in cold regions.

Arctic and alpine tundra and mountain meadows are valued highly for ecosystem services, including wildlife habitat, recreation, mineral resources, fresh water, and summer forage for livestock (Billings, 1979; Johnson, 1979; Körner, 1994; Laycock et al., 1996). These cold habitats, while providing important ecosystem services, are some of the last wilderness areas in North America. They are refuges for nature, often surrounded by large human populations, and are regions important to the global biogeochemical cycling of C and other nutrients (Oechel and Vourlitis, 1994; Körner, 1995; McKane et al., 1997).

Recently, cold-region environments have become the focus of increased attention as they store significant amounts of soil C (Oechel and Billings, 1992) that may be respired as global temperatures rise (Oechel et al., 1993; Chapin and Körner, 1995; Oechel et al., 1995; Michaelson et al., 1996; Jones et al., 1998; Welker et al., 1999). If we are to understand how current human activities affect C cycling, we must be familiar with the processes that govern C fixation, storage,

[1] Research Scientist, Rangeland Ecology and Watershed, Department of Renewable Resources, University of Wyoming, Laramie, WY 82071; e-mail kpovirk@yahoo.com.

[2] Assoc. Prof., Rangeland Ecology and Watershed, Department of Renewable Resources, University of Wyoming, Laramie, WY 82071; e-mail jeff@uwyo.edu.

[3] Professor, Soil Science, Department of Renewable Resources, University of Wyoming, Laramie, WY 82071; e-mail gfv@uwyo.edu.

and release in these high elevation, high latitude, cold region environments (Chapin et al., 1995; Zimov et al., 1993, 1996; Oberbauer et al., 1996).

Land area

Cold-region environments constitute significant portions of the terrestrial biosphere, approaching 25%. These habitats exist on all continents (Heal et al., 1998). Between 60°N and 60°S, the alpine life-zone encompasses approximately 4.5×10^6 km² worldwide. Körner (1995) suggests that 50% of the land area between 60°N and 70°N, and between the altitudes of 600 to 1300 m, should be considered part of the Arctic–alpine life zone (approximately 1.2×10^6 km²) for a total of 5.7×10^6 km² (Fig. 8.1).

Correcting for glaciers, barren landscapes, and mountain deserts, *alpine tundra* represents 3% of the global land area (4×10^6 km²) and at least 1% of the global terrestrial C pool (Körner, 1995). *Arctic tundra* represents 8×10^6 km² of the global surface area, with an NPP (net primary production) of 1,000 kg C/ha/yr, amounting to 180 Pg of soil organic C (SOC) (Scharpenseel and Pfeiffer, 1998). Additionally, tundra represents 13.7% of the global pool of SOC (Scharpenseel and Pfeiffer, 1998).

Mountain meadow ecosystems represent approximately 5% of mountain forested areas and are difficult to define since they are present at both high and low altitudes and along highly variable gradients. Area and C parameters of these small openings in continuous forest are not well defined, as they span multiple soil orders. Within these diverse cold region environments, 7 of the 11 soil orders (prior to the introduction of a 12th soil order, gelisols) are present (Marion et al., 1997).

Importance to carbon flux

Alpine and Arctic tundra and mountain meadow ecosystems may be highly important for their C flux potential (Billings et al., 1982; Oberbauer et al., 1991, 1996; Oechel et al., 1997; Randerson et al., 1997). Because of their vast land area (Bliss and Matveyeva, 1992) and potentially large soil C pool (Oechel and Billings, 1992), these cold region habitats are important in the net exchange of CO_2 between the biosphere and the atmosphere (Oechel and Vourlitis, 1994; Welker et al., 2000).

As a result of global warming, associated with increasing CO_2 concentrations in the atmosphere (Vitousek, 1994; Schlesinger, 1997), it generally is accepted that alpine and Arctic tundra ecosystems are likely to undergo more pronounced changes than ecosystems at lower altitudes and lower latitudes (Körner, 1994; Callaghan and Jonasson, 1995; Chapin et al., 1996). Climatic warming could

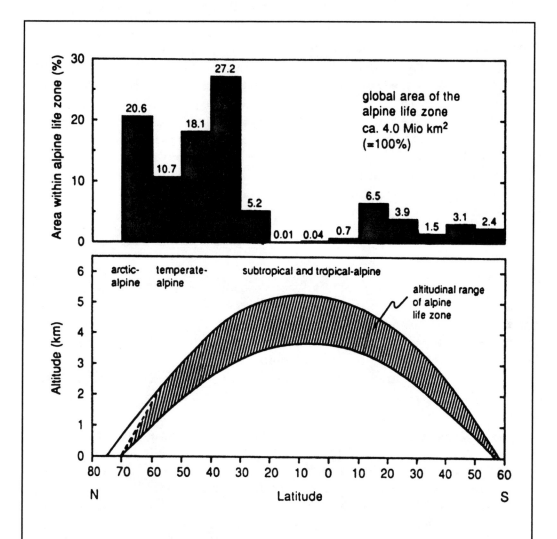

Figure 8.1. An estimation of the latitudinal distribution of the alpine life zone. The lower diagram illustrates the mean lower and upper altitudes of the alpine life zone. The dashed line on the left accounts for the uncertainty between Arctic alpine vegetation and so-called mountain tundra. The upper diagram illustrates the relative contribution of each 10° latitudinal range to the total global area alpine vegetation covers. (Used by permission, Chapin and Körner, 1995.)

cause great alterations of regional biospheric C pools since these cold, wintery ecosystems may be more sensitive to global change (Zimov et al., 1993).

Changes in the balance between photosynthesis and respiration in these areas may be great enough to affect the global pattern of atmospheric CO_2 concentrations (Zimov et al., 1993, 1996). Arctic tundra ecosystems may begin to exhibit greater CO_2 source behavior (Billings et al., 1982; Oechel et al., 1993, 1995) as soil microbial activity is stimulated (Schimel and Clein, 1995), especially since gross ecosystem respiration seems more sensitive to warmer temperatures than gross ecosystem photosynthesis (Oechel et al., 1993; Jones et al., 1998; Shaver et al., 1998; Welker et al., 1999).

In addition, alpine tundra systems at the higher altitudes may respond positively to the documented increase in CO_2 concentrations, as growth currently may be limited (in some scenarios) due to low atmospheric pressure (Körner et al., 1997). Consequently, plant growth and C sequestration in the alpine could be stimulated by increasing CO_2 concentrations (Harrison and Broecker, 1993; Körner, 1995).

Grazing

Ungulates such as domestic livestock, caribou, moose, and other wildlife are the principal herbivores consuming forage and recycling nutrients in cold regions (Bryant et al., 1983; Russell et al., 1993; Jefferies et al., 1994; Walsh et al., 1997). In alpine regions and mountain meadows, livestock and wildlife use tundra and forest openings for summer forage (Bradford, 1998).

In the U.S., these areas are typically portions of federally managed Forest Service lands and are leased to livestock producers and grazed for 1 to 3 months by sheep and/or cattle. Additionally, large, wild herbivores, such as elk, use these areas during the summer. Generally, the forage resources of mountain meadows and some alpine ecosystems are of great enough abundance to support both domestic and wild ungulate forage needs. Although the effects of grazing on soil C vary (Dormaar et al., 1990; Milchunas and Lauenroth, 1993; Frank et al., 1995; Manley et al., 1995), where sheep graze at Libby Flats, WY, the trend is for greater soil C in the grazed than in the ungrazed meadow (Fig. 8.2).

In the Arctic tundra, the Porcupine Caribou migrate from the boreal forest, where they spend the winter consuming lichens, to the Arctic coastal plain, where they calve and consume an array of graminoids, including prevalent species such as *Eriophorum vaginatum* (Russell et al., 1993; Walsh et al., 1997). These animals represent the largest migratory herd of ungulates in North America. They rely on the abundant forage of the coastal plain to recharge energy reserves used during the migration and to support lactating cows. It appears that this migration has been under way for several thousand years, and the entire home range of this herd may encompass 250,000 km² (Russell et al., 1993).

On the opposite side of the Arctic, caribou have been domesticated by the Scandinavians, while those in Lapland (N. Sweden, Norway, and Finland) herd caribou. Raised as livestock, the caribou are herded from the boreal forest to summer ranges on the Arctic Ocean.

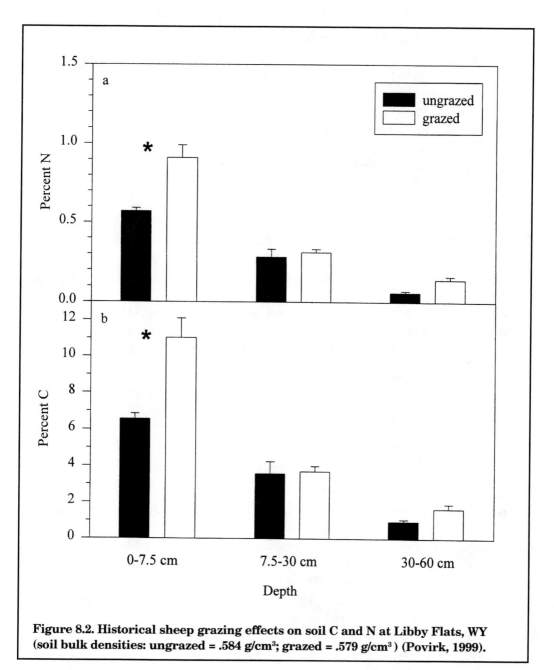

Figure 8.2. Historical sheep grazing effects on soil C and N at Libby Flats, WY (soil bulk densities: ungrazed = .584 g/cm³; grazed = .579 g/cm³) (Povirk, 1999).

Use and Federal Ownership

In the U.S., with the exception of privately owned in-holdings, alpine and Arctic tundra and mountain meadow areas are primarily under federal (U.S. Forest Service, National Park Service, and Bureau of Land Management) ownership and subject to federal management policies, including the Multiple Use Act of 1976. From the late 1860s to the 1880s, before the creation of the Forest Reserves (later the U.S. Forest Service) through the General Revision Act of 1891 (McCarthy, 1977), unregulated livestock (sheep and cattle) grazed many of these areas (Bradford, 1998).

Livestock grazing privileges on the Forest Reserves were given to ranchers who could prove ownership of both livestock and base property sufficient to sustain their livestock throughout the rest of the year (Laycock et al., 1996). In addition, ranchers had to prove prior use, something the ranchers themselves insisted on to retain security of tenure.

Today, many privately owned ranches in the rural western U.S. depend on federal lands for summer livestock forage. Future management decisions, potentially having large financial impacts on rural agricultural communities, should be considered carefully.

Carbon Sequestration Controls and Constraints

Plant carbon assimilation

Higher plants sequester C primarily during the months of May, June, July, and August, months typically snow-free in temperate and high latitude systems (Chapin and Shaver, 1985). During late spring and early summer, rates of C fixation respond rapidly to the daily increases in irradiance of the early morning hours, with maximum assimilation often reached before midday (Körner and Larcher, 1988; Oberbauer et al., 1991, 1996). This is especially the case for C_3 plants in late spring and early summer when water deficits are minimal, though, by late season, C fixation may be reduced.

Although seasonal drought affects rates of C assimilation, air temperature and soil nutrients are the primary limitations to C fixation by plants in cold regions (Welker et al., 1993, 1995; Wookey et al., 1993; Oberbauer et al., 1996). For example, the addition of nutrients to tussock tundra in Alaska increases the rates of leaf photosynthesis, and, in the High Arctic, warmer temperatures result in higher rates of gas exchange and C fixation (Welker et al., 1993; Wookey et al., 1995). In alpine environments, similar limitations to C fixation have occurred, whereby grass and forb C assimilation almost doubled with added N and P fertilizer (Bowman et al., 1995).

Herbivory affects C assimilation in higher plants through both intrinsic and extrinsic mechanisms (McNaughton, 1983; Fahnestock and Detling, 2000). Removing older leaf tissue can keep existing plants in a suspended juvenile stage of growth. These plants often have leaves of higher N content and higher rates of C gain (Jefferies et al., 1994). Additionally, species subject to long-term grazing exposure may have inherently different resource allocation patterns that facilitate higher rates of assimilation.

For instance, Caldwell et al. (1981) found that *Agropyron cristatum* not only had higher rates of photosynthesis following grazing than ungrazed plants, but that greater amounts of C were allocated to leaf blades and sheaths, ensuring future C assimilation. However, this was not the case for *Pseuderoegneria spicata* which did not have an evolutionary history with grazing. In the Pryor Mountain Wild Horse Range, Fahnestock and Detling (2000) found few effects on C fixation between ungrazed and short- and long-term grazing of five cool season grasses.

Herbivory may alter soil properties, and, subsequently, C fixation potential. High densities of animals can result in large inputs of animal N into the grazing landscape. These inputs of N can increase leaf N and result in accelerated rates of leaf C fixation (Jefferies et al., 1994; Wilson and Jefferies, 1996; Tracy and Frank, 1998).

Indirect effects of animals may include changes in soil water properties and individual plant water balance. The removal of leaf area can produce higher water potentials for remaining biomass, further facilitating gas exchanges and C fixation. Studies indicate that browsed *Salix* sp. plants may have higher rates of C fixation (Farquhar et al., 1989) and higher water potentials than unbrowsed plants. However, the loss of leaf tissue still produces less photosynthetically active surface area and potentially less C fixed in soil organic matter (SOM). Both directly and indirectly, it appears that herbivores can affect C fixation, although these responses may be site-, region-, or year-specific and depend on a host of environmental and biotic parameters.

Soil organic carbon (SOC)

Several forms of C comprising SOM pools are broadly categorized as plant litter, microbial biomass, partially decomposed organic remains, dissolved organic substances, and humus materials (Stevenson, 1994). Each of these groupings represent forms of SOC based primarily on solubility and susceptibility to microbial decomposition, which in turn determines C and nutrient availability, C loss (e.g., CO_2 release), and C accumulation in the more resistant SOM pool (Paul and Clark, 1996).

Labile or dynamic SOM pools represent C and nutrient sources that are readily available for decomposition and use by microorganisms, plants, and other dissemination processes, whereas the recalcitrant (stable humus) SOM pool functions as a source of SOC and an organic nutrient reservoir, providing long-term soil stability (Stevenson, 1994; Janzen, 1998; Stevenson and Cole, 1999). While labile SOC components can rapidly immobilize, mineralize, or volatilize, stable SOM significantly influences numerous soil chemical and physical properties (Table 8.1). Many of these properties are important in considering the role of SOC in cold region environments.

Soil C sequestration occurs by converting above- and belowground plant materials, i.e., root and leaf biomass, into SOC pools that preserve C temporarily or long term. Various processes involved in transforming materials into labile or stable SOC constituents depend on inherited soil conditions and on disturbances by natural and human influences.

In order to enhance SOC contents, either increased plant biomass inputs or decreased C losses are required. Conditions described above for the control of C assimilation by plants therefore are important in increasing above- and belowground plant biomass.

Correlations between increased plant residues and SOC accumulations are well documented for agronomic, forestry, and rangeland ecosystems (Campbell and Zentner, 1993; Van Cleve and Powers, 1995; Herrick and Wander, 1998). Environmental and human influences that limit the decomposition of plant biomass inputs, as well as indigenous SOM components, can cause an increase in SOC.

Table 8.1. Some general properties of SOM and their effects on cold region soils.

Property	Comment	Influence
Water retention	Organic matter can hold up to 20 times its weight in water	Helps prevent drying and shrinking. Improves moisture-retaining properties of sandy soils
Color	The typical dark color of many soils is caused by organic matter	May facilitate warming
Mineralization	Decomposition of organic matter yields CO_2, NH_4^+, NO_3^-, PO_4^{3-}, and SO_4^{2-}	Source of nutrients for plant growth
Aggregation	Cements soil particles into structural units called aggregates	Permits exchange of gases, stabilizes structure and increases permeability
Buffer action	Exhibits buffering in slightly acid, neutral, and alkaline ranges	Helps to maintain a uniform reaction in the soil
Cation exchange	Total acidities of isolated fractions of humus range from 300 to 1400 cmoles/kg	Increase cation exchange capacity (CEC) of the soil. From 20 to 70% of the CEC of many soils (e.g., Mollisols) is caused by organic matter

(Adapted from Stevenson, 1994.)

Reducing biological activity responsible for decomposing plant residue and producing CO_2, and protecting SOC physically through increased aggregation, are two mechanisms that can enhance soil C sequestration (Hassink et al., 1993; Lal et al., 1998).

Several processes affect increases (humification, aggregation, sedimentation, and deposition) and decreases (erosion, leaching, decomposition, and volatilization) in SOM contents (Lal et al., 1998). For soil C sequestration, the processes that contribute to SOM losses need to be minimized while processes that result in SOM accumulation and stability need to be maximized.

Mineralogy, soil texture, soil structure, vegetation-soil effects, climate, and the soil's preservation capacity (Bouwman, 1990) influence SOM stability. According to Oades (1988), mineralogy that releases bases (e.g., Ca and Mg) is more favorable for rapid decomposition of plant litter, but it also has a greater tendency to retain SOM and have higher SOC contents. Mineralogy that is acidic can result in slower decomposition of plant litter initially; however, due to the lack of retention properties, the resulting SOM is lost rapidly through more complete degrading processes.

Textural differences also result in greater SOM retention by clay, compared to sandy particle-size materials, which is more apparent in soils with basic mineralogy and alkaline pH (Jenkinson, 1977). Aggregation and structure formation due to clay-organic complexes, both with and without cation (e.g., Ca and Al) bridging, provides a mechanism for the physical protection of SOM. Grassland ecosystems result in greater belowground SOM than forest sites, which have higher aboveground biomass.

Temperature and precipitation play important roles in the rates of plant decomposition and SOM accumulation. Moist, warm conditions are much more favorable for fast breakdown and oxidation processes, whereas cold, wet ecosystems tend to have slow decomposition rates that encourage SOM accumulation; however, plant biomass is often much greater under moist, warm conditions. Microbial activity has a dominant influence on determining decomposition products, rates of reactions, and transformation pathways. In summary, chemical, physical, biological, and hydrological conditions are interrelated factors that determine a soil's capacity to preserve the integrity of its SOM pools.

Decomposition of plant biomass and SOM also depends on environmental factors that influence microbial populations and their respective activities (Paul and Clark, 1996). Over time, the initial plant biomass components reduce as microbial biomass increases and then diminishes with loss of easily degradable plant constituents. Conditions in tundra and mountain meadow ecosystems that contribute to reductions in SOM include both wind and water erosion; low plant biomass production due to the predominant climatic conditions such as cold winters, dry summers, and differences in photoperiod at high altitudes and latitudes; and soils with low fertility status, hydric properties, and weak soil profile development.

Management practices, natural events, and anthropogenically driven climatic changes can influence C loss and/or accumulation by altering the chemical, physical, and biological properties of soil ecosystems. Disturbance of natural ecosystems results in the disruption of soil aggregates that contribute to the stabilization of SOM (Kay, 1998). Exposure of SOM that was physically protected from potential chemical and microbial oxidative changes will lead to rapid decomposition.

Climatic conditions resulting from global warming could increase soil temperatures and thus quicken the rate of SOM decomposition and production of CO_2 (Scharpenseel et al., 1990); but greater plant biomass production with higher atmospheric temperatures could enhance C accumulation, particularly in wetter environments. In addition, reduced precipitation could enhance SOM losses due to higher soil temperatures and increased aerobic conditions, while preventing plant biomass production and residue inputs. However, increased precipitation would have an opposite effect, resulting in C sequestration. Higher atmospheric CO_2 contents may suppress the degradation rates of surface litter, which in some ecosystems is the primary material that contributes to SOM (Bohn, 1990).

Interactions between soil and vegetation

Figure 8.3 shows interrelationships among plants, soils, and herbivores and expected pathways for CO_2 and other greenhouse gases. Of the most concern is the increase in atmospheric CO_2 levels that are direct inputs from plant and soil microbial respiration and represent a loss of C from the soil C pool. Plants convert CO_2 to sugars, physiological products, and other organic constituents. These products can become forage materials for herbivores or become plant residues that, through decomposition, incorporate into SOM pools. Additional greenhouse gases such as N_2O, NO_x, and CH_4 may be released to the atmosphere or incorporated into soils through microbial activity.

In cold regions, increased temperatures, and possibly additional precipitation due to global climate change, may result in greater plant CO_2 assimilation and greater C sequestration into SOM pools. However, this temperature increase also may stimulate microbial communities further and enhance the release of greenhouse gases (mainly CO_2) from these ecosystems.

Alpine Tundra

Physiography

Alpine ecosystems are bands of vegetation at or above the tree line. They include islands of krummholz but do not include forests of normal upright growth

Follett, Kimble, and Lal, editors

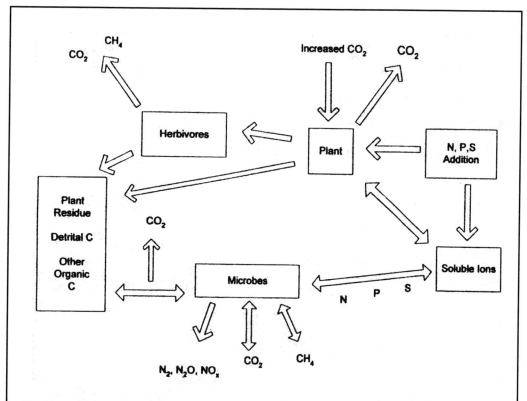

Figure 8.3. Interrelationships among plants, soils, and herbivores and expected pathways for CO_2 and other greenhouse gases (adapted from Follett et al., 1991).

patterns (Bliss, 1985). Alpine ecosystems are highly variable and stressful environments for both plants and animals (Körner, 1994, 1995).

The variability in alpine conditions is due primarily to elevational, latitudinal, and topographic differences, resulting in large differences in photoperiod and maximum radiation, atmospheric pressure, length of growing season, diurnal variation in temperature, precipitation, and soil nutrients, especially N (Billings, 1979; Bliss, 1985; Körner, 1995).

Additionally, regional isolation of floras and habitat fragmentation is common in alpine ecosystems (Körner, 1995). In the U.S., these islands of vegetation are mainly in the West and concentrate in the Rocky Mountains of northern New Mexico, Colorado, Wyoming, Idaho, and Montana and in the Sierra Nevada-Cascade mountain chain of California, Oregon, and Washington (Retzer, 1974; Billings, 1979). However, the geographical distribution of the alpine life-zone is worldwide (Fig. 8.4).

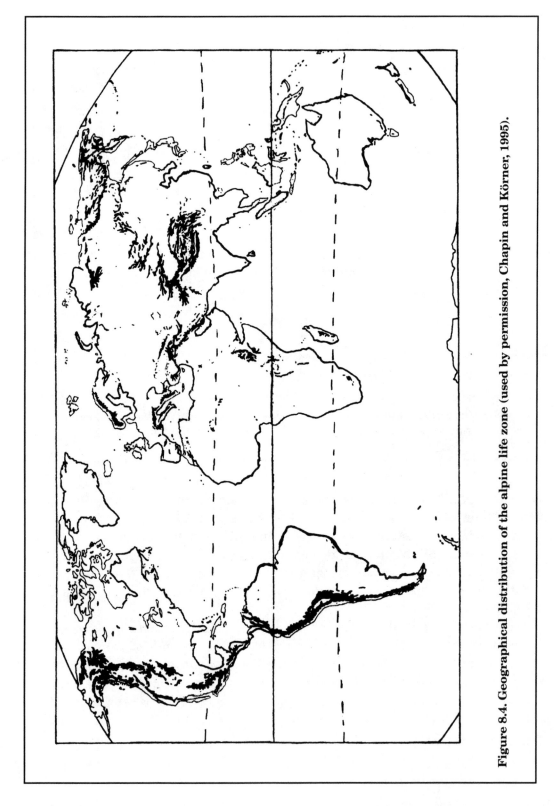

Figure 8.4. Geographical distribution of the alpine life zone (used by permission, Chapin and Körner, 1995).

Vegetation

The vegetation of alpine ecosystems is well adapted to the extremely rigorous physical environment found at high elevations (Körner, 1994; Chapin and Körner, 1995). Environmental attributes, such as exposure or soil properties, control the growth and distribution of alpine plant communities (Bliss, 1956, Billings, 1979).

Plant growth forms characteristic of alpine ecosystems are graminoids, herbaceous dicotyledenous plants, cushion plants, small basal rosette plants, dwarf shrubs (deciduous and evergreen), lichens, mosses, and annual vascular plants (Billings, 1979; Chabot and Mooney, 1985; Knight, 1994). Typically, these species exhibit very conservative growth strategies, even under periods of unusually favorable conditions (Baddley et al., 1994; Wookey et al., 1995; Callaghan and Jonasson, 1995).

Tundra vegetation fixes C at rates typically lower than domesticated species, averaging 15 $\mu mol/m^2/s$ (Bliss, 1985; Chapin and Shaver, 1985; Oberbauer et al., 1996), compared to rates approaching 40 $\mu mol/m^2/s$ (Field and Moony, 1986). Low rates of leaf C fixation are due in part to the inherently slow growth rate of long-lived perennial species common in alpine ecosystems.

These generally low rates of leaf C fixation of course do not apply to all tundra species and settings. For instance, some alpine grasses fix C at relatively high rates, especially when N limitations are not extreme. In the Niwot Ridge alpine ecosystem, *Deschampsia caespitosa* leaves fix C at rates of up to 30 $\mu mol/m^2/s$ under clear sky conditions in mid-summer (pers. comm, Bill Bowman, Univ. of CO). But net C sequestration at the whole system level depends on more than just leaf photosynthesis. It is the difference between whole ecosystem CO_2 fixation (multiple species) and ecosystem respiration (plant, soil, microbial) (Vourlitis et al., 1993; Shaver et al., 1998).

Measurement of whole ecosystem net CO_2 exchange (Vourlitis et al., 1993) in cold region settings has been attempted only recently. To date, many findings from Arctic tundra exist (Oechel et al., 1995; Jones et al., 1998; Shaver et al., 1998), and studies also have been conducted for alpine tundra (Table 8.2).

Winter studies from the Glacier Lakes Ecosystem Experiments station (GLEES) sites, located in the Snowy Range, WY, indicate annual winter efflux (indicating wintertime loss of C) among three alpine sites to average 95 g $C/m^2/yr$ (Sommerfeld et al., 1996). When characterized separately, these sites vary greatly by community type, with the wet alpine meadow being the greatest source of C to the atmosphere (3.57 kg CO_2-C/ha/d). Flux data sets from many sites (e.g., Niwot Ridge, CO, and GLEES) unfortunately are specific to the season (characterizing summer or winter only) and community and therefore do not characterize annual C budgets. However, seasonal data from multiple sites can be combined to form useful measures of annual net ecosystem C flux in a variety of systems (Table 8.2).

Table 8.2. Potential ranges for C flux (kg CO$_2$-C/ha/d) in a host of alpine tundra environments (negative numbers indicate net source, positive numbers indicate net sink of CO$_2$-C). (Data from Brooks et al., 1996; Sommerfeld et al., 1996; Fahnestock et al., 1998; Jones et al., 1998; Povirk, 1999; Welker et al., 1999).

Site		Winter		Summer
Description	Early		Late	
Libby Flats, WY				
alpine <u>Deschampsia</u> meadow				
-grazed	−1.18		−4.17	−5.9
-ungrazed	−0.68		−3.03	+13.9
Niwot Ridge, CO				
dry tundra	−0.55 to 0.90		−8.24	+0.72
Toolik Lake, AK				
dry heath tundra	−.068		−1.71	−3.6 (1996), −0.52 (1997)
tussock tundra	−0.2		−2.6	−4.8 (1996)
GLEES, Snowy Range, WY				
dry tundra	−2.35			
alpine meadow	−1.88			
wet alpine meadow	−3.57		−3.57	

Numbers represent average flux per day over each period or date. Sampling dates and duration vary by site.

Sampling dates: Libby Flats early winter 1 Nov. 1997 - 17 Feb. 1998, late winter 18 Feb. - 20 May, 1998, summer 12 June - 12 Sept. 1998; Niwot Ridge early winter 4 Mar., 1993, late winter 18 May, 1993, summer 5 June - 11 Sept., 1997; Toolik Lake early winter Mar. 3 - 6 1996, late winter May 21-31, 1996, summer 2 June - 3 Sept.,1996 and 23 May - 29 Aug. 1997; GLEES late winter April - June 1993 [dry tundra], early winter Dec. 1993 - Apr.1994 [alpine meadow], Nov. 1992 - June 1993, dates not specified [wet alpine meadow].

Interannual variability is evident from summer measurements at Toolik Lake, AK. During the growing season of 1996, net flux at the dry heath tundra site averaged 3.6 kg CO$_2$-C/ha/d while, in 1997, net flux averaged 0.52 kg CO$_2$-C/ha/d. Results from Libby Flats, WY (3230 m), suggest that ungrazed areas exhibit greater C sink strength than adjacent grazed areas, although these patterns can vary greatly on an interannual basis. Experiments involving simulated warming indicate dry alpine tundra is a net sink for CO$_2$ of about 7 g C/m^2 per summer; however, under a climate warming scenario, dry alpine tundra changes to a C source of almost 10 g C/m^2 per summer (Fig. 8.5).

Quantifying Net CO$_2$ flux can be useful in determining the C balance of grazed and ungrazed areas. However, other parameters, such as soil C characteristics, would be complementary in determining C sequestration potential.

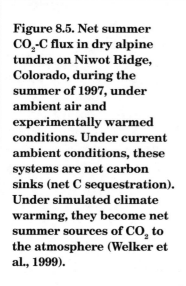

Figure 8.5. Net summer CO_2-C flux in dry alpine tundra on Niwot Ridge, Colorado, during the summer of 1997, under ambient air and experimentally warmed conditions. Under current ambient conditions, these systems are net carbon sinks (net C sequestration). Under simulated climate warming, they become net summer sources of CO_2 to the atmosphere (Welker et al., 1999).

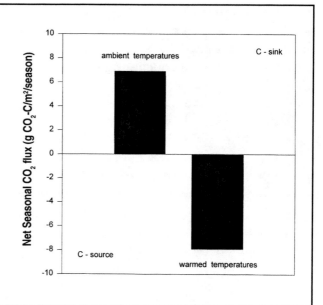

Temperature fluctuations

During the summer, extreme diurnal temperature variations characterize alpine ecosystems, often ranging from 15 to 30°C (Billings, 1979; Chabot and Mooney, 1985; Knight, 1994) during the daytime, to close to 0°C in late summer evenings. Snow is a possibility at virtually any time of the year, but less so in July.

Length of growing season varies yearly, with an average of 70 days/yr (Billings, 1979), as determined by the vigor (depth and length of presence) of the preceding spring's snowpack. Although radiative energy is high, little heat is trapped, due to windy conditions, clear weather, and a lessening of the insulating effect of the atmosphere at higher elevations (Billings, 1979). Wind has a great effect on daytime summer temperatures. On a windy, clear day, temperatures may never rise above 5°C; however, clear and calm weather may bring 25°C days.

Snow

Winter in alpine ecosystems is the longest season (Bliss, 1956). While the alpine system tends to be windblown and free of snow early in winter, snow usually blankets the region in the depth of winter through spring (Billings, 1979; Bliss, 1985). As a result of spring thaw, all of the chemical constituents of a snowpack accumulated over the winter are released during snowmelt.

Snow constituents span all periods and groups of the periodic chart and can have a significant effect on soil microbial processes and spring vegetative growth (Brooks et al., 1997). Snow chemistry is highly variable, depending on the particular material or chemical sources dominating the atmospheric processes. The prolonged presence of snow in these ecosystems is significant, given the thermal conductivity of snow.

Snow and ice act as insulating layers between the soil's surface and the atmosphere, blocking heat conduction from above and below (Sommerfeld et al., 1993, 1996; Fahnestock et al., 1998, 1999). This insulation can have a tremendous effect on microbial activity in soil that is near freezing. Microbial activity occurs down to temperatures of approximately –5°C (Taylor and Parkinson, 1988; Sommerfeld et al., 1993, 1996; Clein and Schimel, 1995; Brooks et al., 1997). Any heat retained in the soil due to the insulating effect of the snowpack could greatly increase microbial respiration rates during the cold winter months (Sommerfeld et al., 1993; Brooks et al., 1997; Fahnestock et al., 1999), possibly creating a source effect. The snowpack also acts as a passive cap controlling CO_2 concentration at the snow-soil interface (Sommerfeld et al., 1996).

Soils

The soils of alpine landscapes are highly variable but, due to climatic factors, generally are not well developed, except in some landscape depressional areas. Physical traits, such as parent material, climate, landform, and geomorphic processes, interact with microflora and vegetation to yield soils spanning multiple orders (Retzer, 1974; Johnson, 1979; Knight, 1994). The harsh climate of alpine environments often limits the chemical weathering of parent material. These soils, often Entisols, Inceptisols, Mollisols (Retzer, 1974), or Spodosols of the Olympic Mountains of Washington (Kuramoto and Bliss, 1970), are generally young, heterogeneous, and poorly formed and have low available nutrients (especially N).

Environmental factors favorable to the retention of high amounts of inorganic soil C are generally not found in alpine ecosystems. Soil C pools in alpine soils are mainly SOC comprised of recalcitrant, long-lived, C-based materials or C of short-term, high-turnover pools. Biologically inert and recalcitrant materials are the dominant components in the SOC pool (Lynch, 1991). Since these humic materials have half-lives often greater than 100 years (Lynch, 1991), the dominant slow C pool cannot be affected radically (increased) in short periods of time. Management techniques can increase or decrease effectively only the short-term or high-turnover SOC.

Alpine ecosystems are subject, by nature, to topographic heterogeneity attributable to erosional processes and glaciation on young landscapes at high altitudes (Brown and Johnston, 1979). Additionally, resistant geologic substrates that weather slowly, such as gneiss, schist, or quartzite (Retzer, 1974), are often parent material sources of alpine soils. The soils formed are usually coarsely textured, low in SOM, low in cation exchange capacity (CEC), and low in nutrients (Brown and Johnston, 1979).

Organic matter (OM) accumulation is slow and depends on the decomposition and recycling of organic compounds. It occurs in the surface horizons which, in concealed, low-lying depressional areas, can create thick, darkly colored A layers (Retzer, 1974) that can be concentrated highly in SOM. The accumulation that does occur mainly arises from root mortality and decomposition.

Cold temperatures that inhibit not only chemical reactions, but also biological enzymatic reactions, limit alpine soil genesis. However, significant microbial activity at –5°C has been observed in tundra, taiga, and aspen woodland ecosystems (Coxson and Parkinson, 1987; Clein and Schimel, 1995), implying that winter microbial respiration and litter decomposition is not as slowed as previously thought.

Since microbial activity may occur most of the winter, if not all season long, C may enter the SOM pool through the decomposition process almost year-round. However, at the same time, loss of C in the form of CO_2 from microbial respiration is also occurring throughout the cooler months (approximately 75% of the season), possibly resulting in a modern net loss of C in these systems, due to increasing global temperatures.

Grazing

The primary grazing pressure in North American alpine ecosystems is through sheep allotments (Bradford, 1998). Although cattle allotments do exist in the alpine, sheep traditionally have ruled the higher altitudes, due to logistical problems, the intolerance of cattle to the thin air of the alpine, and their susceptibility to brisket disease. This disease's often fatal, subacute to chronic syndrome of congenitive heart failure usually affects cattle older than 4 years and cows after calving (Tokarnia et al., 1989).

Because North American alpine plant communities generally have not evolved under large ungulate herbivory, in contrast to some prairie ecosystems, Bliss (1962) suggests that certain alpine plants, such as the alpine sedge, can be destroyed by intensive grazing practices. Introduction of herbivores into these areas can result in losses of sensitive plant species, changes in species composition in response to reduced competition, and physical ecosystem damage (Tieszen and Archer, 1979). Primary producers such as the graminoids, however, have good regrowth characteristics, but they also show reduced CO_2 uptake, depletion of belowground reserves, and a significant reduction in root growth (Tieszen and Archer, 1979).

Alpine grazing also affects the proportion of forb species, with increases in yarrow and cinquefoil and decreases in marsh marigold (Bonham, 1972). Grazing

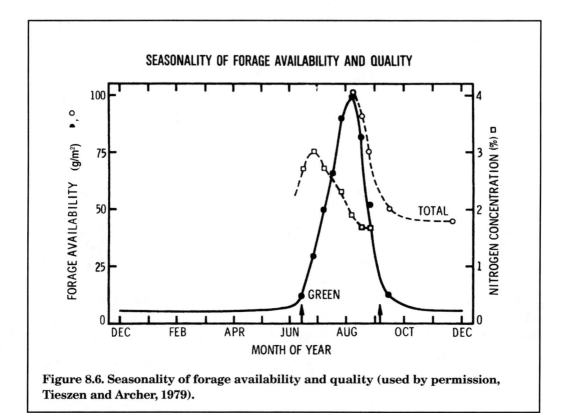

Figure 8.6. Seasonality of forage availability and quality (used by permission, Tieszen and Archer, 1979).

impacts within alpine meadows also may influence the invasion of trees; however, forest fires at the alpine margin have been proposed for enlargement of alpine ecosystems, due to the slow regrowth of trees (Billings, 1979).

Forage quality in the alpine is optimal in late June through July, while availability is best in late July through August (Fig. 8.6). In most years, greening occurs in middle to late June, depending on both the vigor of the preceding year's snowpack and the daily mean temperature. Although forage quality early in the season is high, low availability and concerns over erosion potential on soils that are still saturated by snowmelt waters generally keep turn-on dates to early to mid-July.

Mountain Meadows

Physiography

Mountain meadows, like most montane ecosystems, vary in composition and productivity according to elevation, moisture availability, soil depth, and nutrient availability (Hunt et al., 1988; Knight, 1994). Generally, within mountain meadow ecosystems, lower elevation sites are similar to foothill grasslands. As elevation increases, composition changes, and a greater abundance of forbs and sedges exists in the community (Knight, 1994). Along this elevational gradient, temperatures decrease at a rate of about 6°C per additional 1,000 m (Buol et al., 1989). Although annual snowfall can be highly irregular and inconsistent, it is usually substantial even in years of low precipitation (Klikoff, 1965).

A wide variety of environmental conditions exist and, often, meadows can be characterized as dry, moist, or wet, depending on topographic position. In the lower meadows near Elk City, ID (1220 to 1310 m), average temperature is 4.8°C and average annual precipitation is 79.3 cm (occurring mostly in spring and winter), with a summer high temperature of 16.2°C in July (Leege et al., 1981). In the mid-elevation of Cinnabar Park, WY (2926 m), precipitation is fairly well distributed throughout the year, although the warm season tends to be wetter than the colder months (Vale, 1978). Nearby Fox Park, WY (2748 m), receives an annual average of 435 mm of precipitation, with 392 cm falling as snow (Vale, 1978), while mean temperatures in July are 2.2°C. Existing conditions can be as extreme as those with standing water throughout the summer, to meadows that are quite dry within 2 weeks after snowmelt (Starr, 1974).

Vegetation

C sequestration rates and potentials in mountain meadows are highly variable since community composition can differ greatly among sites. A complexity of environmental factors and gradients control plant communities' patterns, which are driven by snow depth, time of snowpack release, summer moisture availability (Douglas, 1972), and soil type. Mountain meadows also are influenced by the surrounding forest, which can be a combination of Engelman spruce and subalpine fir, Douglas fir, lodgepole pine, and limber pine (Starr, 1974; Knight, 1994) or of the grand fir, western red cedar, or western hemlock habitat types (Leege et al., 1981).

Meadow production varies, but investigations in the Medicine Bow National Forest of southeast Wyoming place production at approximately 53 g/m² (Hunt et al., 1988). In the GLEES sites of southeast Wyoming, three subalpine meadows were measured for their winter C flux potential. Although there was much intersite variation, Sommerfeld et al. (1996) found that these high mountain meadows had an average, equivalent annual wintertime efflux (source of CO_2 to the atmosphere) of 232 g C/m²/yr. These measurements are for winter (potentially 70% of the year [Raich and Schlesinger, 1992]) only and do not represent an annual C budget for the GLEES subalpine meadow sites.

Snow

While long, cold, windy winters characterize the alpine ecosystem, mountain meadows are somewhat protected from the harsh winds and intense ultraviolet light of higher, more exposed elevations. Mountain meadow systems are subject to potentially large amounts of snow and uniform coverage of snow throughout the season. Wet and moist meadows often are covered by water each spring from snowpack runoff, possibly eliminating the invasion of woody plants (Leege et al., 1981).

The combination of long winters, extensive snowpack, and runoff can result in a short growing season during the summer (Leege et al., 1981). Edaphic properties may be the primary control on mountain meadow existence and perseverance.

Soils

Occurring at both low and high altitudes, soils in mountain meadow ecosystems are highly variable, but, along these elevational gradients, some generalizations about soil characteristics are possible. As elevation increases, OM, N, and C:N ratios increase, while pH value, Ca, Mg, and K decrease (Buol et al., 1989).

Snow duration and soil moisture correlate with soil development (Douglas, 1972) in these areas. However, little horizon development is often characteristic of some wet mountain meadow soils, due to soil saturation. Often, poorly developed soils are associated with *Carex* spp. communities where snow duration is long and soils remain moist or saturated well into the season.

Meadows associated with Tsuga-Abies (mature phase) communities often produce soils with more profile development, as snow release is early and soils are more mesic (Douglas, 1972). These soils potentially can contain deep O horizons, consisting mainly of undecomposed litter.

Grazing

Wildlife and livestock grazing of mountain meadows has occurred for years. These habitats favor the production of significant amounts of highly palatable forage, i.e., grasses and forbs. However, livestock grazing can damage these ecosystems, particularly if plants are in their early growth stages and grazing is longterm.

Excessive foraging is detrimental, especially if soils are wet and sensitive to physical damage (Knight, 1994). Studies have shown that heavy continuous grazing can dramatically reduce *Stipa columbiana, Achillea millifolium, Penstemon* sp., *Agoseris* sp., and *Potentilla gracilis,* in addition to others, on dry and moist meadows (Table 8.3). The same study near Elk City, ID, indicated grazing re-

Table 8.3. Frequency values (%) occurring inside and outside exclosures on dry and moist mountain meadow with heavy and light grazing regimes.

	(Dry/Heavy)		(Dry/Light)		(Moist/Heavy)	
Graminoids	Out	In	Out	In	Out	In
Agropyron repens	10	15	80	50		
Agropyron sp.			20	0		
Agrostis alba					100	20
Bromus marginatus			65	35		
Bromus polyanthus			30	40		
Carex microptera					0	45
Carex sp.	25	25			55	85
Dactylis glomerata			5	40		
Danthonia californica					0	20
Danthonia intermedia	70	0				
Deschampsia caespitosa					10	35
Festuca sp.	5	0				
Juncus sp.	0	5	10	10	90	0
Phelum pratense	70	15	30	20	90	45
Poa compressa	25	20				
Poa pratensis	100	100	100	80	50	95
Scirpus columbiana	50	85	0	25		
Forbs						
Achillea millifolium	100	100	95	95	35	80
Agoseris sp.	0	40				
Antennaria rosea	75	75				
Arenaria macrophylla	15	0				
Arenaria seriphylifolia	80	30	20	0		
Aster conspicuous					0	85
Campanula rotundifolia	70	30				
Circium sp.			25	0		
Collomia linearis	60	0	60	0	0	25
Collinsia parvifloria	5	5				
Erigeron sp.	0	5				
Fragaria virginiana	5	70	0	85		
Fraseria albicaulis					0	25
Geranium sp.	5	0				
Hypericum perforatum	0	20				
Lathyrus sp.	0	75	0	5		
Linnea borealis					5	5
Microsteris sp.	50	0				
Penstamon sp.	35	80	5	30		
Perideridia gairdneri	0	75	20	0	0	30
Plantago lanceolata					15	0
Polygonum douglasii	20	0				
Potentilla glandulosa			0	10		
Potentilla gracillis	80	100	70	40	80	70
Prunella vulgaris	0	10	10	5	35	0
Ranunculus uncinatus	30	55	55	10	15	15
Rumex acetosella	30	10	25	95		
Sedum douglasii	55	65				
Taraxacum sp.	70	65	90	50	80	35
Trifolium sp.	20	10	80	30	100	25

* Values indicate percent of twenty 20 x 50 cm plots in which species occurred (modified from Leege et al., 1981)

sulted in an increase in *Phleum pratense, Arenaria seriphylifolia, Campanula rotundifolia, Collumia linearis, Trifolium* sp., and *Danthonia intermedia* (Leege et al., 1981). For these reasons, livestock introduction and length of grazing time are closely monitored, the number of animals allowed on mountain meadows is reduced, and overgrazing is prevented, especially on federal lands.

Mountain meadows are subject to invasion by the surrounding forest. Evidence suggests that the factors that maintain meadows are those that primarily limit tree seedlings' establishment, rather than those that limit tree growth (Dunwiddie, 1977). Abundant tree establishment in meadows has been attributed to moderate cattle grazing, since grazing can reduce competition between tree seedlings and meadow grasses (Dunwiddie, 1977). Intensive grazing may sufficiently open up dense meadow vegetation to permit seedlings' recruitment and establishment.

A potential effect of this forest invasion is the loss of C, in the form of stored humus, from meadow soils. In this case, OM from dense fibrous roots of grasses is replaced by less productive root systems of the surrounding forest. Stored humus then becomes the focus of the vigorous decomposer communities that evolved in the fibrous-root, debris-rich environments (Buol et al., 1989).

Cattle grazed the mountain meadows as early as 1890, grazing most heavily from probably around the turn of the century to the 1930s. To some extent, cattle grazing occurred on most Forest Service lands for at least 10 to 20 years before records were kept (Dunwiddie, 1977). Although the number of cattle on the forests has decreased, it may take total cessation of cattle grazing in some areas to stop the invasion of mountain meadows by *Picea* spp. and *Pinus* spp. seedlings.

Fire suppression also may have contributed to a decline in areas comprising mountain meadows. Tree encroachment and the development of more dense forests are a consequence of restrictions on the burning that occurred during the late 1800s and into the early 1900s (Knight, 1994). Many meadow rangeland areas that were opened for timber harvesting, mining, and recreational activities have begun to revert to their former composition due to restrictions on these types of activities. The effects of climatic change on these ecosystems also may affect the size and extent of mountain meadow environments.

Arctic Tundra

Physiography

Arctic tundra occurs primarily above 66°N latitude in the circumarctic north. This treeless landscape is comprised of evergreen and deciduous dwarf shrubs such as *Dryas* spp., *Betula* sp., graminoids such as *Carex* spp., and herbaceous perennial forbs such as *Polygonum* sp. (Komarkova and Webber, 1980; Chapin and Shaver, 1985a; Walker and Walker, 1991; Walker et al., 1999).

The Arctic is partitioned into two regions — the Low Arctic, encompassing almost all of Alaska, and the High Arctic, typically that area north of about 72° (Chapin and Shaver, 1985a; Bliss and Matveyeva, 1992). This includes most of the Canadian Northwest Territories, Greenland, Svalbard, and portions of northern Eurasia. The Low Arctic is a region where soil C pools are larger on a per area basis (Oechel and Billings, 1992), while the plant biomass and SOM pools are smaller on a per area basis in the High Arctic. However, since the High Arctic constitutes almost 33% of the Arctic, its C pools are globally significant (Oechel and Billings, 1992).

The Arctic is a diverse region with a host of climates, either continental or oceanic, that strongly influence landscape temperature and precipitation patterns (Bliss and Matveyeva, 1992). The Arctic Ocean and the Bering Sea strongly influence the long coast line of Arctic Alaska. During the summer, coastal fog typically limits irradiance, while providing very cold, wet conditions during the growing season (Tieszen, 1978). In contrast, inland Arctic Alaska has a very continental climate, with exceptionally cold winters reaching –40°F in January and February, although, in spring and summer, clear skies can prevail for weeks, enhancing photosynthesis and ecosystem C gain (Oberbauer et al., 1996; Jones et al., 1998). In addition to a diversity of climates, the Arctic region's permafrost occurs to varying depths, controlling patterns of plant root growth and the hydrology of these habitats (Hinzman et al., 1996).

The landscape across the Arctic has been affected significantly by glacial events over the last several 100,000 years, resulting in a mosaic of systems of different ages (Walker and Walker, 1991). Recent findings indicate that this variability results in systems with differing soil acidities. Most systems are slightly acidic in nature (6 to 6.5 pH), while others may be neutral or slightly basic (7 to 7.7 pH). This variability can have major impacts on vegetation, soil biological properties, and trace gas emission properties (Walker et al., 1999).

The Arctic region experiences continual daylight and continual darkness for extended periods each year. These large swings in irradiance have implications for the biogeochemical cycling of C. During the summer, plants assimilate C as soon as temperature and water demands are met, with maximum rates of C sequestration in July (Jones et al., 1998). Long periods of snow cover exist from September to May. Although soils are very cold (–1 to –5°C), C loss from soils occurs from biologically active and respiring microorganisms throughout much of the winter (Clein and Schimel, 1995; Sommerfeld et al., 1996; Brooks et al., 1997; Fahnestock et al., 1998; Welker et al., 2000). During this period, CO_2 typically evolves and mineralization releases soil nutrients (Hobbie, 1996), though it is unclear whether soil microbes or plants assimilate the winter-released N.

The Arctic, like the alpine, has a pulse period of very rapid change in land-scape hydrology (Hinzman et al., 1996). The hydrologic cycle in the Arctic typically begins with snowmelt each spring, which often starts in early to mid-May and lasts until early June (Hinzman et al., 1996). Snow depth and time of thaw vary across the landscape, but thaw often begins with the melting of thin patches on ridge tops and ends in the accumulation areas of valley bottoms.

Typically, within a 3-week period, Arctic landscapes change from being snow covered to almost snow free, initiating the short (<100 days) but intense growing season. With very long and even continuous daylight, CO_2 typically is fixed over almost the entire 24-hour cycle. Rates of CO_2 exchange are determined by:

1. the leaf area index (LAI) of specific communities
2. community composition, as deciduous shrubs and graminoids have higher rates of growth (Chapin and Shaver, 1985b, 1996) and leaf photosynthesis than evergreen plants (Oberbauer et al., 1991, 1996), and
3. the stage of phenologic development and radiation inputs.

Vegetation

Leaf photosynthesis in Arctic plants shows very strong light saturation, between 500 and 1000 mol/m²/s. This may indicate that, even at the low light levels typical of early morning and midnight in the Arctic, plants may be assimilating C at their highest rates (Oberbauer et al., 1996). However, these responses depend somewhat on life-form. Light saturation of leaves on evergreen plants occurs at lower levels, with absolute rates of C sequestration lower by up to 200% (Oberbauer et al., 1996).

In addition to responding to light, leaf photosynthesis responds strongly to temperature in a curvilinear fashion, being low at temperatures less than 10°C, reaching a maximum between 15 and 30°C, and decreasing beyond about 35°C (Oberbauer et al., 1996). Despite this broad range of operation, temperature frequently may limit photosynthesis, since the mean daily temperature is often in the low range of photosynthetic operation.

Other environmental factors that may affect the C sequestration of Arctic plants are water and nutrition (Welker et al., 1993, 1995). However, experimental analysis of whether water limits plant photosynthesis and C gain have shown mixed results. In Alaska, Oberbauer et al. (1996) found that, along a toposequence, water additions had no discernible effects on C gain. In the High Arctic, Welker et al. (1993) found that water additions significantly increased leaf C isotope discrimination, indicating that gas exchange rates were higher in plants that were watered. Water additions in the High Arctic also have been reported to increase leaf biomass.

These responses could be attributed to the fundamental differences in the soil and climate regime between Alaska and Svalbard (High Arctic). This, in conjunction with warmer temperatures and added nutrients, can result in an extended period of ecosystem C gain (Welker et al., 1997).

Many of the processes associated with C sequestration are limited by nutrients, especially by N (Chapin and Shaver, 1996; Shaver et al., 1992, 1998). Often, when soil N availability increases, leaf N concentration and C fixation rates increase accordingly (Field and Mooney, 1986). These nutrient responses and those reported by others (Chapin and Shaver, 1985; Wookey et al., 1993; Wookey et al., 1995; Chapin and Shaver, 1996; Shaver et al., 1998) are likely to be tied to the close association between leaf N content and C fixation rates (Field and Mooney, 1986).

While leaf photosynthesis and its response to abiotic variables may play an important part in the C sequestration of Arctic systems, the balance between gross ecosystem photosynthesis and respiration (soil microbial and plant) actually determines whether an ecosystem is accumulating or losing C (Billings et al., 1982; Oechel et al., 1993; Jones et al., 1998; Welker et al., 2000). Recently, research groups have started to examine whole system C-flux using small chambers placed over entire ecosystems (Vourlitis et al., 1993; Jones et al., 1998). Because of the short stature of these ecosystems, entire systems can be enclosed, and their net C-exchange measured, and then the net flux partitioned into its respective components (Vourlitis et al., 1993).

Several recent findings regarding C sequestration and emission in Arctic ecosystems are important. Oechel et al. (1993) indicate that, over the last 20 years or so, coastal wet sedge systems in Alaska have switched from being C sinks where SOM is accumulating, to C sources where more C is being respired than is being fixed by photosynthesis. This suggests that the Arctic may now be an important C source to the atmosphere, further accelerating the rising CO_2 levels in the atmosphere (Zimov et al., 1993, 1996).

Additionally, experimental warming studies show that warmer summer temperatures may accelerate the amount of CO_2 released from Arctic tundra to the atmosphere (Jones et al., 1998). If warmer summer temperatures are coupled with deeper snow cover in winter, the amount of CO_2 lost to the atmosphere may double from that under ambient conditions at present (Welker et al., 1999).

Temperature fluctuations

The Arctic tundra is a very diverse region with a host of climates that greatly influence temperature and precipitation regionally (Bliss and Matveyeva, 1992). In addition, permafrost to varying depths acts to control patterns of root growth and hydrology in many habitats (Chapin and Shaver, 1995a; Hinzman et al., 1996).

Some parts of the Arctic, however, are experiencing dramatic shifts in weather conditions that have displaced wildlife, initiated the thaw of the permafrost, and increased pollution risks. For example, Arctic indigenous people of the Inuit communities rely on hunting, fishing, and trapping to support their economy and provide food. An increase of 5°C above normal temperature in 1998 resulted in the breakup of important ice field hunting grounds, reduced polar bear weights because of limited food (e.g., seal pups) access, and increased erosion and loss of buildings from thawing of permafrost soils. There are also concerns about contamination from mine wastes, destabilization of roads, bridges, and oil pipelines, and the encroachment of forest on Arctic rangelands that foraging caribou herds use.

Snow

Snow affects Arctic systems in several ways. First, variation in snow depth can affect the temperature of soils over the winter (Fahnestock et al., 1998; Jones et al., 1999). Deep snow cover insulates soils, keeping them relatively warm compared to shallow or snow-free areas. Second, variation in snow distribution and chemistry affects the duration of snowmelt during spring and the amount of water that seeps into the soil and adjoining streams and rivers (Hinzeman et al., 1996).

Considerable attention has been paid to how deep snow causes warmer soil temperatures. The extent to which these warmer soil temperatures (−5 to 0°C) cause significant amounts of heterotrophic respiration during winter is of great concern (Oechel et al., 1997; Fahnestock et al., 1998; 1999; Mast et al., 1998; Welker et al., 2000). This is especially the case when snowfall occurs early in the winter, locking in the heat accumulated over summer and possibly stimulating microbial respiration.

Initially, it was thought that most of the C exchange between the Arctic landscape and the atmosphere took place in summer (Oechel et al., 1993). However, under deep snow, plant roots and/or soil microbes are metabolically active (Schimel and Clein, 1995). Although CO_2 may be released at low rates under the snowpack, it appears that cumulative, season-long, winter CO_2 fluxes may be significant since high levels of CO_2 accumulate at the snow-soil interface and diffuse to the atmosphere (Sommerfeld et al., 1996).

However, strong year-to-year variation in snowfall patterns may exist. Years where early winter temperatures are extremely cold could result in the freezing of soil before substantial snowfall. In this case, even when snow cover eventually becomes deep (mid to later winter), CO_2 efflux may be low for the season (Welker et al., 2000).

Soils

Soils associated with permafrost are common at high latitudes (Marion et al., 1997; Bockheim et al., 1998; Ping et al., 1998). If not completely frozen, Arctic soils are generally wet and cold, limited in N and P, and noted for their high C contents (Billings, 1982; Michaelson et al., 1996; Marion et al., 1997). With the introduction of the Gelisol soil order (*Soil Taxonomy,* USDA-NRCS, 1998), Arctic soils now are classified predominantly as Gelisols, with inclusions of Histisols, Molisols, Inceptisiols, and Entisols (Bockheim et al., 1998; USDA-NRCS, 1998).

In permafrost layers, Arctic soils contain significant amounts of C that originate from cryoturbation. Michaelson et al. (1996) suggested that permafrost C resulted in an underestimation of Arctic soil C, and that, if this C is considered, Arctic soils may comprise as much as 30% of the global soil C pool.

A study of nonacidic and acidic soils in the Kuparuk basin of Arctic Alaska found that C distribution with depth was a function of organic mats and cryoturbation (Bockheim et al., 1998). Nonacidic tundra had organically rich mats, surface horizons with greater C contents, and deeper active layers than acidic tundra. In the upper 1 m, the C and N in active layers of nonacidic and acidic tundra was 60% and 40%, respectively, indicating greater amounts were present in moist nonacidic soils. Figure 8.7 shows a conceptual model of cryoturbation control on soils in Arctic Alaska.

The impact global warming will have on Arctic soils is uncertain. Marion et al. (1997) offer these reasons that Arctic soils in northern ecosystems may play a large role in the future:

1. Arctic soils may be a source or sink to the atmosphere
2. Arctic soils may control energy balance at the earth's surface, and
3. Arctic soils may serve as a source of plant nutrients.

The impact of global warming on Arctic regions may result in higher temperatures and lower precipitation, which in turn could increase the depth of permafrost thaw, enhance SOM decomposition, and possibly change the region from a C sink to a C source of CO_2 to the atmosphere (Billings et al., 1982; Oechel et al., 1993; Ping et al., 1998). Although climate change may affect high latitude environments such as Arctic tundra more, too little is known of the controls and responses of the Arctic regions to forecast or predict global warming's consequences there or the role these regions will play worldwide.

Grazing

In the Arctic, the primary grazers are caribou that migrate from the Boreal Forest to the Arctic in summer (Russell et al., 1993). Because the caribou exceed

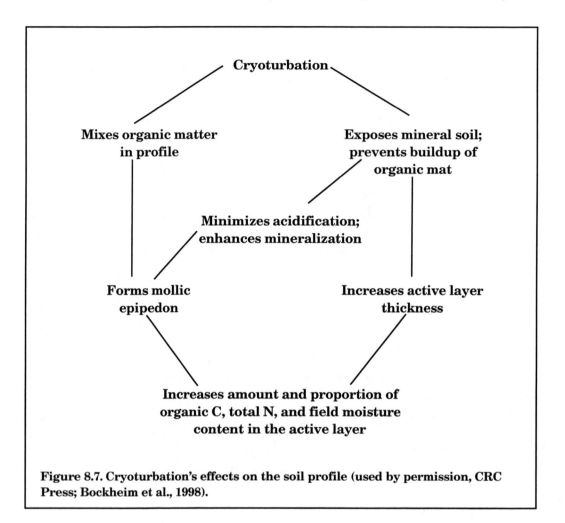

Figure 8.7. Cryoturbation's effects on the soil profile (used by permission, CRC Press; Bockheim et al., 1998).

several hundred thousand, they may alter the vegetation significantly and, subsequently, the C sequestration of these systems.

Herbivores may alter C sequestration through their grazing, which maintains juvenile vegetative tissue that assimilates C at greater rates than mature tissue (McNaughton, 1983). Grazers' feces and urine also may stimulate nutrient cycling in the soil, possibly causing increases in plant nutrient content, growth, and C gain. Unfortunately, few studies have quantified these dynamics in the Arctic. However, in the wet sedge communities of subarctic Canada, where geese have been excluded, plants' growth rates are slower than in adjoining areas (Jefferies et al., 1994).

Federal Management

Most alpine tundra and mountain meadow ecosystems are under federal ownership. Livestock grazing on public lands has come under increasing public scrutiny and criticism from groups concerned with issues such as decreased biodiversity, deteriorated range condition, soil erosion, desertification, depleted watershed and riparian condition, decreased wildlife populations and habitat, and decreased recreational opportunities.

Traditionally, the U.S. Forest Service managed these systems mainly for grazing and logging purposes. More recently, however, the management of alpine and mountain meadow ecosystems has become a mixture of range science and social science as the public becomes more involved in high altitude recreation and subsequently more concerned with land management issues.

Currently, the standards for livestock management differ by forest (district), but mainly they stem from the concept of *range readiness*. This concept has been prominent for the last 20 to 30 years and is based on plant phenology and soil moisture conditions of a given allotment. Additionally, the practices of the residing range conservationist heavily influence the standards for turn-on (the date when privately owned animals are allowed access to forage on federal lands) and further contribute to differences in management techniques across the western U.S.

If areas in cold region environments are to be used for grazing purposes, criteria for interpreting the health of the ecosystem should be evaluated first to determine overall rangeland health (National Research Council, 1994). Three criteria used to determine rangeland health include:

1. degree of soil stability and watershed function
2. integrity of nutrient cycles and energy flow, and
3. presence of functioning recovery mechanisms.

Livestock grazing to some degree affects all these criteria and others, such as biological diversity, riparian area function, wildlife habitat, timber production, and recreation.

For instance, on the Paonia district of the G.M.U.G. (Grand Mesa, Uncompagre, and Gunnison) National Forest, Colorado range conservationist David Bradford (Bradford, 1998) prefers not to use the range readiness standards alone on intensively managed allotments, as the standards are inflexible across landscapes. Bradford prefers to use a set of guidelines which take into consideration vegetation volume, landscape variability, and plant growth and regrowth patterns, along with the traditional standards of phenology and soil moisture conditions.

Unfortunately, the problems that range conservationists encounter in mid- and high-altitude ecosystems are numerous. Elevation gains of 1000 feet in a short distance, or remnant snowbanks, are not uncommon in the alpine or mountain meadow systems. Often corresponding with these abrupt changes in vegetation are changes in soil type and C content (Table 8.4). In these cases, "bottlenecks" of immature vegetation may lag behind the rest of the allotment in development and should remain ungrazed while the adjacent lower elevation vegetation may be ready for grazing by livestock.

Here, using more flexible "guidelines," which may include combinations of holistic, intensive, and rest, deferred or season-long grazing management systems, allow land managers to tailor an appropriate grazing regime for problem areas. In addition to these guidelines, Bradford must incorporate allotment per-

Table 8.4. Major soil orders of cold region environments.

Order	Great Group	Soil Organic Carbon kg C/m³	Subgroup
Entisols	Cryaquents	9.8	Typic Cryaquents
	Cryorthents	6.6-7.2	Typic Cryorthents
			Aquic Cryorthents
			Lithic Cryorthents
Inceptisols	Cryaquepts	20.5	Typic Cryaquepts
			Aeric Cryaquepts
			Aeric Humic Cryaquepts
			Histic Cryaquepts
			Histic Pergelic Cryaquepts
			Humic Cryaquepts
			Humic Pergelic Cryaquepts
			Pergelic Cryaquepts
	Cryochrepts	8.9	Typic Cryochrepts
			Dystic Cryochrepts
			Lithic Cryochrepts
			Pergelic Cryochrepts
	Cryumbrepts	11.0-19.5	Typic Cryumbrepts
			Entic Cryumbrepts
			Lithic Cryumbrepts
			Pergelic Cryumbrepts
Mollisols	Cryaquolls	16.3	Typic Cryaquolls
			Histic Cryaquolls
			Pergelic Cryaquolls
	Cryoborolls	14.1	Typic Cryoborolls
			Aquic Cryoborolls
			Lithic Cryoborolls
			Pergelic Cryoborolls
Histosols	Cryofibrists	~100	Typic Cryofibrists

(Adapted from Retzer, 1974; Davidson and Lefebvre, 1993; Knight, 1994; Kern et al., 1998. Table does not reflect the change in Soil Taxonomy, USDA-NRCS, 1998, which includes a 12th soil order, Gelisols.)

mits and U.S. Forest Service goals, while simultaneously keeping the public informed and satisfied with range conditions (Bradford, 1998). According to Bradford, there is no "one size fits all" management scheme. Although the traditional measure of rangeland condition within a successional paradigm has been discarded, there has been no replacement in terms of structural or functional properties (Ojima et al., 1997).

Social Impacts

Ranchers in the western U.S. depend greatly on the federal natural resource base (Laycock et al., 1996). Federal land management policies can economically affect rural populations that rely on federal lands for summer livestock forage. Approximately 85% (106 Mha) of federally owned land in the West is grazed by domestic livestock all or part of the year. Of the total 124 Mha of federally owned land in the West, 57% is administered by the Bureau of Land Management (BLM) and 43% is administered by the U.S. Forest Service (Laycock et al., 1996). Of this, 90% of BLM and 81% of U.S. Forest Service grazing permits are complement to small to medium cow-calf and cow-calf-yearling operations, which in turn stabilize the western livestock industry and rural communities (Laycock et al., 1996).

Climate change impacts on species composition, productivity, palatability, and forage quality (Ojima et al., 1997) have the potential to greatly affect the already variable year-to-year profits in the western U.S., where beef cattle are the highest or second highest income producer in agriculture (Laycock et al., 1996). Economic significance becomes apparent when the market values of agricultural products are recognized. More than half of the commercial beef cattle operators in the West hold grazing permits.

Conclusion

Soil C sequestration in cold region systems (Arctic and alpine tundra and mountain meadows) has undergone large swings over long time periods. The cold climate and nutrient limitations in these habitats suggest that, in a warmer world, changes in C acquisition and release can be expected. The warmer, drier conditions observed recently in the Arctic may be influencing areas that were previously net C sinks, accumulating soil and plant C. This modern change in net Arctic C flux may happen in alpine tundra under warmer conditions, but grazing history also may influence net C flux and sequestration.

Many studies have shown that herbivores can alter the biogeochemical cycling of C and other nutrients on rangelands. From the existing literature base, we anticipate that past and current grazing practices will alter the balance between C sequestration and C loss from ecosystems. The net effect, however, is less clear. Activities that affect the input of plant residues will affect SOC accumulation; this is well documented in agronomic, forestry, and rangeland ecosystems. Environmental and human influences, such as grazing, that limit the plant biomass inputs, as well as aggregation of indigenous SOM components, may affect C sequestration negatively.

It is also quite possible that grazing may stimulate ecosystem respiration to a greater degree than C assimilation, with heavily grazed areas becoming greater C sources than ungrazed or lightly grazed habitats. However, instances may occur whereby grazed landscapes will have higher rates of C assimilation, leading to long-term C sequestration. For soil C sequestration, the processes that contribute to SOM losses (decomposition, erosion, leaching, etc.) need to be minimized while processes that result in SOM accumulation (humification) and stability (aggregation, sedimentation, etc.) need to be maximized.

Further research on rangeland C balance of rangelands and investigations of climate and CO_2 impacts on rangeland ecosystem responses, such as vegetation, soil, and animal dynamics, are needed. Summer and winter, process-based analyses that identify the mechanisms controlling C flux, and complementary investigations into the mineral nutrition and water relations of rangelands, also would be beneficial.

References

Baddley, J.A., S.J. Woodin, and I.J. Alexander. 1994. Effects of increased nitrogen and phosphorus on the photosynthesis and nutrient relations of three Arctic dwarf shrubs on Svalbard. *Funct. Ecol.* 8:676-685.

Billings, W.D. 1979. Alpine ecosystems of western North America. *In* D.A. Johnson (ed), *Special Management Needs of Alpine Ecosystems, Range Science Series, No. 5*, Society for Range Management. Denver, CO, pp. 6-21.

Billings, W.D., J.O. Luken, D.A. Mortensen, and K.M. Peterson. 1982. Arctic tundra: a source or sink for atmospheric carbon dioxide in a changing environment. *Oecologia* 53:7-11.

Bliss, L.C. 1956. A comparison of plant development in microenvironments of Arctic and alpine plants. *Ecol. Monogr.* 26:303-337.

Bliss, L.C. 1962. Adaptations of Arctic and alpine plants to environmental conditions. *Arctic.* 15:117-144.

Bliss, L.C. 1985. Alpine. *In* B.F. Chabot and H.A. Mooney (eds), *Physiological Ecology of North American Plant Communities*, Chapman and Hall, New York, pp. 41-65.

Bliss, L.C., and N. Matveyeva. 1992. Circumpolar Arctic vegetation. *In* F.S. Chapin, R.L. Jeffries, J.F. Reynolds, G.R. Shaver, and J. Svoboda (eds), *Arctic Ecosystems in a Changing Climate: An Ecophysiological Perspective*, Academic Press, San Diego.

Bockheim, J.G., D.A. Walker, and L.R. Everett. 1998. Soil carbon distribution in nonacidic and acidic tundra of Arctic Alaska. *In* R. Lal, J.M. Kimble, R.F. Follett and B.A. Stewart (eds), *Soil Processes and the Carbon Cycle*, CRC Press, Boca Raton, FL, pp. 143-156.

Bohn, H.L. 1990. Considerations for modeling carbon interactions between soil and atmosphere. *In* H.W. Scharpenseel, M. Schomaker, and A. Ayoub (eds), *Soils on a Warmer Earth*, Elsevier, New York, pp. 391-394.

Bonham, C.D. 1972. Vegetative analysis of grazed and ungrazed alpine hairgrass meadows. *J. of Range Manage.* 25(4):276-279.

Bouwman, A.F. (ed). 1990. *Soil and the Greenhouse Effect*. John Wiley and Sons, New York.

Bowman, W.D., T.A. Theodose, and M.C. Fisk. 1995. Physiological and production responses of plant growth forms to increases in limiting resources in alpine tundra: implications for differential community response to environmental change. *Oecologia* 101:217-227.

Bowman, W.D., T.A. Theodose, J.C. Schardt, and R.T. Conant. 1993. Constraints of nutrient availability on primary production in two alpine tundra communities. *Ecology* 74:2085-2097.

Bradford, D. 1998. Holistic resource management in the West Elks – why it works. *Rangelands* 2(1):6-9.

Brooks, P.D., S.K. Schmidt, and M.W. Williams. 1996. Winter production of CO_2 and N_2O from alpine tundra: environmental controls and relationship to inter-system C and N fluxes. *Oecologia* 110:403-413.

Brown, R.W., and R.S. Johnston. Revegetation of disturbed alpine rangelands. 1979. *In* D.A. Johnson (ed), *Special Management Needs of Alpine Ecosystems, Range Science Series No. 5*, Society for Range Management, Denver, CO, pp. 76-94.

Bryant, J.P., F.S. Chapin III, and D.R. Klein. 1983. Carbon/nutrient balance of boreal plants in relation to vertebrate herbivory. *Oikos* 40:357-368.

Buol, S.W., F.D. Hole, and R.J. McCracken. 1989. *Soil Genesis and Classification.* Iowa State University Press. Ames, IA.

Caldwell, M.M., J.H. Richards, D.A. Johnson, R.S. Nowak, and R.S. Dzurec. 1981. Coping with herbivory: photosynthetic capacity and resource allocation in two semiarid *Agropyron* bunchgrasses Utah. *Oecologia* 50(1):14-24.

Callaghan, T.V., and S. Jonasson. 1995. Implications for changes in Arctic plant biodiversity from environmental manipulation experiments. *In* F. S. Chapin and C. Körner (eds), *Arctic and Alpine Biodiversity: Patterns, Causes and Ecosystem Consequences, Ecological Studies 113*, Springer-Verlag, Berlin, pp. 151-164.

Campbell, C.A., and R.P. Zentner. 1993. Soil organic matter as influenced by crop rotations and fertilization. *Soil Sci. Soc. Am. J.* 57:1034-1040.

Chabot, B.F., and H.A. Mooney (eds). 1985. *Physiological Ecology of North American Plant Communities.* Chapman and Hall. New York.

Chapin, F.S., and C. Körner (eds). 1995. *Arctic and Alpine Biodiversity: Patterns, Causes and Ecosystem Consequences.* Springer-Verlag. Berlin.

Chapin, F.S., and G.R. Shaver. 1985a. Arctic. *In* B.F. Chabot and H.A. Mooney (eds), *Physiological Ecology of North American Plant Communities*, Chapman and Hall, New York, pp. 16-40.

Chapin, F.S., and G.R. Shaver. 1985b. Individualistic growth responses of tundra species to environmental manipulations in the field. *Ecology* 66:564-576.

Chapin, F.S., and G.S. Shaver. 1996. Physiological and growth responses of Arctic plants to a field experiment simulating climatic change. *Ecology* 77:822-840.

Chapin, F.S., G.R. Shaver, A.E. Giblin, K.J. Nadelhoffer, and J.A. Laundre. 1995. Responses of Arctic tundra to experimental and observed changes in climate. *Ecology* 76:694-711.

Chapin, F.S., S.A. Zimov, G.R. Shaver, and S.E. Hobbie. 1996. CO_2 fluctuation at high latitudes. *Nature* 383:585-586.

Clein, J.S., and J.P. Schimel. 1995. Microbial activity of tundra and taiga soils at subzero temperatures. *Soil Biol. Biochem.* 27(9):1231-1234.

Coxson, D.S., and D. Parkinson. 1987. Winter respiratory activity in aspen woodland forest floor litter and soils. *Soil Biol. Biochem.* 19(1):49-59.

Davidson, E.A., and P.A. Lefebvre. 1993. Estimating regional carbon stocks and spatially covarying edaphic factors using soil maps at three scales. *Biogeochem.* 22:107-131.

Dormaar, J.F., and W.D. Willms. 1990. Effect of grazing and cultivation on some chemical properties of soils in the mixed prairie. *J. of Range Manage.* 43:6-9.

Douglas, G.W. 1972. Subalpine plant communities of the western North Cascades, Washington. *Arctic and Alpine Res.* 4(2):147-166.

Dunwiddie, P.W. 1977. Recent tree invasion of subalpine meadows in the Wind River mountains, Wyoming. *Arctic and Alpine Res.* 9(4):393-399.

Fahnestock, J.T., and J.K. Detling. 2000. Morphological and physiological responses of perennial grasses to long-term grazing the Pryor Mountains, Montana. *Am. Mid. Nat.* (in press).

Fahnestock, J.T., M.H. Jones, P.D. Brooks, D.A. Walker, and J.M. Welker. 1998. Winter and spring CO_2 efflux from tundra communities of northern Alaska. *J. Geophys. Res.* 103(D22):29,023-29,027.

Fahnestock, J.T., M.H. Jones, and J.M. Welker. 1999. Wintertime CO_2 efflux from Arctic soils: implications for annual carbon budgets. *Global Biogeochem. Cycles* 13(3):775-779.

Farquhar, G.D., J.R. Ehlringer, and K.T. Hubick. 1989. Carbon isotope discrimination and photosynthesis. *Annu. Rev. Plant Phsiol-Plant. Mol. Biol.* Palo Alto, Calif. v.40:503-547.

Field, C., and H.A. Mooney. 1986. The photosynthesis-nitrogen relationship in wild plants. *In* T.J. Givnish (ed), *On the Economy of Plant Form and Function*. Cambridge University Press, Cambridge, pp. 25-55.

Follett, R.F. 1991. *Strategic Plan for Global Climate Change Research by the Agricultural Research Service: Biogeochemical Dynamics*. USDA. Washington, DC.

Frank, A.B., D.L. Tanaka, L. Hofmann, and R.F. Follett. 1995. Soil carbon and nitrogen of Northern Great Plains grasslands as influenced by long-term grazing. *J. of Range Manage.* 48:470-474.

Harrison, K., and W. Broecker. 1993. A strategy for estimating the impact of CO_2 fertilization on soil carbon storage. *Global Biogeochem. Cycles* 7:69-80.

Hassink, J., L.A. Bouwman, K.B. Zwart, J. Bloem, and L. Brussard. 1993. Relationship between soil texture, physical protection of organic matter, soil biota, and C and N mineralization in grassland soils. *Soil and Tillage Res.* 19:77-87.

Heal, O.W., T.V. Callaghan, C. Cornelissen, C. Korner, and S.E. Lee. 1998. *Global change in Europe's cold regions. Ecosystem Research Report No. 27.* Directorate-General, Science, Research and Development, European Commission. Brussels.

Herrick, J.E., and M.M. Wander. 1998. Relationships between soil organic carbon and soil quality in cropped and rangeland soils: the importance of distribution, composition, and soil biological activity. *In* R. Lal, J.M. Kimble, R.F. Follett, and B.A. Stewart (eds), *Soil Processes and the Carbon Cycle*, CRC Press, Boca Raton, FL, pp. 405-425.

Hinzmann, L., D.L. Kane, C.S. Benson, and K.R. Everett. 1996. Energy balance and hydrological processes in an Arctic watershed. *In* J.F. Reynolds, and J.D. Tenhunen (eds), *Landscape Function and Disturbance in Arctic Tundra*, Springer-Verlag, Berlin, pp. 131-154.

Hobbie, S.E. 1996. Temperature and plant species control over litter decomposition in Alaskan tundra. *Ecol. Monogr.* 66:503-522.

Hunt, H.W., E.R. Ingham, D.C. Coleman, E.T. Elliot, and C.P.P. Reid. 1988. Nitrogen limitation of production and decomposition in prairie, mountain meadow, and pine forest. *Ecology* 69:1009-1016.

Janzen, H.H., C.A. Campbell, E.G. Gregorich, and B.H. Ellert. 1998. Soil carbon dynamics in Canadian agroecosystems. *In* R. Lal, J.M. Kimble, R.F. Follett, and B.A. Stewart (eds), *Soil Processes and the Carbon Cycle*, CRC Press, Boca Raton, FL, pp. 57-80.

Jefferies, R.L., D.R. Klein, and G.R. Shaver. 1994. Vertebrate herbivores and northern plant communities: reciprocal influences and responses. *Oikos* 71(2):193-206.

Jenkinson, D.S. 1977. Studies on the decomposition of plant material in soil. V. The effect of plant cover and soil type on the loss of carbon from 14C (carbon isotope) labeled ryegrass decomposing under field conditions. *J. Soil Sci.* 28:424-434.

Johnson, D.A. *(ed).* 1979. *Special Management Needs of Alpine Ecosystems. Range Science Series No. 5.* Society for Range Management. Denver, CO.

Jones, M.H., J.T. Fahnestock, D.A. Walker, M.D. Walker, and J.M. Welker. 1998. Carbon dioxide fluxes in moist and dry Arctic tundra: responses to increases in summer temperature and winter snow accumulation. *Arctic and Alpine Res.* 30:373-380.

Jones, M.H., J.T. Fahnestock, and J.M. Welker. 1999. Early and late winter CO_2 efflux from Arctic tundra in the Kuparuk river watershed, Alaska. *Arctic, Antarctic, and Alpine Res.* 31(2):187-190.

Kay, B.D. 1998. Soil structure and organic carbon: a review. *In* R. Lal, J.M. Kimble, R.F. Follett, and B.A. Stewart (eds), *Soil Processes and the Carbon Cycle,* CRC Press, Boca Raton, FL, pp. 169-197.

Kern, J.S., D.P. Turner, and R.F. Dodson. 1998. Spatial patterns of soil organic carbon pool size in the Northwestern United States. *In* R. Lal, J.M. Kimble, R.F. Follett, and B.A. Stewart (eds), *Soil Processes and the Carbon Cycle,* CRC Press, Boca Raton, FL, pp. 29-43.

Klikoff, L.G. 1965. Microenvironmental influence on vegetational pattern near timberline in the central Sierra Nevada. *Ecol. Monogr.* 35(2):187-211.

Knight, D.K. 1994. *Mountains and Plains: The Ecology of Wyoming Landscapes.* Yale University Press. New Haven, CT.

Komarkova, V., and P.W. Webber. 1980. Two low Arctic vegetation maps near Atkasook, Alaska. *Arctic and Alpine Res.* 12:447-472.

Körner, C. 1994. Impact of atmospheric changes on high mountain vegetation. *In* M. Beniston *(ed), Mountain Environments in Changing Climates,* Routledge, London, pp. 155-166.

Körner, C. 1995. Alpine plant diversity: a global survey and functional interpretations. *In* F. S. Chapin, and C. Körner (eds), *Arctic and Alpine Biodiversity: Patterns, Causes and Ecosystem Consequences, Ecological Studies 113*, Springer-Verlag, Berlin, pp. 45-62.

Körner, C., M. Diemer, B. Schappi, P. Niklaus, and J.A. Arnone. 1997. The responses of alpine grassland to four seasons of CO_2 enrichment: a synthesis. *Acta Oecol.* 18(3):165-175.

Körner, C., and W. Larcher. 1988. Plant life in cold climates. *Symp Soc. Exp. Biol.* 42:25-57.

Kuramoto, R.T., and L.C. Bliss. 1970. Ecology of subalpine meadows in the Olympic Mountains, Washington. *Ecol. Monogr.* 40(3):317-347.

Lal, R., J.M. Kimble, and R.F. Follett. 1998. Pedospheric processes and the carbon cycle. *In* R. Lal, J.M. Kimble, R.F. Follett, and B.A. Stewart (eds), *Soil Processes and the Carbon Cycle*, CRC Press, Boca Raton, FL, pp. 1-8.

Laycock, W.A., D. Loper, F.W. Obermiller, L. Smith, S.R. Swanson, P.J. Urness, and M. Vavra. 1996. *Grazing on Public Lands. Task Force Report. No. 29.* Council for Agricultural Science and Technology. Ames, IA.

Leege, T.A, J.H. Daryl, and B. Zamora. 1981. Effects of cattle grazing on mountain meadows in Idaho. *J. of Range Manage.* 34(4):324-328.

Lynch, J.M. 1991. Sources and fates of soil organic matter. *In* T.R.G. Gray and W.S. Wilson (eds), *Advances in Soil Organic Matter Research: The Impact on Agriculture and the Environment, G.B. Special Publication no. 90*, Royal Society of Chemistry, Cambridge, pp. 231-237.

Manley, J.T., G.E. Schuman, J.D. Reeder, and R.H. Hart. 1995. Rangeland soil carbon and nitrogen responses to grazing. *J. of Soil and Water Conserv.* 50(3):294-298.

Marion, G.M., J.G. Bockheim, and J. Brown. 1997. Arctic soils and the ITEX experiment. *Global Change Biol.* 3 (suppl. 1):33-43.

Mast, M.A., K.A. Wickland, R.T. Striegal, and D.W. Clow. 1998. Winter fluxes of CO_2 and CH_4 from subalpine soils in Rocky Mountain National Park, Colorado. *Global Biogeochem. Cycles* 12:607-620.

McCarthy, G.M. 1977. *Hour of Trial: The Conservation Conflict in Colorado and the West, 1891-1907.* University of Oklahoma Press. Norman, OK.

McClung, C. 1997. *Environmental Assessment for Libby Flats Sheep Allotment.* U.S.F.S, Medicine Bow-Routte National Forest. Laramie, WY.

McKane, R.B., E.B. Rastetter, G.R. Shaver, K.J. Nadelhoffer, A.E. Giblin, J.A. Laundre, and F.S. Chapin, III, 1997. Climatic effects on tundra carbon storage inferred from experimental data and a model. *Ecology* 78:1188-1198.

McNaughton, S.J. 1983. Physiological and ecological implications of herbivory. Ecological interactions, plant chemistry and responses to herbivory, plant-herbivore symbiosis. *Encycl. Plant Physiol. New Ser.* Springer. Berlin, W. Germany. V. 12c:657-677.

Michaelson, G.J., C.L. Ping, and J.M. Kimble. 1996. Carbon content and distribution in tundra soils in Arctic Alaska. *In* R. Lal, J.M. Kimble, R.F. Follett, and B.A. Stewart (eds), *Soil Processes and the Carbon Cycle*, CRC Press, Boca Raton, FL.

Milchunas, D.G., and W.K. Lauenroth. 1993. Quantitative effects of grazing on vegetation and soils over a global range of environments. *Ecol. Monogr.* 63:327-366.

National Research Council. 1994. *Rangeland Health: New Methods to Classify, Inventory, and Monitor Rangelands.* National Academy Press. Washington, DC.

Oades, J.M. 1988. The retention of organic matter in soils. *Biogeochemistry* 5:35-70.

Oberbauer, S.F., W. Cheng, C.T. Gillespie, B. Ostendorf, A. Sala, R. Gebauer, R.A. Virginia, and J.D Tenhunen. 1996. Landscape patterns of carbon dioxide exchange in tundra ecosystems. *In* J.F. Reynolds and J.D. Tenhunen (eds), *Landscape Function and Disturbance in Arctic Tundra*, Springer-Verlag, Berlin, pp. 223-256.

Oberbauer, S.F., J.D. Tenhunen, and J.F. Reynolds. 1991. Environmental effects on CO_2 efflux from water track and tussock tundra in Arctic Alaska. *Arctic and Alpine Res.* 23:162-169.

Oechel, W.C., and W.D. Billings. 1992. Effects of global change on the carbon balance of Arctic plants and ecosystems. *In* F.S. Chapin, III, R.L. Jeffries, J.F. Reynolds, G.R. Shaver, and J. Svoboda (eds), *Arctic Ecosystems in a Changing Climate: An Ecophysiological Perspective*, Academic Press, San Diego, CA, pp. 139-168.

Oechel, W.C., S.J. Hastings, G.L. Vourlitis, M.A. Jenkins, G. Riechers, and N. Grulke. 1993. Recent changes of Arctic tundra ecosystems from a carbon sink to a source. *Nature* 361:520-523.

Oechel, W.C., and G.L. Vourlitis. 1994. The effects of climate change on land-atmosphere feedbacks in Arctic tundra regions. *Trends Ecol. Evol.* 9:324-329.

Oechel, W.C., G.L. Vourlitis, and S.J. Hastings. 1997. Cold season CO_2 emission from Arctic soils. *Global Biogeochem. Cycles* 11:163-172.

Oechel, W.C., G.L. Vourlitis, S.J. Hastings, and S.A. Bochkarev. 1995. Change in Arctic CO_2 flux over two decades: effects of climate change at Barrow, Alaska. *Ecol. Appl.* 5:846-855.

Ojima, D.S., W.E. Easterling, and C. Donofrio. 1997. Climate Change Impacts on the Great Plains. *Workshop Report. 27-29 May, 1997.* Colorado State University. Fort Collins, CO.

Paul, E.A., and F.E. Clark. 1996. *Soil Microbiology and Biochemistry.* 2nd ed. Academic Press. San Diego, CA.

Ping, C.I., G.J. Michaelson, W.M. Loya, R.J. Chandler, and R.L. Malcolm. 1998. Characteristics of soil organic matter in Arctic ecosystems of Alaska. *In* R. Lal, J.M. Kimble, R.F. Follett, and B.A. Stewart (eds), *Soil Processes and the Carbon Cycle,* CRC Press, Boca Raton, FL.

Povirk, K. 1999. *Carbon and nitrogen dynamics of an alpine grassland: effects of grazing history and experimental warming on CO_2 flux and soil properties.* M.S. thesis, Department of Renewable Resources, University of Wyoming. Laramie, WY.

Raich, J.W., and W.H. Schlesinger. 1992. The global carbon dioxide flux in soil respiration and its relationship to vegetation and climate. *Ser. B.* 44:81-99.

Randerson, J.T., M.V. Thompson, T.J. Conway, I.Y. Fung, and C.B. Field. 1997. The contribution of terrestrial sources and sinks to trends in the seasonal cycle of atmospheric carbon dioxide. *Global Biogeochem. Cycles* 11:535-560.

Retzer, J.L., 1974. Alpine Soils. *In* J.D. Ives and R.G. Barry (eds), *Arctic and Alpine Environments.* Methuen and Co., Ltd, London, pp. 771-802.

Russell, D.E., A.M. Martell, and W.A.C. Nixon. 1993. *Rangifer: Range Ecology of the Porcupine Caribou Herd in Canada. Special Issue NO.* Nordic Council for Reindeer Research (NOR). Harstad, Norway.

Scharpenseel, J.W., and E.M. Pfeiffer. 1998. Carbon turnover in different climates and environments. *In* R. Lal, J.M. Kimble, R.F. Follett, and B.A. Stewart (eds), *Soil Processes and the Carbon Cycle,* CRC Press, Boca Raton, FL.

Scharpenseel, H.W., M. Schomaker, and A. Ayoub (eds). 1990. *Soils on a Warmer Earth.* Elsevier. New York.

Schimel, J.P., and J.S. Clein. 1995. Microbial response to freeze-thaw cycles in tundra and taiga soils. *Soil Biol. Biochem.* 28 (8):1061-1066.

Schlesinger, W.H. 1997. *Biogeochemistry.* 2nd ed. Academic Press. San Diego.

Shaver, G.R., L.C. Johnson, D.H. Cades, G. Murray, J.A. Laundre, E.B. Rastetter, K.J. Nadelhoffer, and A.E. Giblin. 1998. Biomass accumulation and CO_2 flux in three Alaskan wet sedge tundras: responses to nutrients, temperature and light. *Ecol. Monogr.* 68:75-99.

Sommerfeld, R.A., A.R. Mosier, and R.C. Musselman. 1993. CO_2, CH_4 and N_2O flux through a Wyoming snowpack and implications for global budgets. *Nature* 361:140-142.

Sommerfeld, R.A., W.J. Massman, and R.C. Musselman. 1996. Diffusional flux of CO_2 through snow: spatial and temporal variability among alpine-subalpine sites. *Global Biogeochem. Cycles* 10:473-482.

Starr, C.H. 1974. *Subalpine meadow vegetation in relation to environment at Headquarters Park, Medicine Bow Mountains, Wyoming.* Thesis, University of Wyoming. Laramie, WY.

Stevenson, F.J. 1994. *Humus Chemistry: Genesis, Composition, Reactions.* John Wiley and Sons, Inc. New York.

Stevenson, F.J., and M.A. Cole. 1999. *Cycles of Soils: Carbon, Nitrogen, Phosphorus, Sulfur, Micronutrients.* 2nd ed. John Wiley and Sons, Inc. New York.

Taylor, B.R., and D. Parkinson. 1988. Does repeated freezing and thawing accelerate decay of leaf litter? *Soil Biol. Biochem.* 20:657-665.

Tieszen, L.L. 1978. *Vegetation and Production Ecology of an Alaskan Arctic Tundra.* Springer. Berlin.

Tieszen, L.L., and S. Archer. 1979. Physiological responses of plants in tundra grazing systems. *In* D.A. Johnson *(ed)*, *Special Management Needs of Alpine Ecosystems, Range Science Series, No. 5,* Society for Range Management, Denver, CO, pp. 22-42.

Tokarnia, C.H., A. Gava, and P.V. Peixoto. 1989. "Swollen brisket disease" (edema of sternal region) in cattle in the state of Santa Catarena, Brazil. *Pesq. Vet. Bras.* 9(3-4):73-83.

Tracy, B.F., and D.A. Frank. 1998. Herbivore influence on soil microbial biomass and nitrogen mineralization in a northern grassland ecosystem: Yellowstone National Park. *Oecologia* 114(4):556-562.

U.S.D.A.-N.R.C.S. 1998. *Keys to Soil Taxonomy.* 8th ed. U.S. Government Printing Office, Washington D.C.

Vale, T.R. 1978. Tree invasion of Cinnabar Park in Wyoming. *Am. Mid. Nat.* 100(2):277-284.

VanCleve, K., and R.F. Powers. 1995. Soil carbon, soil formation, and ecosystem development. *In* W.W. McFee and J.M. Kelly (eds), *Carbon Forms and Functions in Forest Soils*, Soil Sci. Soc. Amer., Inc., Madison, WI, pp. 155-200.

Vitousek, P.M. 1994. Beyond global warming: ecology and global change. *Ecology* 75:1861-1876.

Vourlitis, G.L., W.C. Oechel, S.J. Hastings, and M.A. Jenkins. 1993. A system for measuring in situ CO_2 and CH_4 flux in unmanaged ecosystems: an Arctic example. *Funct. Ecol.* 7:369-379.

Walker, D.A., and M.D. Walker. 1991. History and pattern of disturbance in Alaskan Arctic terrestrial ecosystems: a hierarchical approach to analyzing landscape change. *J. Appl. Ecol.* 28:244-276.

Walker, D.A., N.A. Auerbach, J.G. Bockheim, F.S. Chapin III, W. Eugster, J.Y. King, J.P. McFadden, G.J. Michaelson, F.E. Nelson, W.C. Oechel, C.L. Ping, W.S. Reeburgh, S. Regli, N.I. Shiklomanov, and G.L. Vourlitis. 1998. Energy and trace gas fluxes across a soil pH boundary in the Arctic. *Nature* 394:469-472.

Walker, M.D., D. A. Walker, A.M. Arft, T. Bardsley, P.D. Brooks, J.T. Fahnestock, M.H. Jones, M. Losleben, A. N. Parson, T.R. Seastedt, P.L. Turner, and J.M Welker. 1999. Long-term experimental manipulation of winter snow regime and summer temperature in Arctic and alpine tundra: An integrated ecosystem approach. *Hydrol. Proc.* 13:2315-2330.

Walsh, N., T. McCabe, A.N. Parsons, and J.M. Welker. 1997. Experimental manipulations of snow depth: effects on mineral nutrition content of caribou forage. *Global Change Biol.* 3 (suppl. 1):158-164.

Welker, J.M., K.A. Brown, and J.T. Fahnestock. 1999. CO_2 flux in Arctic and alpine dry tundra: comparative field responses under ambient and experimentally warmed conditions. *Arctic, Antarctic, and Alpine Res.* 31:272-277.

Welker, J.M., J.T. Fahnestock, and M.H. Jones. 2000. Annual CO_2 efflux from moist and dry tundra in AK: Response to field increases in summer temperatures and deeper snow. *Climatic Change* 44:139-150.

Welker, J.M., T.H.E. Heaton, B. Sprio, and T.V. Callaghan. 1995. Indirect effects of winter climate on the $d^{13}C$ and the dD characteristics of annual growth segments in the long-lived. Arctic clonal plant *Cassiope tetragona*. *Palaeoclimate Res.* 15:105-120.

Welker, J.M., U. Molau, A.N. Parsons, C. Robinson, and P.A. Wookey. 1997. Response of Dryas octopetala to ITEX manipulations: a synthesis with circumpolar comparisons. *Global Change Biol.* 3 (suppl. 1):61-73.

Welker, J.M., P.A. Wookey, A.N. Parson, M.C. Press, T.V. Callaghan, and J.A. Lee. 1993. Leaf carbon isotope discrimination and vegetative responses of *Dryas octopetala* to temperature and water manipulations in a high Arctic polar semi-desert, Svalbard. *Oecologia* 95:463-469.

Wilson, D.J., and R.L. Jefferies. 1996. Nitrogen mineralization, plant growth and goose herbivory in an Arctic coastal ecosystem. *Ecology* 84(6):841-851.

Wookey, P.A., A.N. Parsons, J.M. Welker, J.A. Potter, T.V. Callaghan, J.A. Lee, and M.C. Press. 1993. Comparative responses of phenology and reproductive development to simulate environmental change in sub-Arctic and high Arctic plants. *Oikos* 67:490-502.

Wookey, P.A., C.H. Robinson, A.N. Parsons, J.M. Welker, M.C. Press, T.V. Callaghan, and J.A. Lee. 1995. Environmental constraints on the growth, photosynthesis and reproductive development of *Dryas octopetala* at a high Arctic polar semi-desert, Svalbard. *Oecologia* 12(4):478-489.

Zimov, S.A., G.M. Zimova, S.P. Daviodov, A.I. Daviodova, Y.V. Voropaev, Z.V. Voropaeva, S.F. Prosiannikov, O.V. Prosiannikova, I.V. Semiletova, and I.P. Semiletov. 1993. Winter biotic activity and production of CO_2 in Siberian soils: a factor in the greenhouse effect. *J. Geophys. Res.* 98:5017-5023.

Zimov, S.A., S.P. Davidov, Y.V. Voropaev, S.F. Prosiannikov, I.P. Semiletov, M.C. Chapin, and F.S. Chapin. 1996. Siberian CO_2 efflux in winter as a CO_2 source and cause of seasonality in atmospheric CO_2. *Clim. Change* 33:111-120.

SECTION **3**

Managerial and Environmental Impacts on U.S. Grazing Land

CHAPTER **9**

Soil Erosion and Carbon Dynamics on Grazing Land

R. Lal[1]

Introduction

Grazing land is comprised of rangeland and pasture. *Rangeland* is uncultivated land that provides the necessities of life for grazing and browsing animals. *Grazing* implies the animals' consumption of standing forage (edible grasses and forbs) by livestock or wildlife, and *browsing* refers to their consumption of edible leaves and twigs from woody plants (trees and shrubs) (Holechek et al., 1989).

Rangeland includes a wide range of ecosystems, e.g., native, perennial grassland, annual grassland, desert shrub, and forest. Depending on the rainfall regime and the vegetation, there are at least five types of rangeland in the U.S. (Table 9.1). Rangelands offer several products for humans, including forage for livestock, fish and wildlife, timber and firewood, open space and recreation, and water, along with other miscellaneous products. *Range management* involves manipulating the ecosystem to optimize the *sustainable* output of products.

In contrast to rangelands, pastures periodically are cultivated to enhance productivity of introduced forage species. In addition, pastures receive agronomic inputs (e.g., irrigation, fertilizers, and manures) and often are fenced for managed/controlled grazing.

[1] Prof. of Soil Science, School of Natural Resources, The Ohio State University, Columbus, Ohio, e-mail lal.1@osu.edu.

Table 9.1. Types of rangelands (modified from Holechek et al., 1989).

Type	Precipitation (mm/yr)	Soils	Description
1. Grassland	250-900	Mollisols, deep, loamy textured	Free of woody plants, and dominated by grasses (sedges and rushes), forbs, and shrubs
2. Desert shrublands	< 250	Aridisols and Entisols, sandy to loamy textured	Woody plants (<2 m tall) with sparse herbaceous understory
3. Savanna woodlands	300-500	Alfisols	Scattered low growing trees (<12 m tall) with productive herbaceous understory
4. Forests	>500	Alfisols, Ultisols, Oxisols	Trees (>12 m tall) and closely spaced
5. Tundra	250-500	Cryosols with permafrost	Treeless plains in Arctic or high elevation (alpine)

Table 9.2. Estimates of land cover in the U.S. (USDA, 1997).

Land Cover		Area (Mha)
Total surface area		785.4
Federal land		165.2
Nonfederal land		600.4
(a) Developed		37.4
(b) Rural	563.0	
(i) Cropland	154.8	
(ii) Pastureland	51.0	
(iii) Rangeland	161.5	
(iv) Forest land	159.9	
(v) Minor cover	35.8	
Total grazing land (pastureland + rangeland) = 212.5 MHa		

Status of U.S. Grazing Land

Area

Total area under grazing land (rangeland plus pasture) in the U.S. is great, several hundred million hectares (Mha) (Tables 9.2 and 9.3). Exact estimates vary because of the different definitions involved.

The FAO (1982) estimated the land area under cropland at 164.9 Mha (21% of the total land area), permanent pasture at 204.2 Mha (26%), and forest and woodland at 243.5 Mha (31%). The USDA (1997) estimates the area of privately owned

Range Type		Area (Mha)	Forage Production (kg/ha of dry matter)
Grasslands			
(i)	Tallgrass prairie	15	1500-3500
(ii)	Southern mixed prairie	20	1000-2500
(iii)	Northern mixed prairie	30	1000-2000
(iv)	Short grass prairie	20	600-1000
(v)	California annual grassland	3	400-3500
(vi)	Palouse prairie	3	300-1000
Desert shrublands			
(i)	Hot desert	26	150-500
(ii)	Cold desert	73	100-500
Woodlands			
(i)	Pinon-juniper woodland	17	100-600
(ii)	Mountain shrubland	13	800-2500
(iii)	Western coniferous forest	59	300-1000
(iv)	Southern pine forest	81	2000-4000
(v)	Eastern deciduous forest	100	1200-3500
(vi)	Oak woodland	16	700-2500
Tundra			
(i)	Alpine	4	500-700
		Total 480	

Table 9.3. Types of rangeland in the U.S. (modified from Holechek et al., 1989).

grazing lands (sum of pasture and rangeland as Table 9.2 shows) at 212 Mha. Holechek et al. (1989) estimated the total land area under different types of rangelands at about 480 Mha (Table 9.3). The forage production potential varies widely, depending on the vegetation and the rainfall regime.

Soil erosion

Soil erosion risks on U.S. grazing land are less than on cropland but more than on forest/woodland. The magnitude of soil erosion on grazing land depends on numerous factors (Fig. 9.1). Numerous factors also affect inter-rill erosion on rangelands (Wilcox and Wood, 1989). Climatic (rainfall and runoff) erosiveness and soil erodibility are characteristics of an ecoregion, and little can be done to alter these. In addition, vegetative and other possible factors in some ecosystems (e.g., dryer and warmer climates such as in the U.S. Southwest) are more vulnerable to soil erosion that may result from grazing, burning, or other practices than are ecoregions with greater precipitation and/or lower temperature regimes. However, other factors affecting soil erosion can be altered to regulate its rate and severity.

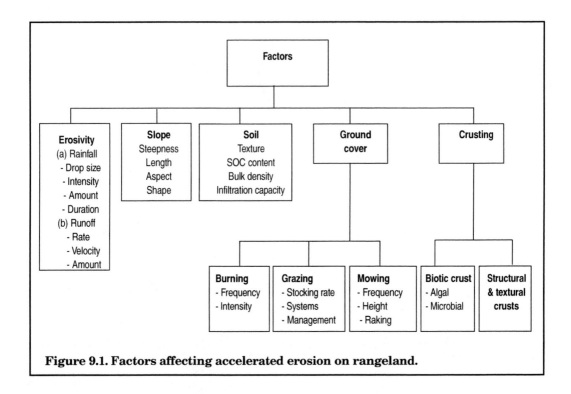

Figure 9.1. Factors affecting accelerated erosion on rangeland.

Slope and soil erosion

A significant percentage of semiarid rangeland has steep slopes, often >25%. Soil erosion increases exponentially with increase in slope gradient (Zingg, 1940):

$$E = aS^b \qquad \text{(Equation 1)}$$

where E is soil erosion, S is slope gradient (%), and a and b are empirical constants. The value of b usually ranges from 1.35 to 2.0. Increasing slope gradient by a factor of 4 (e.g., from 2% to 8%) increases velocity of flow by about 2 times, eroding power, 4 times, material carried, 32 times, and size of the material carried, 64 times. However, erosion increases with slope up to a certain point, after which erosion decreases with further increase in slope. Renner (1936) observed that susceptibility of rangeland to erosion was not influenced by slope gradient beyond 70%. Further, the effect of slope gradient on rangeland is less than on cropland.

Eqation 2 shows another model of erosion McCool et al. (1987) suggested for inter-rill erosion.

$$E = a \, Sin^b \, Q + \; C \qquad \text{(Equation 2)}$$

where Q is the slope angle in degrees and a, b, and C are empirical constants. Wilcox and Wood (1989) observed that, in the semiarid regions of New Mexico, sediment concentration increased exponentially with an increase in slope gradient (r = 0.80). However, inter-rill erosion correlated slightly less (r = 0.6 for dry run, r = 0.73 for wet run) with slope gradient. Moreover, inter-rill erosion decreased exponentially with an increase in litter biomass and rock cover.

Ground cover

Ground cover moderates the effects of climatic erosiveness. Ground cover consists of organic cover or biomass as well as inorganic cover comprised of rocks, stones, and gravel. The organic cover (live and dead) is highly labile and depends on seasons (rainfall and temperature regimes) and management (controlled fire, grazing, water management, nutrient application). Management of the riparian zone can enhance sediment filtration and improve water quality (Frasier et al., 1998; Pearce et al., 1989a, b)

Grazing and infiltration rate

Grazing affects soil erosion by altering the vegetative cover and soil structure (Milchanus and Lauenroth, 1993). Grazing intensity also can affect the infiltration rate through its effect on nutrient balance and soil quality (Dormaar and Willms, 1990; Dormaar et al., 1990; Frank et al., 1995). The infiltration rate depends on the degree of ground cover, as influenced by management practices. In South Dakota, Rauzi and Hanson (1966) observed that heavy grazing decreased the plant cover and the infiltration rate. The equilibrium infiltration rate was 2.7, 4.3, and 7.5 cm/hr for heavy, moderate, and light grazing, respectively.

In Texas, Pluhar et al. (1986) observed that the infiltration rate increased and sediment production declined as vegetative standing crop and cover increased, soil bulk density decreased, and soil organic matter and aggregate stability increased. Sediment production was the least and the infiltration rate highest for the moderate stocking treatment.

Livestock grazing also influences runoff and erosion from forest land. Stoeckeler (1959) reported infiltration rates of an ungrazed oak woodland to be 150 times greater than for an adjacent, heavily grazed woodland. Duvall and Linnartz (1967) observed that infiltration rates for heavily grazed, moderately grazed, and ungrazed longleaf pine (*Pinus palustris*) – bluestem (*Andropogon* spp., *Schizachyrium* spp.) pasture were 20, 30, and 46 mm/hr, respectively.

Thurow et al. (1988) monitored the effects of grazing intensity on the infiltration rate and inter-rill erosion for rangeland on Edwards Plateau, Texas. The infiltration rate decreased and inter-rill erosion increased more on heavily stocked than on moderately stocked pastures. Total organic cover was the most important factor determining infiltration rate.

Wilcox et al. (1988) studied the impact of sheep grazing on steep slopes (average 50%) in the Guadalupe Mountains of southeastern New Mexico. Infilterability on the grazed slopes was 12% to 17% lower than on the ungrazed slopes. Sediment concentration and sediment production by the runoff from lightly grazed slopes was significantly higher than from the ungrazed slopes for the dry run (Table 9.4). These data did not indicate that steep slopes (30% to 70%) in semiarid regions are any more hydrologically sensitive to light grazing than are moderate slopes (<10%).

Grazing and erosion

In addition to affecting ground cover, grazing affects soil quality (Bauer et al., 1987; Berg, 1988; Branson et al., 1962). Soil erosion risks, both by wind and water, increase with a decrease in vegetative cover (Thurow et al., 1986) and with an increase in grazing intensity (Wood et al., 1989).

Branson et al. (1970, 1981) observed a linear increase in the annual runoff amount with an increase in percent bare soil in 17 watersheds of western Colorado. Soil erosion increases with an increase in runoff rate, and with an increase in grazing intensity. Grazing's intensity also decreases the infiltration rate, despite the perceived positive loosening effect of the grazing animals' hoof action on compacted/crusted soils (Warren et al., 1986a, b, c).

Pluhar et al. (1986) observed that soil erosion was greater on heavily grazed than on moderately grazed plots. Dunford (1949) measured the impact of grazing on soil erosion on pine-bunchgrass range in Colorado for a 6-year period (1937-1942) before and after grazing (1942-1948). Soil erosion increased from about 80 kg/ha before to 180 kg/ha after heavy grazing, 75 kg/ha before to 100 kg/ha after moderate grazing, and 110 kg/ha before to 90 kg/ha after no grazing on the ungrazed control.

Many studies have shown that moderate grazing increases erosion over the ungrazed condition (Thurow et al., 1986; Pluhar et al., 1986). However, the differences are usually small and practically unimportant. Several experiments also have shown that short-term grazing increases soil erosion, compared to moderate continuous grazing (Thurow et al., 1986; Warren et al., 1986a, b; Pluhar et al., 1986).

Table 9.4. Grazing's effects on cumulative infiltration and sediment production for steep slopes in southeastern New Mexico (modified from Wilcox and Wood, 1988).

Parameter	Run	Ungrazed	Grazed	Grazed:Ungrazed
I. Cumulative infiltration (cm)				
	Dry	6.52	5.78	0.89
	Wet	4.31	3.56	0.83
II. Sediment production (kg/ha)				
	Dry	883	1390	1.57
	Wet	719	949	1.32

Burning and erosion

Prescribed burning is used widely to manage vegetation on rangeland (Wright, 1974). Fire on rangeland, natural or as a management tool, removes vegetative cover, influences soil properties (e.g., wetability) (De Bano et al., 1970; 1976), and accelerates runoff and soil erosion.

Emmerich and Cox (1992) studied the effects of burning on runoff and erosion in southeastern Arizona. Immediately after a burn, runoff and erosion did not change significantly. However, significantly greater surface runoff and sediment production occurred in the fall than in the spring. Other studies have reported increases in runoff and erosion after burning (Wright et al., 1976, 1982).

Late spring burn leads to a decreased infiltration rate. Hester et al. (1997) evaluated burning's effects on hydrologic characteristics for the Edwards Plateau, Texas. They observed that equilibrium infiltration rates of unburned areas were significantly greater on sites dominated by oak (*Quercus virginiana*) (200 mm/hr) or Juniper (*Juniperus ashei*) (183 mm/hr) than on sites dominated by bunchgrasses (146 mm/hr) or short grass (105 mm/hr). Equilibrium infiltration rates on burned areas were reduced significantly on sites dominated by bunchgrass (110 mm/hr) and short grass (76 mm/hr) and on oak sites that were cut and burned (129 mm/hr).

Before burning, inter-rill erosion was much lower under the tree sites (oak = 2 kg/ha; juniper = 34 kg/ha). After burning, inter-rill erosion significantly increased for all vegetation types (short grass = 5,766 kg/ha; bunchgrass = 4,463 kg/ha; oak = 4,500 kg/ha; juniper = 1,926 kg/ha) (Table 9.5). Total organic cover (r = 0.74) and bulk density of 0 to 3 cm (r = 0.46) correlated most strongly with inter-rill erosion.

Table 9.5. Burning's effect on sediment yield for a rangeland in Edwards Plateau, Texas, using simulated rainfall (modified from Hester et al., 1997).

Vegetation	Burned	Unburned	Burned:Unburned
	------------------------kg/ha------------------------		
Oak	4500a	2b	2500
Juniper	1926b	34b	57
Bunchgrass	4463a	300b	15
Short grass	5766a	1299b	4

Figures in the row followed by the same letter are statistically similar.

Measurement and Prediction of Erosion

Depending on the objectives, one can assess runoff and soil erosion from rangeland at different scales. One can measure runoff and soil erosion experimentally using surface runoff plots of 5 to 50 m^2 (Williams and Buckhouse, 1991; Wilcox, 1994) or large watersheds (Baker, 1982; Gray et al., 1996). Small runoff plots are useful for studying soil processes affecting runoff and soil erosion, e.g., disturbance regime and runoff initiation (Wilcox, 1994). Experimental assessment of erosion is an essential but slow and expensive undertaking. Therefore, simulation modeling is used widely instead. The usefulness and accuracy of these models vary.

The Water Erosion Prediction Process (WEPP) model (Lane and Nearing, 1989) and hydrological modeling techniques (Osborn and Simanton, 1990) have been used to simulate the major processes that affect soil erosion by water (Wilcox et al., 1992). Simanton et al. (1991) evaluated the application of WEPP for nine rangeland sites throughout the western U.S. They concluded that canopy cover had little effect on runoff, infiltration, initial abstractions, and erosion rate under simulated rainfall conditions.

Saubi et al. (1995) compared measured surface runoff with the simulated results of the WEPP model and observed good agreement between accurate depictions of texture and accurate predictions of erosion. WEPP poorly predicted runoff from a small sagebrush watershed (1 ha) (Wilcox et al., 1990) unless the infiltration parameters (Green and Ampt parameters) were determined from field measurements (Wilcox et al., 1992).

Osborn and Simanton (1990) observed that the KINEROS model can be useful for evaluating the impact of rangeland treatments on water yield. The KINEROS is a nonlinear, deterministic parameter model (Rovey et al., 1977). However, when Wright and Hansen (1991) used SPUR (Simulation of Production and Utilization of Rangelands) and ERNYM (Ekalaka Rangeland Hydrology and Yield Model) to evaluate runoff potential, the synthetic precipitation record failed to simulate extreme precipitation events that significantly reduced forecasted runoff values.

Models are used effectively for large-scale assessment of soil erosion, e.g., USLE and Wind Erosion Equation. USDA-NRI (1992) estimated the highly erodible land area to be 76.6 Mha under forest land, 99.8 Mha under rangeland, and 22.9 Mha under pasture (Table 9.6). Total land area of highly erodible grazing land thus is estimated at 122.7 Mha. Highly erodible lands are prone to losing topsoil by wind and water erosion. The data in Table 9.7 show that annual sediment loss from grazing land is estimated at 1.76 billion tons by wind erosion and 0.61 billion tons by water erosion. Thus the total annual sediment loss from grazing land is about 2.37 billion tons/yr.

Table 9.6. The estimated area of highly erodible land (USDA-NRI, 1992).	
Land Type	Area (Mha)
Forest	76.6
Rangeland	99.8
Pastureland	22.9

Table 9.7. Soil erosion on grazing land (USDA-NRI, 1992).		
Land Type	Wind Erosion	Water Erosion
	-----------MMT/yr-----------	
Forest	0.0	0.0
Rangeland	1750±117.3	4820±14.6
Pastureland	10.0±1.8	126.0±2.9
Total	1760	608

Soil Erosion and Carbon Dynamics on Rangeland

Because a large proportion of soil organic C (SOC) is concentrated near the soil surface, it is highly vulnerable to soil erosion processes. In addition, since SOC is a light fraction, wind and water runoff transport SOC content preferentially. The enrichment ratio of eroded sediments for SOC is always >1 and may be as much as 5. Zobeck and Fryrear (1986) observed that the enrichment ratio for SOC ranged from 1.7 to 2.9 for windblown sediments collected at Big Spring, Texas. Moreover, the SOC content of the sediment ranged from 2.0% to 3.5% (Table 9.8).

Two approaches can estimate the impact of erosion on SOC. The first approach involves estimating the C displaced in gross soil erosion. The data in Table 9.9 use this approach to estimate the C emission to the atmosphere from the SOC displaced by erosion. We based the calculations on three assumptions:

1. Estimates of soil erosion account for gross values and do not consider deposition.
2. The SOC content in sediment is 2.0% in windblown material and 3.0% in water entrained sediment.
3. 20% of the SOC displaced by wind and water erosion is emitted into the atmosphere as CO_2, through mineralization and/or oxidative processes.

The second approach considers the sediment transported to the ocean. Lal et al. (1998) estimated that, of the 336 MMT/yr of suspended load transported by U.S. rivers to the ocean, cropland contributes 250 MMT/yr (or 75%) . The remainder, 86 MMT/yr, is from grazing land. Assuming a delivery ratio of 10%, SOC content of 3%, and gaseous emission of 20%, the C emission due to water erosion from U.S. grazing land is 5.2 MMTC/yr. To this must be added the component due to wind erosion. With gross wind erosion of 1760 MMT/yr, the SOC emission (with 2% SOC content and 20% emission) is 7.0 MMT/yr. Therefore, the total C emission due to erosion by water and wind from rangelands is 12.2 MMT/yr.

These preliminary calculations show that a total of 10.7 to 12.2 MMTC/yr may be emitted into the atmosphere through accelerated soil erosion on grazing land. These calculations do not account for the soil inorganic C (SIC).

Table 9.8. SOC dynamics influenced by accelerated erosion on U.S. grazing land.

Parameter	Wind Erosion	Water Erosion
Soil erosion (MMT/yr)	1760	610
Mean SOC content (%)	2000	3000
Total SOC displaced by erosion (MMT/yr)	35200	18300
Mean emission to the atmosphere (%)	20000	20000
Carbon emission from exposed calciferous layer	2100-2400	1200-1400
Total C emission to the atmosphere (MMT/yr)	**7000-8200**	**3700-4000**

Table 9.9. Enrichment ratio of soil organic matter for windblown sediments over a height of 0.15 to 2.0 m in 1984 at Big Spring, TX (recalculated from Zobek and Fryrear, 1986).

Date	Storm Rating	SOC (g/kg)	Enrichment Ratio
4/11	6	3.5	2.9
4/21	5	2.1	1.8
4/26	4	2,11.8	
4/27	5	2.2	1.9
4/30	7	2.5	2.0
5/2, 3, 8, 11	3	2.3	1.9
6/6, 8	6	2.0	1.7

In addition to releasing, displacing, and redistributing C in the sediments, exposure of the calciferous and petrocalcic horizon to the surface (Figure 9.2) can lead to CO_2 emission. The petrocalcic horizon thus exposed is subject to acidifying reactions from acid rains, use of fertilizers, and other anthropogenic activities. No experimental data estimate the amount of C released through this reaction, but it may approximate 33% of that released through the sediment or 3.3 to 3.8 MMTC/yr.

As a crude estimate, therefore, the total amount of C released due to erosion from grazing land may be 14 to 16 MMTC/yr. This amount is comparable to that released by erosion from U.S. cropland (Lal et al., 1998). Soil erosion management, through the use of appropriate erosion controls, can decrease these emissions into the atmosphere.

Restoring Grazing Land to Sequester C

Some rangelands are severely degraded, and the extent and severity of degradation varies among ecoregions. In the western U.S., only about 40% of the 160 Mha of rangeland is in good condition for forage production (USDA, 1987). Degra-

Figure 9.2. The calciferous horizon exposed by anthropogenic disturbance can cause CO$_2$ emissions.

dation of grazing land may be due to accelerated soil erosion, compaction, drought, and salinization.

Because of loss in biomass production, soil degradation leads to a decrease in the SOC pool. The SOC content of undisturbed grasslands is generally high (Tables 9.10 and 9.11). Surface soils in some grasslands may contain as much as 6.0% SOC content. Soil erosion and other degrading processes may decrease SOC content by as much as 25%.

Restoring degraded grazing lands can lead to C accretion in soil (Henzell et al., 1967; Manley et al., 1995). There are several options for restoring the soil quality of degraded grazing lands. The basic strategy is to enhance the vegetative cover, improve soil structure, and decrease runoff.

Runoff management

Several engineering devices/techniques are used to manage runoff. These include contour trenches and furrows, pitting, ripping, water harvesting, water spreading, etc. It is important to identify critical points for gully erosion, and control measures should be sited on these critical areas (Heede, 1970).

Improving vegetative cover

About 70 to 90% of ground cover is needed to effectively reduce risks of soil erosion. This may translate into 4 to 6 tons/ha of herbage and litter for effective erosion control. Thurow et al. (1988) observed an exponential decrease in inter-rill erosion with increase in litter accumulation. Soil erosion decreased from 8000 kg/ha for no ground cover to 10 kg/ha for ground covered with litter at the rate of 2500 kg/ha (Eq. 3).

$$Y = 7881 \times 5(10)^{-0.0015x}, r^2 = 0.94) \qquad \text{(Equation 3)}$$

where Y is soil erosion in kg/ha and x is litter accumulation in kg/ha. It is generally believed that spots of bare soil should be no larger than 10 cm in diameter in the wheatgrass type and 5 cm in the annual type (Heady and Child, 1994).

Maintaining riparian zones

In montane riparian communities of northern Colorado, Frasier et al. (1998) observed that the organic layer on the soil surface exhibited signs of water repellency. The experiments in the montane riparian communities in Colorado showed more time generally required for the runoff to begin in the sedge community than in the grass community. There were also lower equilibrium runoff percentages from dry runs in the sedge community than in the grass community.

Table 9.10. SOC content of Konza prairie soils on interfluves (modified from Ransom et al., 1996).

(a) Summit		(b) Summit-depression		(c) Shoulder	
Depth (cm)	SOC (%)	Depth (cm)	SOC (%)	Depth (cm)	SOC (%)
0-10	4.39	0-9	5.77	0-9	3.40
10-17	2.68	9-20	2.43	9-18	2.37
17-33	1.50	20-31	1.62	18-28	1.85
33-41	1.14	31-42	1.07	28-37	1.60
41-58	0.70	42-57	0.70	37-66	1.41
58-71	0.54	57-73	0.45	66-79	0.76
71-86	0.31	73-77	0.44	79-90	0.93
86-97	0.32	77-80	—	90-93	—
97-107	0.31				
107-127	0.32				
127-148	0.35				
148-150	—				

Table 9.11. SOC content of Konza prairie soils on side slopes, footslopes, terraces, and floodplains (modified from Ransom et al., 1996).

Side Slope (Tuttle)		Side Slope (Benfield)		Low Terrace Floodplain		High Terrace	
Depth (cm)	SOC (%)	Depth (cm)	SOC (%)	Depth (cm)	SOC (%)	Depth (cm)	SOC (%)
0-8	5.10	0-9	6.06	0-12	5.94	0-8	6.14
8-17	4.40	9-22	4.12	12-33	2.99	8-25	3.35
17-32	3.78	22-42	2.70	33-43	2.00	25-46	1.70
32-57	2.62	42-58	1.06	43-53	1.09	46-77	0.84
57-70	0.86	58-80	0.71	53-68	0.69	77-85	0.54
70-109	0.40	80-98	0.39	68-86	0.63	85-102	0.46
109-154	0.17	98-111	0.28	86-121	0.45	102-125	0.59
		111-135	0.15	121-140	0.71	125-137	0.43
		135-181	0.10	140-155	0.63	137-152	0.25
		181-190	0.06	155-156	—	152-182	0.13
						182-194	0.12
						194-206	0.08
						206-222	0.05

No differences in time to runoff initiation for either dry or wet runs could be attributed to the vegetation's height for either plant community. Sediment yield also was not affected by the vegetation's height (Pearce et al., 1998a, b). However, sediment retention depended on physical characteristics, e.g., texture. Important variables that influenced sediment's movement downslope included % surface vegetative cover, aboveground biomass, % shrubs, surface roughness coefficient, texture of sediment, % bare ground, and vegetative spp. and density.

Potential to Sequester SOC in Grazing Land by Restoring Eroded Soils

The total area of highly erodible grazing land is 123 Mha (Table 9.6). Controlling soil erosion and restoring eroded soils by improving biomass production may sequester SOC in soil at the rate of 100 to 200 kg/ha/yr. Thus the total potential to sequester SOC by restoring eroded soils is 12 to 25 MMTC/yr. This finite potential may be realized over a 25- to 50-year period.

Adopting effective erosion control measures also may reduce C emissions from eroded sediments redistributed over the landscape. Annual C emission to the atmosphere caused by erosion is estimated at 14 to 16 MMTC/yr. Therefore, adopting effective erosion control measures and restoring eroded grazing lands can sequester 26 to 41 MMTC/yr.

References

Baker, M.B., Jr. 1982. *Hydrologic regime of forested areas in the Beaver Creek Watershed*. USDA Forest Service, Rocky Mtn. For and Range Exp. Sta. Fort Collins, CO.

Bauer, A., C.V. Cole, and A.L. Black. 1987. Soil property comparisons in virgin grasslands between grazed and nongrazed management systems. *Soil Sci. Soc. Am. J.* 51:176-182.

Berg, W.A. 1988. Soil nitrogen accumulation in fertilized pastures of the southern Plains. *J. Range Manage.* 41:22-25.

Branson, F.A., R.F. Miller, and R.S. McQueen. 1962. Effects of contour furrow grazing intensities, and soils on infiltration rates, soil moisture and vegetation near Fort Peck, Montana. *J. Range Manage.* 15:151-158.

Branson, F.A., and J.R. Owen. 1970. Plant cover, runoff and sediment yield relationship on Mancos shale in western Colorado. *Water Resour. Res.* 6:783-790.

Branson, F.A., G.F. Gifford, K.G. Renard, and R.F. Hadley. 1981. *Rangeland Hydrology. Range Science Series #1*. Soc. Range. Manage. Denver, CO.

De Bano, L.F., L.D. Mann, and D.A. Hamilton. 1970. Translocation of hydrophobic substances in soil by burning. *Soil Sci. Soc. Am. Proc.* 34:130-133.

De Bano, L.F., S.M. Savage, and D.A. Hamilton. 1976. The transfer of heat and hydrophobic substances during burning. *Soil Sci. Soc. Am. J.* 40:779-782.

Dormaar, J.F., and W.D. Willms. 1990. Effect of grazing and cultivation on some chemical properties of soils in the mixed prairie. *J. Range Manage.* 43:456-460.

Dormaar, J.F., S. Smoliak, and W.D. Willms. 1990. Distribution of nitrogen fractions in grazed and ungrazed grassland. Ah horizons. *J. Range Manage.* 43:6-9.

Dunford, E.G. 1949. *Relation of grazing to runoff and erosion on bunchgrass ranges.* USDA-FS. Note RM-7.

Duvall, V.L., and L.E. Linnartz. 1967. Influence of grazing and fire on vegetation and soil of longleaf pine-bluestem range. *J. Range Manage.* 20:241-247.

Emmerich, W.E., and J.R. Cox. 1992. Hydrologic characteristics immediately after seasonal burning on introduced and native grasslands. *J. Range Manage.* 45:476-479.

FAO 1982. *Production Yearbook.* FAO. Rome, Italy.

FAO 1995. *Production Yearbook.* FAO. Rome, Italy.

Frank, A.B., D.L. Tanaka, L. Hoffman, and R.F. Follett. 1995. Soil carbon and nitrogen of Northern Great Plains grasslands as influenced by long-term grazing. *J. Range Manage.* 48:470-474.

Frasier, G.W., M.J. Trlica, W.C. Leininger, R.A. Pearce, and A. Fernald. 1998. Runoff from simulated rainfall in 2-montane riparian communities. *J. Range Manage.* 51:315-322.

Gray, L.J., G.L. Macpherson, J.K. Koelliker, and W.K. Dodds. 1996. Hydrology and aquatic ecology. *In* A.K. Knapp, J.M. Briggs, D.C. Hartnett, and S.L. Collins (eds), *Grassland Dynamics: Long-Term Ecological Research in Tallgrass Prairie*, LTER Publications Committee, New York, pp. 159-176.

Heady, H.F., and R.D. Child. 1994. *Rangeland Ecology and Management.* Westview Press. Boulder, CO.

Heede, B.H. 1970. Morphology of gullies in the Colorado Rocky Mountains. *Bull. Int. Assoc. Sci. Hydrol.* XV 2:79-89.

Henzell, E.F., I.F. Fergus, and A.E. Martin. 1967. Accretion studies of soil organic matter. *J. Aust. Inst. Agric. Sci.* 33:35-37.

Hester, J.W., T.L. Thurow, and C.A. Taylor, Jr. 1997. Hydrologic characteristics of vegetation types as affected by prescribed burning. *J. Range Manage.* 50:199-204.

Holechek, J.L., R.D. Pieper, and C.H. Herbel. 1989. *Range Management: Principles and Practices.* Regents/Prentice Hall. Englewood Cliffs, NJ.

Lal, R., J.M. Kimble, R.F. Follett, and C.V. Cole. 1998. *The Potential of U.S. Cropland to Sequester C and Mitigate the Greenhouse Effect.* Ann Arbor Press. Chelsea, MI.

Lane, L.J., and M.A. Nearing (eds). 1989. *USDA-Water Erosion Prediction Project: Hill slope Profile Version. NSERL Rep. 2.* Nat. Soil Erosion Res. Lab., USDA-ARS. W. Lafayette, IN.

Manley, J.T., G.E. Schuman, J.D. Reeder, and R.H. Hart. 1995. Rangeland soil C and N responses to grazing. *J. Soil Water Conserv.* 50:294-298.

McCool, D.K., L.C. Brown, G.R. Foster, C.K. Mutchler, and L.D. Meyer. 1987. Revised slope steepness factors for the USLE. *Trans. Am. Soc. Agric. Eng.* 30:1387-1396.

Milchanus, D.G., and W.K. Lauenroth. 1993. Quantitative effects of grazing on vegetation and soils over a global range of environments. *Ecol. Monogr.* 63:327-366.

Osborn, H.B., and J.R. Simanton. 1990. Hydrologic modeling of a treated rangeland watershed. *J. Range Manage.* 43:474-481.

Pearce, R.A., M.J. Trlica, W.C. Leininger, D.E. Mergen, and G. Frasier. 1998a. Sediment movement through riparian vegetation under simulated rainfall and overland flow. *J. Range Manage.* 51:301-308.

Pearce, R.A., G.W. Frasier, M.J. Trlica, W.C. Leininger, J.D. Stednick, and J.L. Smith. 1998b. Sediment filtration in a montane riparian zone under simulated rainfall. *J. Range Manage.* 51:309-314.

Pluhar, J.J., R.W. Knight, and R.K. Heitschmidt. 1986. Infiltration rates and sediment production as influenced by grazing systems in the Texas rolling plains. *J. Range Manage.* 40:240-243.

Ransom, M.D., C.W. Rice, T.C. Todd, and W.A. Wehmueller. 1996. Soils and soil biota. *In* A.K. Knapp, J.M. Briggs, D.C. Hartnett, and S.L. Collins (eds), *Grassland Dynamics: Long-Term Ecological Research in Tallgrass Prairie,* LTER Publications Committee, New York, pp. 48-66.

Rauzi, F.C., and C.L. Hanson. 1966. Water intake and runoff as affected by intensity of grazing. *J. Range Manage.* 19:351-356.

Renner, F.G. 1936. *Conditions influencing erosion on the Boise River Watershed.* USDA Tech. Bull. 528. Washington, D.C.

Rovey, E.W., D.A. Woolhiser, and R.E. Smith. 1977. *A distributed kinematic model of upland watersheds. Hydrology Pap. 93.* Colorado State Univ., Fort Collins, CO.

Savbi, M.R., W-J. Rawls, and R.W. Knight. 1995. Water erosion prediction project (WEPP) rangeland hydrology component evaluation on a Texas range site. *J. Range Manage.* 48:535-541.

Simanton, J.R., M.A. Weltz, and H.D. Larsen. 1991. Rangeland experiments to parametrize the water erosion prediction project model: Vegetation and canopy cover effects. *J. Range Manage.* 44:276-282.

Stoeckler, J.H. 1959. *Trampling by livestock drastically reduces infiltration rate of soil in oak and pine woods in southwestern Wisconsin.* USDA For. Serv. Lake State Forest. Exp. Sta. Tech. Note 556.

Thurow, T.L., W.H. Blackburn, and C.A. Taylor, Jr. 1986. Hydrologic characteristics of vegetation types as affected by livestock grazing systems, Edwards Plateau, Texas. *J. Range Manage.* 39:505-508.

Thurow, T.L., W.H. Blackburn, and C.A Taylor, Jr. 1988. Infiltration and inter-rill erosion responses to selected livestock grazing strategies, Edwards Plateau, Texas. *J. Range Manage.* 41:296-302.

USDA 1987. *The second RCA appraisal. Soil, water, and related resources on nonfederal land in the United States.* USDA. Washington, DC.

USDA 1997. *Statistical Abstracts of the United States.* Economics and Statistics Administration. Washington, D.C.

USDA-NRCS 1992. *National Resource Inventory, Data Base.* USDA. Washington, DC.

Warren, S.D. W.H. Blackburn, and C.A. Taylor, Jr. 1986a. Effect of season and stage of rotation cycle on hydrologic condition of rangeland under intensive rotation grazing. *J. Range Manage.* 39:486-491.

Warren, S.D., T.L. Thurow, W.H. Blackburn, and N.E. Garza. 1986b. The influence of livestock trampling under intensive rotation grazing on soil hydrologic conditions. *J. Range Manage.* 39:49-496.

Warren, S.D., W.H. Blackburn, and C.A. Taylor, Jr. 1986c. Soil hydrologic response to number of pastures and stocking density under intensive rotation grazing. *J. Range Manage.* 39:500-505.

Wilcox, B.P. 1994. Runoff and erosion in intercanopy zones of pinyon-juniper woodlands. *J. Range Manage.* 47:285-295.

Wilcox, B.P., and M.K. Wood. 1988. Hydrologic impact of sheep grazing on steep slopes in semi arid rangelands. *J. Range Manage.* 41:303-306.

Wilcox, B.P., M.K. Wood, and J.M. Tromble. 1988. Factors influencing infiltrability of semi-arid mountain slopes. *J. Range Manage.* 41:197-206.

Wilcox, B., and M.K. Wood. 1989. Factors influencing inter-rill erosion from semi-arid slopes in New Mexico. *J. Range Manage.* 42:66-70.

Wilcox, B.P., W.J. Rawls, D.L. Brakensiek, and J.R. Wright. 1990. Predicting runoff from rangeland catchments: a comparison of two models. *Water Resour. Res.* 26:2401-2410.

Wilcox, B.P., M. Sbaa, W.H. Blackburn, and J.H. Milligan. 1992. Runoff prediction from sagebrush rangelands using water erosion prediction project technology. *J. Range Manage.* 45:470-474.

Williams, J.D., and J.C. Buckhouse. 1991. Surface runoff plot design for use in watershed research. *J. Range Manage.* 44:411-412.

Wood, J.C., W.H. Blackburn, H.A. Pearson, and T.K. Hunter. 1989. Infiltration and runoff water quality response to silvicultural and grazing treatments on a long-leaf pine forest. *J. Range Manage.* 42:378-381.

Wright, H.A. 1974. Range burning. *J. Range Manage.* 27:5-11.

Wright, H.A., F.M. Churchill, and W.C. Stevens. 1976. Effects of prescribed burning on sediment, water yield and water quality from dozed juniper lands in central Texas. *J. Range Manage.* 29:294-298.

Wright, H.A., F.M. Churchill, and W.C. Stevens. 1982. Soil loss, runoff and water quality of seeded and unseeded steep watersheds following prescribed burning. *J. Range Manage.* 35:382-385.

Wright, J.R., and C.L. Hanson. 1991. Use of stochastically generated weather records with rangeland simulation models. *J. Range Manage.* 44:282-285.

Zingg, A.W. 1940. Degree and length of land slope as it affects soil loss in runoff. *Agric. Eng.* 21:59-64.

Zobeck, T.M., and D.W. Fryrear. 1986. Chemical and physical characteristics of wind-blown sediment. II. Chemical characteristics and total soil and nutrient discharge. *Trans. ASAE* 29:1037-1041.

CHAPTER **10**

The Physical Quality of Soil on Grazing Lands and Its Effects on Sequestering Carbon

R. Lal[1]

Introduction

This chapter reviews how managing grazing lands (rangelands and pastures) affects soil's physical quality, with particular reference to soil organic C (SOC) content and its effect on soil. Grazing and cultivation also influence the soil chemical properties of rangelands (Dormaar and Willums, 1989), but this chapter focuses only on soil's physical quality.

Seasonal alterations in soil structure are mostly due to differences in moisture and temperature regimes and to changes in organic cover on the soil surface. Soil aggregation increases with the increase in SOC content, and the latter varies with the quantity and quality of biomass on the soil surface and the proportion of bare areas exposed to climatic elements. The quantity (mass per unit area and the % of ground cover) and quality (C:N ratio and the lignin content) of the biomass depend on management. Important management factors that influence soil structure and the C pool in the soil of grazing land are the grazing system, biomass burning, soil and water conservation, soil fertility management, and predominant forage species.

Basic Definitions

Soil quality relates to "inherent attributes of soils that are inferred from soil characteristics" (SSSA, 1987) and refers to "the capacity of a soil to function

[1] Prof. of Soil Science, School of Natural Resources, The Ohio State University, Columbus, Ohio, e-mail lal.1@osu.edu.

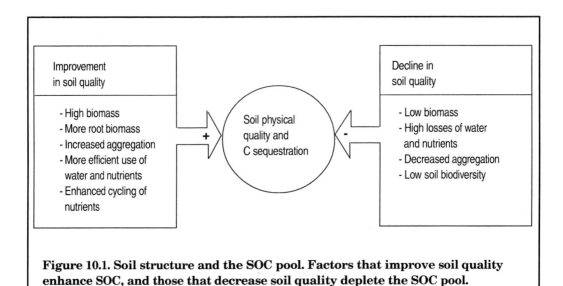

Figure 10.1. Soil structure and the SOC pool. Factors that improve soil quality enhance SOC, and those that decrease soil quality deplete the SOC pool.

within ecosystem boundaries to sustain biological productivity, maintain environmental quality, and promote plant and animal health" (Doran and Parkin, 1994). Soil structure also relates to soil resilience, the "ability of a soil to restore its productivity and environmental regulatory capacity following a perturbation" (Lal, 1994).

Soil's physical quality relates to the *status* of those physical properties that influence biomass productivity and the environment. Indicators of soil's physical quality include bulk density, soil structure, infiltration capacity, pore size distribution, and water retention characteristics. SOC content affects all these index properties, especially soil structure. The latter is a complex property, often difficult to define, and involves a wide range of attributes including aggregation, susceptibility to crusting and surface seal formation, and total porosity and pore size distribution. In addition to productivity, soil structure also affects C sequestration (Fig. 10.1) in soil, both directly and indirectly.

Direct effects of soil structure include those on biomass production, conversion of biomass to humus, and incorporation of C as humic substances and organo-mineral complexes within stable aggregates. Indirect effects of structure on C sequestration are due to changes in other soil properties that influence biomass production, e.g., nutrient cycling, water and energy balances, water and nutrient use efficiencies, and diversity of soil fauna's activity and species (i.e., earthworms, termites).

Soil structure refers to the form, stability, resilience, and vulnerability of soil aggregates and voids (Kay, 1996). *Structural form* refers to properties that describe the arrangement of void and solid space, e.g., porosity, pore size distribution, and aggregate/ped shape and size. *Structural stability* refers to soil's ability to maintain its arrangement of void and solid space against natural or anthropogenic disruptive forces. *Structural resilience* refers to soil's ability to recover its form following a perturbation, while *structural vulnerability* is its inability to recover. Considering the complexity of structural attributes, and scale measurement ranging from microscopic to field, it is difficult to identify a structural index that is climate-sensitive and universally applicable for all soils, land uses, and ecological environments.

Soil structure is a dynamic property, with large spatial and temporal variations. While soil structure directly and indirectly affects C sequestration as just mentioned, soil organic matter (SOM) content and soil moisture and temperature regimes strongly influence soil structure, and changes in SOM content affect soil structure directly and indirectly.

Direct effects of SOM content are those related to the development of aggregates through the formation of organo-mineral complexes, leading to changes in:

1. total aggregation
2. aggregate size distribution
3. aggregate stability against disruptive forces of wind and water
4. aggregate strength and resistance to compacting forces of vehicular traffic or farm equipment.

Indirect effects of SOM content are those related to soil biodiversity and to the activity and species diversity of soil fauna. Earthworms, termites, and other macro- and microfauna have drastic influence on soil structure. Activities of soil fauna related to structural properties are:

1. burrowing
2. mixing and soil turnover
3. biomass decomposition
4. fecal ejecta and body fluids that bind and enrich soil particles.

Soil fauna thrive best under moist/humid environments within a temperature range of 20 to 25°C.

Indicators of the Physical Quality of Soil

Assessment of soil's physical quality should be quantitative. It involves evaluating numerous properties, including bulk density, crusting, aggregation, wettability, infiltration rate, moisture retention characteristics, and pore size distribution.

The key properties that determine soil's physical quality differ among soils and ecoregions. We need to identify these properties and formulate them in an appropriate index which we then can use to indicate rangeland's productivity and its capacity to moderate the environment.

Soil bulk density

Weight per unit volume is an important indicator of soil's physical quality. Data on soil bulk density are needed for evaluating elemental pools (e.g., C, N, P, S, etc.) and water balances in soil in relation to natural (temporal and spatial) and anthropogenic (management) perturbations.

Bulk density of the surface (0 to 50 mm) layer, important in determining the surface hydrologic process, is difficult to determine (Frasier and Keiser, 1993). It is thus important to standardize methods for specific soil types, e.g., coarsely textured soils, gravelly soils, stony soils, organic soils, etc.

Bulk density of the subsoil (150 to 250 mm) is important to the root system's development, incorporation of biomass into the deeper layers, and activity of soil fauna (e.g., termites, earthworms). Because of large temporal and spatial variations, it is difficult to obtain a reliable index of soil's bulk density that represents the entire field, landscape, or soilscape.

Crusting

Crust or *surface seal* refers to a thin, laminated, hard, and relatively impermeable soil layer that restricts movement of water and air into the soil. Soils prone to crusting have a low infiltration capacity (i_c) and a high runoff rate. A soil crust may have high strength rather than high bulk density. Brakensiek and Rawls (1983) observed that a stable soil crust can reduce the saturated hydraulic conductivity (K_s) by 75 to 90% of that observed on noncrusted soil.

Susceptibility to crusting is high in soils containing high silt content, low SOC content, and a predominance of low activity clays. Management practices that accentuate crusting include heavy grazing and burning of biomass. Trampling by cattle, although it seems to break the crust, accentuates the adverse effects of crusting because it compacts the subsoil.

Aggregation

The proportion of stable aggregates in the soil is an important indicator of soil's physical quality. *Aggregation* refers to the arrangement, size distribution, and stability of secondary particles. Secondary particles are formed through the cementing effect of organo-mineral complexes. *Aggregate stability* may refer to the action of water (wet sieving) or wind (dry sieving) relevant to susceptibility to erosion by water and wind. Aggregate stability also influences susceptibility to compaction, crusting, anaerobiosis, and drought.

Wettability

Wettability refers to ease of wetting and is governed by the contact angle of the water at the soil/air interface. Some soils are more wettable than others, and non-wettability leads to a low infiltration rate and high risks of runoff and erosion.

Non-wettable rangeland soils are numerous, especially in western states (De-Bano and Krammer, 1966; DeBano et al., 1967; DeBano, 1981). Their low wettability or hydrophobicity may be due to microorganisms or chemical substances. The presence of algal or microbial crust on soils of arid and semiarid regions may accentuate hydrophobicity. Fire intensifies non-wettability, especially when the soil temperature reaches 200 to 425°C, even for a short duration.

Coatings of organic materials that condense on the surfaces of soil particles cause water repellency. These materials are mainly allophatic hydrocarbons that cause severe repellency problems in coarsely textured soils with a low surface area (Scholl, 1975), and they increase the contact angle of the soil-water interface after burning. DeBano et al. (1970) observed that intense heat during burning can lead to downward diffusion of volatile organic materials, exacerbating the non-wettability of subsoil.

Infiltration capacity

Infiltration is a key hydrologic process that affects the water balance, soil water reserve, runoff, and erosion. The *infiltration capacity* or i_c refers to the equilibrium rate of water entry into the soil at the soil-air interface and designates the infiltration flux resulting when water at atmospheric pressure is made freely available at the soil surface (Hillel, 1971). The i_c is an important soil physical property because it integrates the effect of numerous other properties, e.g., texture, bulk density, aggregation, wettability, pore size distribution, and pore continuity. In fact, i_c is an important index of soil's physical quality.

Because of its importance to all hydrological processes, the infiltration characteristics of grazing land soils have been studied extensively. Several researchers (Hutton and Gifford, 1988; Rawls et al., 1989; Kidwell et al., 1997) have used the Green-Ampt infiltration model to predict the infiltration capacity of western rangeland soils. Equation 1 describes the Green and Ampt (1911) model:

$$i = K_s(1 + f_a \, y \, / \, I \,) \qquad \text{(Equation 1)}$$

where i is infiltration rate (cm/hr), K_s is hydraulic conductivity of the wetted zone, f_a is aeration porosity (%), y is suction parameter (cm), and I is cumulative infiltration. Equation 1 shows that the infiltration rate increases with an increase in K_s and decreases with an increase in I (or time after the infiltration began).

Naeth et al. (1991a) used the Kostiakov (1932) model to describe the infiltration process (Eq. 2):

$$i = at^b \qquad\qquad \text{(Equation 2)}$$

where i is infiltration rate, t is time, and a and b are empirical constants. Constants a and b are influenced by management, e.g., grazing intensity, burning, forage species, etc. Naeth et al. (1991b) observed that the parameter "a" declined with grazing's intensity and earliness in the growing season.

The i_c approaches K_s of the wetted layer at infinite (long) time. Several experiments have shown that K_s decreases with short-duration heavy grazing. Dormaar et al. (1997) observed that mean K_s of the 0 to 3 cm layer was 5.2 ± 0.7 cm/hr for ungrazed fescue grassland, compared with 3.2 ± 1.5 cm /hr for a short-duration grazed treatment. Similar values for 3- to 6-cm depth were 3.6 ± 0.9 cm/hr for an ungrazed control versus 3.0 ± 0.3 cm/hr for the grazed treatment. Soil compaction and a decrease in macroporosity caused a decline in soil structure. All other factors remaining the same, K_s depends on land use and ground cover. Structural degradation of the bare soil leads to reduction in the K_s.

Moisture retention characteristics

Knowledge of the quantitative relationship between soil moisture content and soil moisture potential (energy status) or the pF curves is needed to assess these factors and their alterations with management: the available water holding capacity; total porosity; pore size distribution; and changes in pore size distribution (soil structure). For soils of the same textural class, pF curves are influenced strongly by changes in aggregation and SOC content.

Those grazing regimes and forage species that enhance SOC content also influence pF curves (Naeth et al., 1991a, b, c). Soil water potential also is influenced by precipitation, weather parameters affecting evapotranspiration, and surface soil characteristics (Herbel and Gibbens, 1989). Determining pF curves on undisturbed soils or under *in situ* conditions is a challenging task. An appropriate strategy is to determine pF curves *in situ* under field conditions.

A relevant index of the available water reserve in an arid environment is the least limiting water range (LLWR) (de Silva et al., 1994). It is defined as the water content at which aeration, water potential, and resistance to penetration reach values that are critical to or limiting to plant growth (de Silva et al., 1994). The LLWR is a modification of the original concept of the nonlimiting water range

Letey (1985) proposed. The LLWR can be related to other soil physical properties through development of appropriate pedotransfer functions (Eq. 3).

LLWR = f(SOC, WSA, clay, bulk density)   (Equation 3)

Physical Quality Index

Considering all the indicators of soil's physical quality, it may be useful to establish critical limits or threshold values of key soil properties, beyond which rangeland productivity declines. Examples of threshold values are:

Soil property	Threshold value
Clay content (%)	10-15
Bulk density (Mg/m^3)	1.3-1.5
Crust strength (kg/cm^2)	2.0-3.0
Infiltrability (cm/hr)	0.2-0.5
Aggregation (WSA %)	10-20
Available water capacity	10-15

All these properties strongly correlate with the SOC content. Quantifying them and establishing their temporal trends in relation to management can provide important guidelines for managing soil's physical quality.

It is also important to establish a physical quality index based on these properties. Equation 4 shows some examples of such indices, which are specific to soils and ecoregions.

S_q = f(clay, SOC)

S_q = f(bulk density, i_c)   (Equation 4)

S_q = f(WSA, i_c, AWC)

S_q = f(i_c, AWC, SOC)

Establishing these pedotransfer functions provides an important management tool.

Infiltration Capacity

Soil's infiltration capacity governs water losses due to runoff. Management of rangeland that influences i_c can regulate runoff (Tromble, 1976). The size distribution, continuity, and stability of pores or voids determine the i_c. In turn, soil properties and managerially induced changes in the organic cover influence these structural attributes. The grazing system, biomass burning, and seasonal i_c influence the organic cover, due to differences in soil moisture and temperature regimes, because drier vegetation, soils, and other possible factors in some ecosystems (for example, in the U.S. Southwest) are more vulnerable than other ecoregions to the effects of grazing, burning, or other practices. Management moderates the effects of soil properties on i_c.

Grazing and infiltration capacity

Grazing affects i_c by decreasing total organic cover on the soil surface, and by exacerbating soil compaction through the trampling effect. Experiments conducted by Pluhar et al. (1987) at the Texas Experimental Ranch near Throckmorton, Texas, showed that i_c increased as vegetative standing crop and cover increased, soil bulk density decreased, and SOM and stability increased (Tables 10.1 and 10.2). Grazing caused a significant decline in i_c by reducing vegetative cover and standing crop and increasing the bare ground. At Edwards Plateau, Texas, Thurow et al. (1988) observed that the infiltration rate strongly correlated with total organic cover (Eq. 5).

$$i \text{ (mm/hr)} = -12.4 + 0.37 \text{ (\% organic cover)}, R^2 = 0.96 \qquad \text{(Equation 5)}$$

They also observed that i_c decreased and inter-rill erosion increased in the heavily stocked pastures. Further, infiltration rates were seasonally cyclic in the heavily stocked pastures, with no significant seasonal trend for the medium stocking treatment.

In Alice, Texas, Weltz and Blackburn (1995) observed that the K_s was least for the bare soil and highest under the shrub cluster. The relative K_s was in the order 1:1·14:1·91 for horizon 1; 1:1·11:1·58 for horizon 2; and 1:1·06:1·56 for horizon 3 (Table 10.3) for bare soil, grass interspace, and shrub site, respectively. The effects of ground cover on K_s were significant in all of the six horizons studied. For the sixth horizon, K_s in the shrub site was 3 times that of the bare soil.

Burning and infiltration capacity

In addition to decreasing organic cover, burning also influences soil properties that affect water retention and transmission characteristics. Effects of burning on soil structure are complex and influenced by numerous interacting factors. The extent of biomass removed and the surface area exposed depend on the intensity/severity of burning. The magnitude of changes in soil properties depends on the soil temperature during and after burning. A bare soil without vegetative cover and with a dark color may have greater diurnal fluctuations in soil temperature than a covered soil surface.

Experiments conducted at Edwards Plateau, Texas, by Hester et al. (1997) showed changes in i_c of the soil due to burning (Table 10.4). Fire reduced the infiltration rate on the oak and juniper vegetation types. Burning made the soil of the

Table 10.1. The effect of grazing treatments on structural and hydrological properties of soils in Texas (modified from Pluhar et al., 1986).

Parameters	Moderately Stocked Continuous Grazing	Heavily Stocked Rotational Grazing			
		14-Paddock		42-Paddock	
		A	B	A	B
Vegetation cover (%)	92a	80b	74c	88a	77bc
Soil bulk density (MT/m³)	1.1a	1.2a	1.1a	1.2a	1.2a
Soil organic matter (%)	6.5a	6.3ab	4.4c	6.7a	5.7b
Aggregate stability (%)	72a	68a	65a	68a	70a
Infiltration rate (mm/hr)	89a	87ab	77b	85ab	78ab

A = before grazing; B = after grazing. Figures in the row followed by the same letter are statistically similar.

Table 10.2. The effects of grazing treatments on the soil infiltration rate in Throckmorton, Texas (modified from Pluhar et al., 1987).

Grazing Treatment	Mid-grass			Short Grass		
	Before	After	After:Before	Before	After	After:Before
	——mm/hr——		---------------			
14-Paddock rotation	95a	64b	0.67	75a	55b	0.73
42-Paddock rotation	81ab	85a	1.05	86a	79a	0.92
4-Pasture, 3-herd deferred rotation	86a	81ab	0.94	80a	68ab	0.85
Moderately stocked continuous	89a	85a	0.96			
Ungrazed exclosure	88a	—	—	—	—	

Figures in the column followed by the same letter are statistically similar.

Table 10.3. Vegetative cover's effects on soil hydraulic conductivity at La Capita Research Area, Alice, Texas (Weltz and Blackburn, 1995).

Soil Horizon	Bare Soil	Grass Interspace	Shrub Cluster
	------------------------	——mm/s——	------------------------
1	0.58	0.67	1.12
2	0.32	0.35	0.50
3	0.27	0.28	0.42
4	0.20	0.25	0.37
5	0.07	0.12	0.22
6	0.07	0.12	0.20

Table 10.4. The effect of rangeland burning on soil properties at Edwards Plateau, Texas (Hester et al., 1997).

Soil Parameters	Oak			Juniper			Bunch Grass			Short Grass		
	B	U	B:U	B	U	B:U	B	U	B:U	B	U	B:U
Infiltration rate (mm/hr)	129de	202a	0.64	162bc	183ab	0.89	110e	146cd	0.75	76f	105ef	0.72
Soil organic matter (%)	12.3a	11.9ab	1.03	10.6b	10.4b	1.02	5.7c	6.9c	0.83	5.3c	5.9cd	0.90
Aggregate stability (%)	81.2a	84.5a	0.96	86.5a	81.5a	1.06	66.0b	67.0b	0.99	62.3b	67.1b	0.93
Bulk density 0-30 mm (MT/m³)	0.64b	0.51c	1.25	0.70b	0.76b	0.92	0.97a	1.00a	0.97	0.95a	0.97a	0.98
Bulk density 50-80 mm (MT/m³)	0.88ab	0.66c	1.33	0.81b	0.97a	0.84	0.99a	0.96a	1.03	1.01a	0.98a	1.04

Figures in the row followed by the same letter are statistically similar. B = Burned. U = Unburned.

oak sites hydrophobic. Thus the soil had a low infiltration rate, despite having good structural properties. Consequently, infiltration rate poorly correlated with variables that reflect soil structure on the burned sites (soil organic matter r = 0.58, aggregate stability r = 0.57, bulk density at 0 to 30 mm r = 0.40).

Seasonal effects on infiltration capacity

Vegetative cover has a major effect on i_c (Wilcox et al., 1988); and seasonal differences in vegetative cover, due to differences in soil temperature and moisture regimes, influence i_c and other soil structural attributes. Thurow et al. (1988) conducted time series analyses and observed a seasonally cyclic pattern in the i_c of the heavily stocked treatment. The infiltration rate was low during the drought period, due to the low vegetation/organic cover.

Soil properties and infiltration capacity

The presence of skeletal or coarse materials within and on the soil surface can influence i_c. Wilcox et al. (1988) observed that i_c negatively correlates with rock cover. While protecting the soil from raindrop impact, development of a stone carapace tends to inhibit infiltration and increase runoff (Thornes, 1980).

Tromble et al. (1974) also found i_c to correlate negatively with gravel (<10 mm) cover in the semiarid rangeland of Arizona. The i_c of stone pavement is low because of the lack of interconnected macropores in the surface horizon. Because of its strong impact on the infiltration rate (i_c), Rawls et al. (1988) developed empirical equations to predict skeletal fraction, as % by weight of gravel or soil rock in the top 7.6 cm, from ground cover measurements made outside the plant canopy.

In addition to stone, tree canopy also can influence the i_c of rangelands and pastures (Frost and Edinger, 1991).

The Physical Quality of Soil and C Sequestration

Data from several experiments in North America show that light and moderate grazing has no effect on SOC content compared with ungrazed exclosures (Dormaar et al., 1989, 1990, 1994, 1997; Naeth et al., 1991c; Frank et al., 1995; Berg et al., 1996). In some cases, even intense/heavy grazing caused no significant changes in the SOC pool. Berg et al. (1997) observed that 50 years of moderate grazing by cattle had no effect on the SOC content of the surface 5 cm of a sandy soil.

However, grazing systems can influence soil water regime (Sturges, 1992), vegetative species (Dormaar et al., 1997), and ground cover (Clary, 1995; Watters et al., 1995). Therefore, on a long-term basis, heavy grazing may have adverse impacts on SOC content, especially for soils of low inherent fertility. For fertile soils where heavy grazing may induce change of species, however, an increase in the SOC pool over time is possible (Frank et al., 1995).

Schacht et al. (1995) studied the influence of mowing and burning treatments on soil's quality in Nebraska. They assessed soil quality by measuring i_c, bulk density, pH, electrical conductivity, and SOC and N contents. The data showed that changes in physical and chemical properties of grassland soils in response to burning were minor and short-lived. Annual or cyclic (4 years) burning did not affect the SOC content. The principal effect of repeated burning and mowing, however, was to reduce the i_c.

Evaluating the effects of land use on soil quality requires considering two specific issues. Converting natural ecosystems to managed pastures can deplete the SOC pool. The magnitude of reduction by conversion to pasture may be less than that to arable land use. Converting abandoned cropland to native rangeland may improve SOC content. Dormaar et al. (1990) observed that recovery of abandoned cropland, in order to form the type of OM that occurs in native chernozemic soils, may take 75 to 150 years.

Potential to Sequester C by Restoring Degraded Grazing Lands

Adopting recommended grazing systems and restoring degraded grazing land can enhance SOC content. The data in Table 10.5 show that heavily stocked pastures lost 4.8 MTC/ha/30 cm for 14-paddock treatment and 2.1 MTC/ha/30 cm for 42-paddock treatment over a 1-year period (MTC = metric ton of C). Compared with the ungrazed exclosure, the moderately stocked and continuously grazed

Table 10.5. Grazing's effects on the SOC pool in the top 30-cm layer over a 1-year period (recalculated from Pluhar et al., 1986).

Treatment	SOC Pool	D SOC
	————MTC/ha————	
Moderately stocked continuous grazing	12.62	—
Heavily stocked		
(i) 14-Paddock before grazing	13.35	—
(ii) 14-Paddock after grazing	8.55	−4.80
(iii) 42-Paddock before grazing	14.19	—
(iv) 42-Paddock after grazing	12.06	−2.13

treatment lost about 1.2 MTC/ha/30 cm. Therefore, the loss of C due to grazing ranged from 1.2 to 4.8 MTC/ha/yr.

These calculations are based on the assumption that the loss of SOC occurred in the top 30-cm layer. However, if the reduction of SOC were confined to the top 10-cm layer, the loss of C due to grazing ranged from 0.4 to 1.6 MTC/ha/yr. Therefore, adoption of moderate continuous rather than heavy grazing can lead to SOC sequestration at the rate of 0.18 to 1.36 MTC/ha/yr.

The data in Table 10.6 from Alberta show the SOC loss of 1.1 to 2.2 MTC/ha due to grazing. In contrast, the data of Frank et al. (1995) from Mandan, ND, showed little effect of grazing on SOC content (Table 10.7). Compared to the exclosure, the moderately grazed pasture contained less SOC. Heavy grazing did not reduce SOC compared to the exclosure, and it increased SOC compared to the moderately grazed treatment, probably as a result of an increase in species composition to blue grama (*Bouteloua gracilis* [HBK] Lag. ex Griffiths), a species with a dense, shallow root system. Berg et al. (1997) also reported little effect of grazing on the SOC pool of grazing land in Oklahoma.

The data in Table 10.8 show the impact of burning on the SOC pool in the top 8-cm layer of a soil in the Edwards Plateau, Texas. Burning increased the SOC content under oak, probably due to the addition of C in ash and as charcoal (black C). In contrast, burning in other treatments caused a drastic decline in SOC content — 2.7 MTC/ha in short grass, 5.2 MTC/ha in juniper, and 5.5 MTC/ha in the bunchgrass treatment. These data show that biomass burning can lead to severe depletion of the SOC pool, and the magnitude of depletion depends on the type of vegetation, intensity of burning, soil type, and other conditions at the time of burning.

Experiments conducted by Schacht et al. (1996) showed that burning decreased SOC content at site 1. Annual consumption of litter by fires left the soil

Table 10.6. Effect of short-term grazing on the SOC pool in the top 6 cm of a Chernozemic soil in Macleod, Alberta (recalculated from Dormaar et al., 1989).

Date	Ungrazed	Grazed	SOC
	----------------MTC/ha----------		
7 May, 1985	24.4	26.4	+2.0
15 Oct., 1985	22.8	20.6	−2.2
30 April, 1986	24.4	23.2	−1.2
30 Sept., 1986	26.5	25.4	−1.1

Table 10.7. Effect of grazing intensity on the SOC pool of a prairie soil in Mandan, ND (recalculated from Frank et al., 1995).

Depth (cm)	Exclosure	Moderately Grazed	Heavily Grazed
	--------------------MTC/ha--------------------		
0-30.4	7.2a	6.4b	7.4a
0-106.7	14.1a	11.7b	14.0a

surface relatively bare and removed a significant amount of organic matter that might otherwise have been incorporated into the soil. The mean (average of all burning treatments) SOC content of all burning treatments was lower than that of the unburned control by 2.2 MTC/ha for site 1 (Table 10.9). An annual burn in the fall reduced the SOC content at both sites, with a decrease of 1.5 MTC/ha for site 1 and 3.0 MTC/ha for site 2.

The data in Table 10.9 show differences among treatments, and among sites for the same treatment. For site 1, the SOC pool was the maximum for the 4-year burn in summer or fall, medium for the 4-year burn in the annual burn treatment during spring, and least for the 4-year burn in summer. Therefore, the recommended frequency and season of burning depend on the site characteristics (e.g., rainfall, temperature, vegetation, biomass, drainage, and soil characteristics). Data from long-term experiments are needed to develop site-specific recommendations.

The data in Table 10.10 show the effect of mowing on SOC content. Mowing without raking caused an increase in SOC content. However, mowing with raking caused a drastic decrease in the SOC content.

The calculations in Tables 10.5 through 10.10 show the potential to sequester C in rangeland soils by adopting an improved grazing system and eliminating or

Table 10.8. Burning's effects on the SOC pool in the 8-cm layer (calculated from Hester et al., 1997).

Treatment	Before Burning	After Burning	DSOC
	MTC/ha		
Oak	33.0	44.0	+11.0
Juniper	42.6	37.4	−5.2
Bunchgrass	31.8	26.3	−5.5
Short grass	27.1	24.4	−2.7

Table 10.9. The mean SOC pool in the top 30-cm layer of a silt loam grassland in Nebraska (recalculated from Schacht et al., 1996).

Treatment	Site 1	Site 2
	MTC/ha	
Control	59.4 ± 9.8	59.8 ± 4.8
Annual burn, fall	57.9 ± 4.5	56.8 ± 7.1
Annual burn, spring	59.2 ± 6.5	68.4 ± 3.9
Annual burn, summer	51.4 ± 9.1	60.9 ± 4.1
4-year burn, fall	63.5 ± 6.5	65.1 ± 6.9
4-year burn, spring	47.8 ± 9.1	64.6 ± 2.9
4-year burn, summer	63.4 ± 2.2	55.4 ± 4.5
Mean of all burn treatments	57.2 ± 6.4	61.8 ± 5.1

The data of SOM presented by authors was divided by 1.7 to convert to SOC content.

Table 10.10. The effect of mowing on the SOC pool in the top 30-cm layer of a silt loam grassland in Nebraska (recalculated from Schacht et al., 1996).

Treatment	SOC Pool MTC/ha
Control	59.4 ± 9.8
Mow, fall	66.5 ± 6.1
Mow, spring	64.9 ± 9.8
Mow, summer	63.4 ± 10.1
Mow/rake, fall	56.6 ± 13.9
Mow/rake, spring	44.8 ± 11.1
Mow/rake, summer	63.1 ± 11.1

decreasing the frequency of burning. Burning once in 4 years, in spring or fall, may be better than annual burning. The time of mowing/raking may be another factor that needs assessment for different soils and ecological conditions.

Conclusions

Soil's physical quality affects the SOC pool through its influence on biomass production, soil-water balance, nutrient dynamics, and soil fauna activity and species diversity. Especially in more arid/semiarid ecosystems, grazing land management (e.g., fire, grazing intensity, forage species, nutrient input) can have a strong influence on soil's physical quality and on the SOC pool.

The impact of stocking rates (e.g., heavy vs. light grazing) on the SOC pool depends on the soil, vegetation, and ecoregional characteristics. For soils of low inherent fertility, heavy grazing for a long period can adversely affect the SOC pool through a consequent decline in soil's physical quality, change in soil water balance, and depletion of soil fertility. If heavy grazing induces a species shift from cool season grasses (C_3) to warm season grasses (C_4), an increase in SOC may occur, especially in soils of high inherent fertility.

The effects of burning on the SOC pool depend on several factors, including the species, the biomass quantity, the intensity and duration of the fire, and the season of burning. In the long run, burning may lead to depletion of the SOC pool. Infrequent and controlled burning (once in 4 years) may enhance the SOC pool.

Mowing may enhance the SOC pool by increasing biomass production. The clippings, however, should be retained and recycled on the soil.

We need to develop an appropriate index of soil's physical quality. Such an index will involve key soil physical properties that may vary among soil types and ecoregions. Important soil physical properties for judicious management of rangeland include bulk density, infiltration capacity, clay content, and available water capacity.

References

Berg, W.A., J.A. Bradford, and P.L. Sims. 1997. Long-term nitrogen and vegetation change on sandhill rangeland. *J. Range Manage.* 50:482-486.

Brakensiek, D.L., and W.J. Rawls. 1983. Agricultural management effects on soil properties, Part II. Green-Ampt parameters for crusting soils. *Trans. ASAE* 26:1753-1757.

Clary, W.P. 1995. Vegetation and soil responses to grazing simulation on riparian meadows. *J. Range Manage.* 48:18-25.

da Silva, A.P., B.D. Kay, and E. Perfect. 1994. Characteristics of the least limiting water range of soils. *Soil Sci. Soc. Am. J.* 58:1775-1781.

DeBano, L.F. 1981. *Water repellent soils: a state-of-the-art.* USDA Pacific Southwest Forest and Range Exp. Sta. Gen. Tech. Rep. PSW-46.

DeBano, L.F., and J.S. Krammer. 1966. Water repellent soils and their relation to wild fire temperatures. *Int. Assoc. Sci. Hydrol. Bull.* XI Annee 2:14-19.

DeBano, L.F., J.F. Osborn, J.S. Krammes, and J. Letey, Jr. 1967. *Soil wettability and wetting agents – our current knowledge of the problem. USDA Forest Serv. Res. Paper.* Pacific Southwest Forest and Range Exp. Stn. Berkeley, CA. PSW-43.

DeBano, L.F., L.D. Mann, and D.A. Hamilton. 1970. Translocation of hydrophobic substances into soil by burning organic litter. *Soil Sci. Soc. Am. Proc.* 34:130-133.

Doran, J.W., and T.B. Parkin. 1994. Defining and assessing soil quality. *Defining Soil Quality for a Sustainable Environment. Special Publication 29.* SSA. Madison, WI, pp. 61-89.

Dormaar, J.F., S. Smoliak, and W.D. Willms. 1989. Vegetation and soil responses to short duration grazing on fescue grasslands. *J. Range Manage.* 42:252-256.

Dormaar, J.F., and W.D. Willms. 1990. Effect of grazing and cultivation on some chemical properties of soils in the mixed prairie. *J. Range Manage.* 43:456-460.

Dormaar, J.F., S. Smoliak, and W.D. Willms. 1990. Soil chemical properties during succession from abandoned cropland to native range. *J. Range Manage.* 43:260-264.

Dormaar, J.F., M.A. Naeth, W.D. Willms, and D.S. Chanasyk. 1994. Effect of native prairie, crested wheatgrass and Russian wild rye on soil chemical properties. *J. Range Manage.* 48:258-263.

Dormaar, J.F., B.W. Adams, and W.D. Willms. 1997. Impact of rotational grazing on mixed prairie soils and vegetation. *J. Range Manage.* 50:647-651.

Frank, A.B., D.L. Tanaka, L. Hofmann, and R.F. Follett. 1995. Soil carbon and nitrogen of Northern Great Plains grasslands as influenced by long-term grazing. *J. Range Manage.* 48:470-474.

Frasier, G.W., and J. Keiser. 1993. Thin layer measurement of soil bulk density. *J. Range Manage.* 46:91-93.

Frost, W.E., and S.B. Edinger. 1991. Effects of tree canopies on soil characteristics of annual rangeland. *J. Range Manage.* 44:286-288.

Green, W.H., and G.A. Ampt. 1911. Studies on soil physics: a flow of air and Water through soils. *J. Agric. Sci.* 4:1-24.

Herbel, C.H., and R.P. Gibbens. 1989. Matric potential of clay loam soils on arid rangelands in southern New Mexico. *J. Range Manage.* 42:386-392.

Hester, J.W., T.L. Thurow, and C.A. Tayor, Jr. 1997. Hydrologic characteristic of vegetation types as affected by prescribed burning. *J. Range Manage.* 50:199-204.

Hillel, D. 1971. Soil and Water: Physical principles and processes. *J. Range Manage.* 50:290-299.

Hutten, N.C., and G.F. Gifford. 1988. Using the Green and Ampt infiltration equation on native and plowed rangeland soils. *J. Range Manage.* 41:159-161.

Kay, B.D. 1996. Soil structure and organic carbon:a review. *Proc. Int'l Symposium "Carbon Sequestration in Soils," 22-26 July 1996.* The Ohio State University. Columbus, OH.

Kidwell, M.R., M.A. Weltz, and P. Guertin. 1997. Estimation of Green-Ampt effective hydraulic conductivity for rangeland. *J. Range Manage.* 50:290-299.

Kostiakov, A.N. 1932. On the dynamics of the coefficient of water percolation in soils and on the necessity of studying it from a dynamic point of view for purpose of amelioration. *Trans. 6ᵗʰ Comm. Int. Soc. Soil Sci. (Russian).* Part A:17-21.

Lal, R. 1994. Sustainable land use systems and soil resilience. *In* D.J. Greenland and I. Szabolcs (eds), *Soil Resilience and Sustainable Land Use*, CAB Int., Wallingford, UK, pp. 41-68.

Letey, J. 1985. Relationship between soil physical properties and crop production. *Adv. Soil Sci.* 1:277-294.

Naeth, M.A., A.W. Bailey, D.S. Chanasyk, and D.J. Pluth. 1991a. Water holding capacity of litter and soil organic matter in mixed prairie and fescue grassland ecosystems of Alberta. *J. Range Manage.* 44:13-17.

Naeth, M.A., D.S. Chanasyk, and A.W. Bailey. 1991b. Applicability of the Kostiakov equation to mixed prairie and fescue grassland of Alberta. *J. Range Manage.* 44:18-21.

Naeth, M.A., A.W. Bailey, D.J. Pluth, D.S. Chanasyk, and R.T. Hardin. 1991c. Grazing impact on litter and soil organic matter in mixed prairie and fescue grassland ecosystems of Alberta. *J. Range Manage.* 44:7-12.

Pluhar, J.J., R.W. Knight, and R.K. Heitschmidt. 1987. Infiltration rate and sediment production as influenced by grazing systems in the Texas Rolling Plains. *J. Range Manage.* 40:240-243.

Rawls, W.J., D.L. Brakensiek, J.R. Simanton, and C.L. Hanson. 1988. Prediction of soil cover and soil rock for rangeland infiltration. *J. Range Manage.* 41:307-308.

Rawls, W.J., D.L. Brakensiek, and M.R. Savabi. 1989. Infiltration parameter of rangeland soils. *J. Range Manage.* 42:139-142.

Schacht, W.H., J. Stubbendieck, T.B. Bragg, A.J. Smart, and J.W. Doran. 1995. Soil quality response of reestablished grasslands to mowing and burning. *J. Range Manage.* 49:458-463.

Scholl, D.G. 1975. Soil wettability and fire in Arizona Chaparral. *Soil Sci. Soc. Am. Proc.* 39:356-361.

Soil Science Society of America. 1987. *Glossary of Soil Science Terms.* Madison, WI.

Sturges, D.L. 1993. Soil-water and vegetation dynamics through 20 years after big sagebrush control. *J. Range Manage.* 46:161-175.

Thornes, J.B. 1980. Erosional process of running water and their spatial and temporal controls: a theoretical view point. *In* M.J. Kirby and R.P.C. Morgan (eds), *Soil Erosion*, J. Wiley & Sons, New York, pp. 129-182.

Thurow, T.L., W.H. Blackburn, and C.A. Taylor, Jr. 1988. Infiltration and inter-rill erosion responses to selected live-stock grazing strategies, Edwards Plateau, Texas. *J. Range Manage.* 41:296-302.

Tromble, J.M. 1976. Semi-arid rangeland treatment and surface runoff. *J. Range Manage.* 29:251-255.

Tromble, J.M., K.G. Renard, and A.P. Thatcher. 1974. Infiltration for 3 rangeland soil-vegetation complexes. *J. Range Manage.* 27:318-321.

Watters, S.E., M.A. Weltz, and E.L. Smith. 1995. Evaluation of a soil conservation rating system in southeastern Arizona. *J. Range Manage.* 49:277-284.

Weltz, M.A., and W.H. Blackburn. 1995. Water budget for south Texas rangelands. *J. Range Manage.* 48:45-52.

Wilcox, B.P., M.K. Wood, and J.M. Tromble. 1988. Factors influencing infiltrability of semi arid mountain slopes. *J. Range Manage.* 41:197-206.

CHAPTER **11**

The Dynamics of Soil Carbon in Rangelands

G.E. Schuman,[1] J.E. Herrick,[2] and H.H. Janzen[3]

Introduction

The terrestrial biosphere contains large reserves of carbon (C) — about 1500 Pg of C in the surface meter of soil (Batjes, 1996; Eswaran et al., 1995) and another 600 Pg of C in vegetation (Houghton, 1995; Schimel, 1995). Together, these pools contain three times as much C as the atmosphere and, consequently, a small change in C storage in plants or soils has important implications for atmospheric CO_2. This relationship has gained attention with the recognition that atmospheric CO_2 content is increasing and with the consequent reevaluation of ways to increase C storage in the biosphere and thereby reduce atmospheric CO_2 levels.

Grazing lands occupy about 3 billion ha of land worldwide, about twice the area devoted to cultivated agriculture (Buyanovsky and Wagner, 1998; Bronson et al., 1997). These grazing land soils contain large reserves of C, especially in temperate regions, where soil C in the surface meter accounts for 12.5 to 18.4 kg C/m^2 (Paustian et al., 1997). The U.S. has about 336 Mha of grazing lands, which include about 161 Mha of rangeland, most of which is in the Great Plains region of the country (Sobecki et al., Ch. 2).

Highest rates of C gain in rangeland soils occur early in soil formation. With time, rates of accumulation diminish as the soils approach a new equilibrium or steady-state level (Schlesinger, 1990, 1995; Chadwichk et al., 1994). Computer simulation suggested that much of the C accumulation in a rangeland soil occurred within the first 5000 years of development (Parton et al., 1988). Consequently, most soils in U.S. rangelands may now be past the stage of rapid C accrual.

---

[1] Soil Scientist, Rangeland Resources Research, High Plains Grasslands Research Station, USDA, ARS, 8408 Hildreth Road, Cheyenne, WY 82009.

[2] Soil Scientist, Jornada Experimental Range, USDA, ARS, P.O. Box 3003, MSC 3JER, New Mexico State University, Las Cruces, NM 88003-8003.

[3] Research Scientist, Lethbridge Research Centre, Agriculture and Agri-Food Canada, P.O. Box 3000, Lethbridge, AB T1J 4B1 Canada.

But that does not mean the soil C is static. Changes in soil C may still occur in response to a wide range of management and environmental factors. Given the sheer size of the C pool in U.S. rangelands, we need to understand these potential changes to best manage these systems for C storage and continued productivity.

C dynamics of rangelands is a very complex issue which climate, soils, plant community, and management affect. Research has reported very mixed and inconsistent responses to these variables because of our limited understanding of their interaction and because of the complexity and heterogeneity of rangeland ecosystems. Carbon dynamics of croplands and their response to management systems or changes in cultural practices are quite simple in comparison to rangelands, because cultivated soil is more homogeneous than noncultivated systems, and its C inputs are larger and managed more uniformly.

This chapter discusses the effects of erosion and grazing on C dynamics and also tries to estimate the potential of this large land resource as a C sink.

Effects of Erosion

Soil erosion is one of the most visible drivers of C sequestration related to management in many systems. As Chapter 9 by Lal details, soil erosion can both increase and reduce carbon sequestration. Soil erosion exposes fixed organic carbon (OC) to higher oxidation rates by breaking the bonds which physically protect soil organic matter (SOM) (Gregorich et al., 1997). Organic matter (OM) which is not oxidized, however, may be deposited in another part of the landscape in which SOM turnover rates are lower than in the source location (Gregorich et al., 1998). This effectively increases C sequestration rates.

Much of the western U.S. drains into closed basins. Runoff from these rangelands collects in lower parts of the landscape in ephemeral lakes or playas. The soils in these areas often have higher SOM contents (see Table 5.1 in Ch. 5, Bird et al.). It is not clear, however, if this is due to the higher rates of net primary production (Huenneke, 1995), to differences in turnover rates due to soil texture, temperature, or moisture regime, or to redeposition of SOM from surrounding rangeland.

The moisture regime of the depositional area, in particular, has a significant impact on C storage. Rabenhorst (1995) showed that C density was eight times higher on a poorly drained soil than on a well drained soil formed from the same parent material. In hydrologically open systems, the SOM removed from rangelands may be deposited in marine sediments where the likelihood of reoxidation is extremely low (Fan et al., 1998).

The C movement associated with soil erosion in native rangelands tends to be proportionally higher than that in croplands due to the fact that OM is more highly concentrated in the top few millimeters of the rangeland soils and the steady-state level of SOM is higher in rangeland than cropland, given the same soil. The effects of erosion in high altitude and high latitude regions are magnified by the accumulation of an organic horizon at the soil surface, which frequently is underlain by permafrost.

Livestock management affects soil erosion in at least four ways (Fig. 11.1):

1. Hoof action generates surface disturbances which increase erodibility both directly and indirectly.
2. Hoof action incorporates surface litter and standing dead material into the soil, potentially increasing SOM and reducing soil erodibility.
3. Grazing reduces canopy cover directly.
4. Grazing can lead to changes in species composition.

Soil surface disturbance

Disturbance of physically and biologically crusted soils significantly reduces resistance to both wind and water erosion. Belnap and Gillette (1998) found that simulated disturbance by cattle significantly reduced threshold friction velocities on a variety of rangeland soils and that the effect of disturbance varied as a function of texture and of the development of the biological crust (Table 11.1).

A recently completed experiment conducted on a similar suite of soils showed that a disturbed soils' recovery of resistance to wind erosion is a function of rainfall intensity (J.E. Herrick, unpublished data). Consequently, the time of disturbance relative to aeolian and precipitation events may be as important as the type or intensity of the disturbance. Soil surface disturbance during grazing also increases susceptibility to water erosion (Weltz and Wood, 1986; McIvor et al., 1995).

Incorporation of aboveground material

The effect on production of incorporating litter and standing dead material into the soil, relative to the destruction of physical and biological crusts, has been debated hotly but has received very little scientific attention. Those who argue for a net benefit of animal impact also tend to view any increase in soil erosion as either minimal, benign, or both (Savory, 1988). Others (e.g., Belnap, 1995) argue that the costs of increases in erodibility far outweigh the benefits of incorporating OM, particularly in systems in which biological crusts largely stabilize the soil surface. Very few experiments have been designed to separate the effects of increased OM inputs to the soil surface from other changes associated with grazing.

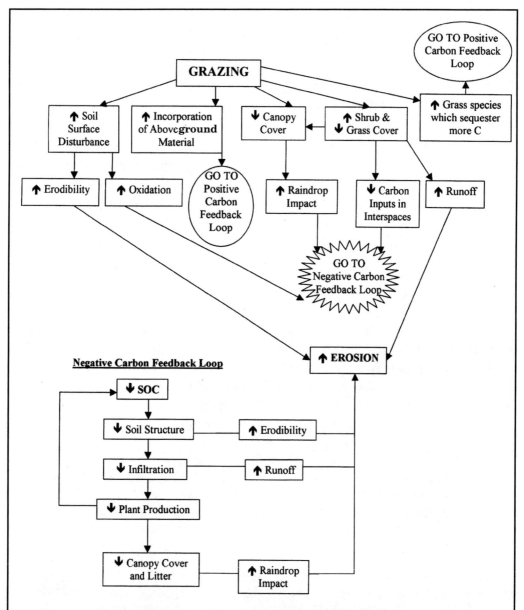

Figure 11.1. Grazing management's effects on soil erosion and interactions with soil carbon. The figure shows a negative carbon feedback loop. It applies directly to the "soil surface disturbance" and a "grass-to-shrub change in composition of species" pathway. For the "incorporation of aboveground material" pathway, a positive feedback loop applies; simply reverse the arrows within the box(es).

Table 11.1. Percent reduction of threshold friction velocities (TFV) following trampling by cattle for four sites in the northern Chihuahuan Desert (from Belnap and Gillette, 1998).

Site	Crust Type	% Sand	% Silt	% Clay	TFV Reduction
Sand	Well Developed Cyanobacterial	72.4	14.3	13.3	33-81
Gravel	Weakly Developed Cyanobacterial	68.7	11.6	19.6	22
Playa	Mineral	33.1	49.3	17.6	86
Silt	Well-Developed Cyanobacterial & Lichen	30.1	40.6	29.3	60

Canopy cover reduction

A number of studies have demonstrated that, in arid and semiarid rangelands in which vegetative cover rarely reaches 100%, total ground cover is the single most important factor affecting sediment production by water (Wood et al., 1987; Blackburn and Pierson, 1994). Reductions in canopy cover by grazing can approach 100%, depending on the management system (Savory, 1988). Other management activities also can increase the susceptibility of soil to wind erosion, through a combination of canopy cover reductions and increases in erodibility. Stout and Zobeck (1998) found that threshold wind speeds still were increasing 6 months after a fire burned a grass-dominated rangeland near Lubbock, Texas.

Changes in species composition

While the factors discussed above can have significant short-term effects on erosion and subsequent loss of C from rangelands, grazing's effects of changing the plant community structure and composition are more important over the long term. Using isotopes, researchers have correlated episodic erosional periods, recorded in soil profiles in the Chihuahuan Desert, with the replacement of C_4 grasses by C_3 shrubs (Monger et al., 1998).

This prehistoric record is reinforced by historic increases in soil erosion associated with a transition from a predominantly grass-dominated (*Bouteloua* spp., *Sporobolus* spp. and *Aristida* spp.) landscape to one dominated by mesquite (*Prosopis glandulosa*) and creosotebush (*Larrea tridentata*) (Buffington and Herbel, 1965). Follett et al. (1997), using stable C isotopes, also reported a shift from a predominately warm season (C_4) plant community to a more cool season (C_3) shrub plant community near Lawton, Oklahoma, and Big Springs, Texas.

The current transition results from an interaction between drought and the change in the disturbance regime associated with high levels of livestock grazing beginning at the end of the 19th century (Schlesinger et al., 1990). Although grazing by livestock can trigger these transitions, grazing management alone, or even the complete removal of livestock, is often insufficient to reverse the changes once they have begun. A total of 4.6 cm of soil was lost from one pasture over a 45-year period during which there was no grazing by domestic livestock (Gibbens et al., 1983). At the beginning of the study, the pasture was covered by a mix of black grama grass (*Bouteloua eriopoda*) and mesquite, with some areas already dominated by mesquite. By the end of the study, the complete area (over 250 ha) was covered by coppice sand dunes formed around mesquite plants.

These increases in erosion induced by vegetative changes are not necessarily associated with a reduction in C storage and, in fact, actually may increase C sequestration, at least over the short term. Over longer periods, however, exposure of the calcic horizon at the surface by erosion may lead to a net reduction in C storage, due to increased losses of carbonate (Ch. 4, Monger). This redistribution of SOM and other resources to create patches of different sizes has been discussed for other rangeland ecosystems (e.g., Tongway and Ludwig, 1990).

Net impact on C sequestration

The impact of soil erosion on C sequestration in source areas of sediment is clearly negative. In severely degraded areas in which both net primary productivity and plant canopy cover are reduced, C sequestration is obviously negative at the landscape scale. The net impact of erosion on C sequestration in less highly degraded rangeland systems is less clear and cannot be separated from feedbacks within the plant community (Fig. 11.1). The net impacts depend on the fate of the displaced C, on subsequent changes in the plant community, and on net primary productivity at both source and deposition areas.

In a global analysis, Lal (1995) assumed that up to 20% of displaced SOM mineralizes each year and is lost as CO_2. Few quantitative estimates of mineralization rates are available, and no landscape level studies on the net impacts of soil erosion have included burial and ecosystem level changes in net primary productivity. Net changes in C sequestration associated with plant community changes, however, have been measured. These figures integrate all sources of change in sources and sinks of C, including erosion.

Effects of Grazing

Evaluations of grazing's effects on soil C changes have varied in results. Milchunas and Lauenroth (1993) evaluated 236 data sets compiled worldwide that compared the effects of grazing on species composition, aboveground net primary production, root biomass, and soil nutrients. Of the many studies they reviewed, only 34 compared the effects of grazed vs. ungrazed (exclosure) areas on soil C. Of these 34, about 40% showed an increase in soil C in response to grazing and about 60% showed a decrease or no response to grazing.

These 34 studies represent research reported from 1947 to 1990 from throughout the world. Burke et al. (1997) further reviewed the biogeochemistry of rangelands in central North America and highlighted the belowground C's varied and inconsistent response to management.

Current research

Significant research has been completed since 1990, and it also shows considerable variance in C's response to grazing. Most of this research has compared systems with various grazing intensities and strategies to small ungrazed exclosures.

Bauer et al. (1987) compared relict rangelands and nearby grazed rangelands and reported that differences in soil C across soil texture and sampling depths were not consistent among grazing treatments. However, when averaged across soil textures and depths, the relict rangelands had 1.27 kg C/m² more in the 0.46 m soil depth than in the grazed areas, but soil N was higher in the grazed areas. They concluded that, since fencing of the rangelands in that part of the Northern Great Plains probably occurred about 75 years before their study, the differences would account for about 165 kg C/ha/yr of C loss due to grazing.

C vs. N response

The difference in C and N response to grazing is puzzling, since grazing has been shown to increase carbon and nitrogen cycling in rangeland systems. Grazing has been shown to increase the amount of medium litter, fine OM, and very fine OM in mixed grass prairie and parkland fescue sites (Naeth et al., 1991). Shariff et al. (1994) also reported that moderate grazing resulted in higher decomposition and soil N mineralization and lower N releases via decomposition than that observed on the long-term, nongrazed, and heavily grazed treatments. Litter and root decomposition averaged 55% on the moderately grazed treatment, 13% on the nongrazed, and 19% on the heavily grazed treatments. They suggested "that moderate grazing may lead to a greater conservation of N since this treatment had a higher level of N retention in OM (litter and dead roots)."

Enhanced N status has been shown to increase the rate of C and N mineralization and the total OC in the surface 2.5 and 7.5 cm of the soil in grasslands reestablished on marginal, highly erodible croplands (Reeder et al., 1998a). Frank et al. (1995) reported that moderate grazing (2.6 ha/steer) reduced the soil OC by 17% in the 107-cm soil profile, compared to that of the nongrazed exclosure they studied. However, they, like Dormaar and Willms (1990) and Smoliak et al. (1972), attributed the increase in soil C under heavy grazing to a change in plant community composition, which translated to a shallower and more robust root system of the C_4 species which replaced the more deeply rooting C_3 species.

Change in plant community

Grazing intensities that cause dramatic changes in a plant community's composition generally are identified as heavy or excessive and should be avoided to prevent reducing the quality of the resource, reducing production, reducing seasonality of the forage, and potentially reducing the sustainability of the system. Season-long grazing also can harm certain components of the plant community. Season-long grazing and/or heavy grazing reduce, or nearly eliminate, desirable C_3 species and shrubs from rangeland ecosystems (Schuman et al., 1999; Manley et al., 1997; Schuman et al., 1990; Smoliak et al., 1972).

Dormaar and Willms (1998) demonstrated that grazing at light stocking intensities (1.2 AUM/ha) did not have any effect on the SOC content after 44 years of grazing; however, heavy (2.4 AUM/ha) and very heavy (4.8 AUM/ha) grazing significantly reduced the SOC in the foothills of southwestern Alberta. They believe the heavy grazing intensities "jeopardized the sustainability of the ecosystem by reducing the fertility and water holding capacity."

Grazing intensity and strategy

As noted earlier, grazing does not always reduce SOC. Berg et al. (1997) evaluated the effects, in a native sandhill rangeland in western Oklahoma, of 50 years of grazing on vegetation and soil C and N. They evaluated pastures that were grazed at a moderate stocking intensity (0.29 yearling steer/ha/yr or 0.14 cow-calf pairs/ha/yr) on a year-long basis. This grazing intensity was defined as the level of grazing that left 1/3 of the average forage production at the end of the grazing period. They also evaluated three soil sampling procedures to determine their relative effects on the evaluation of soil C and N. They found that grazing significantly increased soil C mass in the surface 5 cm of the profile compared to that in the nongrazed exclosures but felt that a significant portion of this increase was due to the difference in bulk density (grazed 1.35 g/cm^3 vs. ungrazed 1.19 g/cm^3).

When sampling to a constant mass (5 cm deep or less) or sampling to a constant 5-cm depth and expressing the soil C on a concentration basis, the differences exhibited between grazed and ungrazed pastures were not significant at P≤

0.05. In both cases, the concentration of SOC was still 10 and 17% greater in the grazed than in the nongrazed treatments, whether sampled to a constant mass or to a constant depth.

Berg et al. (1997) concluded that long-term grazing at the moderate stocking rate resulted in significant changes in plant community composition (based on frequency measurements) and that, as Hobbie (1992) suggested, "these changes are both a cause and effect of differences in nutrient cycling." Dormaar et al. (1984) also reported higher OC in the grazed than in the ungrazed treatments at Manyberries (mixed grass association) and Stavely (fescue grassland association), Alberta, with the exception of late winter/early spring at the Stavely location.

In southeast Wyoming at the USDA-ARS, High Plains Grasslands Research Station, SOC was significantly greater in the surface 30 cm of mixed grass rangeland pastures grazed season-long for 12 years at a light stocking rate (22 steer-days/ha) than in the nongrazed exclosures (Manley et al., 1995). SOC also was higher in pastures heavily grazed (67 steer days/ha) using grazing strategies of continuous season-long, rotationally deferred, and short duration, than in the nongrazed exclosures, when evaluating only the surface 15 cm of the soil (Manley et al., 1995).

In collateral studies, Schuman et al. (1999) evaluated the C balance of this mixed grass prairie site and concluded that the C mass in the plant-soil (30 cm depth) system was significantly greater in the continuous, season-long, light and heavy grazing treatments than in the ungrazed exclosures. Evaluation of the 0- to 60-cm soil-plant system showed that 89% to 93% of the system C was stored in SOM within the soil profile. Less than 10% of the C was found in the vegetative components (above- and belowground). The heavy stocking rate altered the plant community's composition, which could account for a portion of the change in the distribution of carbon among the system's components.

Schuman et al. (1999) concluded that the livestock's hoof action helped break down and incorporate the standing dead biomass and litter into the soil and enhanced biological decomposition, thereby reducing the potential loss of C from the system via photochemical oxidation. Blue grama, with a typically dense but shallow rooting system, showed an increase in root biomass under heavy grazing, while the western wheatgrass showed a decrease; however, no differences in root biomass were observed between the nongrazed, lightly grazed, and heavily grazed treatments. Therefore, root biomass differences could not be used to explain the increase in C.

Further studies were completed at this site to delineate the reasons for the greater C observed in grazed pastures than in ungrazed exclosures. Research evaluating the effects of livestock grazing on CO_2 exchange rates (CER) was completed using a closed chamber system (LeCain et al., 2000). The CER values were adjusted to account for soil respiration. Higher early season CER was observed on the grazed pastures than on the ungrazed exclosures.

The CER on the grazed pastures was 1.5 to 2 times higher from April through June than on the exclosures. This higher early season CER related to earlier spring greening in the grazed pastures. Schuman et al. (2000) note that this could account for a portion of the increased soil C noted in the grazed treatments because, by the end of June, 60% to 80% of the aboveground production has occurred. April through June also is the period of optimum soil moisture.

Studies similar to those by Schuman et al. (1999) also were conducted at the USDA-ARS Central Plains Experimental Range in northeastern Colorado on the short grass steppe and found greater SOC under a heavily grazed pasture than in an adjacent exclosure after 50 years of grazing (Reeder et al., 1998b). However, a similar response to grazing was not evident on the lightly grazed treatment and its adjacent control. The lack of consistency in this case was attributed to the fact that, even though the soil series was carefully selected to be the same, the clay contents in the A and B horizon were significantly greater on the lightly grazed pasture and its exclosure.

Derner et al. (1997) also evaluated the effect of grazing on the short grass steppe in northeastern Colorado and found that SOC content was 43% and 55% higher in the 0- to 5- and 5- to 15-cm depth on the grazed than on the ungrazed treatment. However, they did not find this same relationship with grazing in the tall grass and midgrass communities they studied.

Grazing vs. mowing

Hassink and Neeteson (1991) compared the effects of grazing and mowing on the SOC status of the soil. After 4 years of treatment, they found that the amount of soil C was significantly greater in the top 5 cm of the soil under grazing than under mowing, for both a sandy and loamy soil. They concluded that, under grazing conditions, more organic material returns to the soil than under mowing, and dung and urine also return to the soil and result in more rapid cycling of C. They also concluded that, through the livestock's hoof action, grazing increases the amount of aboveground herbage incorporated into the soil.

Explanations for varied responses to grazing

The effects of grazing on SOC seem to be inconsistent and variable, and numerous explanations have been offered. Schuman et al. (1999) and Berg et al. (1997) identified factors that require careful consideration in this discussion.

Methodology, soil, and climate

Schuman et al. (1999) pointed out that consistent sampling procedures, including the handling of surface litter, comparable soil series/characteristics, similar baseline plant communities, and consistent laboratory methodologies are essential to ensuring sound treatment evaluations. Berg et al. (1997) expanded the list to include reporting on a concentration basis and ensuring similar and appropriate sampling depths and valid field replication.

Considerable discussion has occurred over reporting concentration, rather than correcting for differences in bulk density, to report quantity (Skene, 1966; Henzell et al., 1967). Berg et al. recommended comparing adjacent pastures, to show contrasts in plant communities induced by grazing, as a good approach to determining grazing's effects on soil C.

Rauzi et al. (1968) used a similar approach to characterize infiltration rates in the Great Plains. Dormaar et al. (1977) also found that total C and N, C:N ratios, and numerous other SOM characteristics depend greatly on the season of sampling. Precipitation variation among years also has as great or greater effects on a plant community's response and annual net primary productivity as do grazing treatments (Michunas and Lauenroth, 1993).

Plant response to grazing

Many physiological characteristics of plants also may help explain the differences various researchers observed. Caldwell et al. (1981) and Dormaar et al. (1995) point out that species differ greatly in their response to grazing and, hence, soil C's responses may differ greatly. Crested wheatgrass (*Agropyron desertorum*) has shown greater flexibility of resource allocation for C following defoliation, with more going to the root system with curtailed root growth, whereas root growth in *A. spicatum* continued at the same rate after as before defoliation (Caldwell et al., 1981).

Photosynthesis of grazed plants

Much direct and indirect evidence has shown that grazing or simulated defoliation influences the rate of photosynthesis and that the actual interaction between the grazer and the plant is important (Painter and Detling, 1981; Dyer and Bokhari, 1976; Detling et al., 1979). Painter and Detling (1981) simulated a moderate and heavy defoliation of western wheatgrass plants and found that net photosynthesis was greater on those plants than on the control plants.

Others have reported the enhancement of photosynthesis in the undamaged leaves of partially defoliated plants, which is thought to result from decreased mesophyll's resistance to CO_2's diffusion in new leaves. A single, simulated grazing of blue grama resulted in greater net photosynthesis, and it also appears that the increased photosynthesis results from increased photosynthetic capacity of the tissues that remained or were produced after defoliation (Detling et al., 1979).

Plant growth

Detling et al. (1979) reported that total C losses from root and crown respiration were less in the clipped plants because of their reduced biomass. Hodgkinson and Baas Becking (1977) reported that, when shoots are grazed, root mortality frequently increases and root extension and branching decreases. This root mortality could result in significant increases in SOC over a relatively short time. Stevenson and White (1941) reported greater root production under grazed than ungrazed prairie, even though many grazing studies simulated in greenhouses have shown decreased root growth (Johnston, 1961; Wilson, 1960).

The physical and biochemical effects of grazing by livestock and insects have greater effects on regrowth than simulated grazing. Dyer and Bokhari (1976) reported that letting grasshoppers feed on hydroponically grown blue grama resulted in much greater regrowth potential than simply clipping plants, partially because of the increase in tiller production.

Schuman et al. (1990) also showed increased rhizome production of western wheatgrass under grazing. Dyer and Bokhari (1976) reported that grasshoppers' grazing resulted in an increase in belowground respiration and root exudation. Grasshoppers' grazing increased the rate of change in pH of the root medium by 4% per day, and they believe this is the result of increased respiration of the roots (release of CO_2) and the production of organic acids in the root exudates. They state that this increase "could well affect the total metabolism and total biomass of the plants to a far greater extent than direct consumption or litter production." The observed effects on tiller production could have a significant impact on later growth potentials.

Jameson (1964) and Reardon et al. (1972, 1974) believe that cattle's grazing on grass also stimulates metabolic activities through the mechanical action and absorption of saliva into the grass plant. Reardon et al. (1974) believe that the thiamine in saliva may be important because of coevolutionary considerations between the herbivore and the producer. Dormaar (1988) reported that the growth of blue grama and rye (*Secale cereale*) significantly affected the soil pH, and that blue grama had a slightly greater effect than the rye.

Follett, Kimble, and Lal, editors

Complexity of issues

The variable responses of soil C to grazing is complex and cannot be attributed solely to methodological variations and constraints. Variations in soil C in response to management alternatives undoubtedly reflect the complexity of the soil C balance and the many, poorly understood, indirect effects of grazing (Table 11.2).

Table 11.2 summarizes the literature this chapter cites that evaluates the effects of grazing on soil C. As noted earlier, erosion induced by grazing, even in very limited magnitude, could have major impacts on soil C because of the surface soil material's importance in C storage and activity and its susceptibility to erosion.

Table 11.2. Responses of grassland soil C to grazing.

Carbon Response	Reference	Comment(s)
Increase in C	Milchunas & Lauenroth, 1993	14 sites, grazed
	Frank et al., 1995	Heavy grazing, PCC
	Dormaar & Willms, 1990	Heavy grazing, PCC
	Schuman et al., 1999	Light & heavy grazing
	Manley et al., 1995	Light & heavy grazing
	Berg et al., 1997	Moderate grazing
	Dormaar et al., 1984	Grazed, PCC
	Reeder et al., 1998b	Heavy grazing, PCC
	Derner et al., 1997	Grazed, short grass
	Smoliak et al., 1972	Heavy grazing, PCC
Decrease/No Change in C	Milchunas & Lauenroth, 1993	20 sites, grazed
	Bauer et al., 1987	Moderate grazing, increased N
	Frank et al., 1995	Moderate grazing, increased N
	Dormaar & Willms, 1998	Light grazing (no change) heavy & very heavy grazing (reduced soil C)
	Reeder et al., 1998b	Light grazing (no change)
	Derner et al., 1997	Grazed, no change tall- & mid-grass

PCC = plant community change with increased root biomass resulting.

Rangelands' Potential to Sequester Carbon

Most rangeland soils are probably at or near an equilibrium state, if the vegetative community and conditions have been consistent for many years. They are an important C storage zone, but their potential for enhanced C storage may be limited under these equilibrium conditions.

Many of the proposed effects of management on C storage in rangelands assume constant environmental conditions. In fact, we know that climate and environment are not static, and they may be changing at an accelerated rate because of human activity. One of the most profound changes is the increase in atmospheric CO_2. Current concentrations are already about 30% higher than in preindustrial times, and concentrations are increasing at a rate of 1.5 ppmv per year (Houghton, 1997).

Increases in atmospheric CO_2 can affect rangeland C storage in several ways. First, elevated CO_2 can accelerate rates of photosynthesis directly, thereby increasing biomass production (Owensby et al., 1994; van Ginkel and Gorisson, 1998). Increases in assimilated C may be allocated preferentially below ground (Morgan et al., 1994). Increases in atmospheric CO_2 also may enhance yields by improving the efficiency of water use (Owensby, 1993; Morgan et al., 1994).

Elevated atmospheric CO_2 also may slow the rate of decomposition, which in turn would increase C in the system (Owensby, 1993). Research has shown that roots of grasses grown under enriched CO_2 decompose more slowly than those of plants grown under ambient CO_2 levels (Gorrison et al., 1995; van Ginkel et al., 1996; van Ginkel and Gorrison, 1998). Consequently, increases in atmospheric CO_2 may enhance SOC both by increasing the amount of plant litter returned to the soil through increases in net primary production, and by slowing its decomposition within the soil. Owensby (1993) suggests that doubling the atmospheric CO_2 over 50 to 70 years might elicit a 12% to 15% gain in SOM under tall grass prairie.

The effect of elevated CO_2 cannot be evaluated independently of other concomitant changes. Increases in temperature may accelerate microbial decomposition of OM (Jenkinson et al., 1991; Raich and Schlesinger, 1992; Trumbore et al., 1996), thereby offsetting some of the possible gains in stored C from increased atmospheric CO_2. A further confounding factor is possible accelerated N mineralization with higher temperatures, which could enhance productivity and ameliorate the loss of C (Schimel, 1995). Therefore, the net impact of a changing atmosphere will reflect the interaction of numerous factors (Lashof et al., 1997).

Enhanced deposition of humanly induced N changes will have a significant effect on C storage in rangelands. The amount of N_2 fixed by human activities already rivals or exceeds that from natural sources (Vitousek et al., 1997), and further increases are expected (Jeffries and Maron, 1997). On rangelands, most of which are limited in N, deposition of supplemental N can promote productivity and enhance C storage (Wedlin and Tilman, 1996). The magnitude of gains in terms of C storage will vary with the plant community, which the degree of N deposited also will affect.

N deposition has been suggested as an important mechanism of terrestrial C gain (Schindler and Bayley, 1993). Some short-term fertilization studies have suggested possible effects of N deposition on C dynamics (Reeder et al., 1998a); however, the long-term impact of low-level increases in N on a wide range of rangeland communities remains uncertain.

Estimate of Potential C Gains

It should be obvious from the above discussion that estimating potential C gains is more complicated for rangeland soils than for cultivated lands. Though grazing lands occupy about twice the area of croplands in the U.S., comparatively few long-term studies have been made of soil C's dynamics in grazing land. Impacts of grazing land management on soil C tend to be less abrupt than those on cultivated land. Grazing land soils have inherently high C densities, and these C stocks change relatively slowly because management's effects, like changes in grazing strategy, often are subtle and imposed gradually.

Finally, rangeland ecosystems are much more complex than most croplands. Whereas plant communities of single species are imposed on croplands, rangeland communities are diverse and evolved over time in response to climate, soils, and management practices. Consequently, a shift in management may induce a gradual change in species composition, and the eventual impact of that change and its various feedbacks on soil C may not be measured until years or decades later.

Because of these complexities, the potential rates of C sequestration on rangelands can only be qualitative estimates. Nevertheless, given the vast area and large C pool involved, it is important to at least derive some preliminary estimates of possible changes in C storage.

Assumptions

To estimate the potential gain of C in U.S. rangelands, we make the following assumptions:

1. Under consistent management and environment, soil C reserves in any ecosystem eventually approach a steady-state value, beyond which, rates of soil C gain or loss are negligible from the standpoint of the atmospheric CO_2 pool (Odum, 1969; Johnson, 1995).

2. Change in the current trend of net C exchange depends on a shift in management or environmental conditions. Thus, the C content of the soil currently at a steady state can be altered appreciably only by a shift in management or climate. Similarly, a soil currently losing C, approaching a new, lower steady state, can be shifted from that course only by a change in management or other external factor(s).

3. From the perspective of reducing atmospheric CO_2 gains, an avoided loss of soil C is as important as a gain in soil C (Izaurralde et al., 1999). A management choice that avoids losing 1 MMT of soil C has exactly the same benefit to atmospheric CO_2 as a management choice that gains 1 MMT of soil C.

Estimated benefits

These assumptions make it possible to derive at least conceptual estimates of the benefits of good rangeland management for mitigating increases in atmospheric CO_2 (Table 11.3).

According to recent estimates, about one-third of the U.S. rangeland area (54 Mha) has no serious ecological or management problem (USDA, NRCS, 1998; USDI, BLM, 1998; David Wheeler, USDA, FS, Lakewood, CO, personal communication, 1999). If we assume that the soil C content of these rangelands is at steady state, then these soils have little potential for further C storage gains (assumption 1). But the remaining two-thirds of the U.S. rangelands, presumably, have some constraints which limit productivity and, hence, C storage.

The potential C gain in these lands varies widely, but any improvements in soil C may be gradual, perhaps in the area of 0.1 MTC/yr. Much faster rates of C storage gain are possible in previously cultivated lands reseeded to grass. Bruce et al. (1999) estimated that "set aside" lands under the Conservation Reserve Program (CRP) are gaining C at a rate of about 0.6 MTC/yr, yielding a total rate of C sequestration of about 8 MMTC/yr. Thus, improved management may sequester up to 11 MMTC/yr in permanent rangelands and another 8 MMTC/yr in recently reestablished grasslands (Table 11.3).

Equally as important as the potential C gains are any averted losses (assumption 3) (Table 11.3). Because rangelands represent such a large pool of C, preserving existing reserves is especially important. One way to avoid losses is to maintain or establish optimal grazing strategies. Recent studies have shown that properly grazed rangeland can gain C at a rate of about 0.3 MTC/yr, relative to that in a corresponding ungrazed exclosure (Schuman et al., 1999; Manley et al., 1995). Consequently, maintaining good grazing strategies on all well managed rangelands, rather than ceasing grazing, as some advocate, would yield an avoided loss of about 16 MMTC/yr.

Table 11.3. Estimated potential benefits to the mitigation of atmospheric CO_2 from the adoption of improved management of grasslands and potential avoided losses from grasslands in the U.S.

Land Use	Area[4] (Mha)	Rate[5] (MTC/ha/yr)	Rate[5] (MMTC/yr)
Potential Mitigation Gains			
Well managed grasslands[1]	54	0	0
Poorly managed grasslands[1]	107	0.1	11
CRP grasslands[2]	13	0.6	8
		Total Gain	19
Potential Avoided Losses			
Well managed grasslands[1]	54	0.3	16
Poorly managed grasslands[1]	107	0.2	21
CRP grasslands[3]	13	0.3	4
		Total Avoided Loss	41

[1] *According to USDA-NRCS (1998), 33% (54 Mha) of the private U.S. rangelands are reported to have no serious ecological or management problems, and 67% (109 Mha) would benefit from enhanced management or restoration. Of federally managed rangelands under the authority of the Bureau of Land Management, 37% (2 Mha) are in good to excellent condition and 63% (4 Mha) are in fair to poorer condition (USDI-BLM 1998). 80% (0.8 Mha) of the National Grasslands in the Rocky Mountain Region managed by the USDA-FS meet the forest plan objectives, and 20% (0.2 Mha) are considered moving toward those objectives, not meeting or moving toward them, or undetermined (David Wheeler, Rock Mountain Regional USDA-FS, Lakewood, CO, personal communication, 1999).*

[2] *CRP = Conservation Reserve Program; area and rate of potential C gain from Bruce et al. (1999).*

[3] *Rate of avoided loss (0.3 MTC/ha/yr) based on rate measured by Doran et al. (1998) at site converted from established grassland to a no-till wheat-fallow system.*

[4] *Land area values are those compiled by Sobecki et al., Ch. 2, and are used to provide consistency throughout this volume. However, the values in footnote #1 above are those the respective agencies published in their inventory reports.*

[5] *Potential rates apply to possible changes in the period soon after a management change (or in the case of CRP lands, to current rates on 'existing' lands). Rates of accrual will diminish with time (perhaps after a few years or decades) as soil C approaches a new steady state.*

This value assumes the net gain relative to losses incurred if grazing were discontinued on all well managed rangelands, so it represents a maximum value. Similarly, avoided losses by maintaining current grazing practices on "poorly managed" rangelands might amount to as much as 21 MMTC/yr.

Finally, preventing cultivation of existing CRP lands can prevent potentially large losses of stored C. Assuming that conversion of CRP lands to cropland would result in C loss at a rate of 0.3 MTC/ha/yr (Doran et al., 1998), preserving all CRP lands in rangelands avoids a potential loss of 4 MMTC/ha/yr. This avoided loss is in addition to the rates of C accrual at these sites, assumed to be 0.6 MTC/ha/yr. Potential avoided loss from all these sources therefore totals about 41 MMTC/yr. This value does not yet include any losses avoided by preventing further cultivation of existing, reestablished, or native rangelands.

These estimates are just that, estimates, and perhaps better viewed as conceptual illustrations to show the magnitude of potential C sequestration or avoided losses. Nevertheless, they demonstrate that the vast C pool held in U.S. rangelands cannot be ignored in assessing the link between agricultural management and atmospheric CO_2. If nothing else, they illustrate the importance to atmospheric CO_2 of preventing disturbance of these vast C-rich systems.

Conclusion

Carbon dynamics of rangeland soils are very complex, and only limited detailed research has been conducted to assess the role of management and grazing strategies on soil C changes and the potential for enhanced C sequestration. Soil erosion also can affect the soil C pools in rangeland ecosystems, but much of the erosion does not result in loss of C from the landscape but rather a repositioning on the landscape. However, a significant portion of this redeposited SOM may undergo mineralization and subsequently be lost from the system.

The U.S. rangelands are already a large repository of C because of their high C density and vast land area. Improved practices on poorly managed rangelands could further increase this C store; our estimate suggests a theoretical accumulation rate of 11 MMTC/yr, though this might be a theoretical maximum. A further gain of 8 MMTC/yr might accrue from maintaining existing CRP lands. But perhaps the most important management opportunity, from the standpoint of atmospheric CO_2, is the preservation of current stores of C in U.S. rangelands. For example, conservation practices might prevent losses amounting to as much as 41 MMTC/yr.

All of these estimates perhaps are viewed best as conceptual illustrations rather than quantitative values. Nevertheless, they illustrate the significance of U.S. rangelands as a C storehouse, and the importance of management to preserve or even enhance their rich C reserves. As more directed research specifically addresses the effects of management on C storage, and as CO_2 flux measurements become available, a more thorough and accurate estimation of C sequestration by rangelands will be possible.

References

Batjes, N.H. 1996. Total carbon and nitrogen in the soils of the world. *Eur. J. of Soil Sci.* 47:151-163.

Bauer, A., C.V. Cole, and A.L. Black. 1987. Soil property comparisons in virgin grasslands between grazed and nongrazed management systems. *Soil Sci. Soc. of Am. J.* 51:176-182.

Belnap, J. 1995. Surface disturbances: their role in accelerating desertification. *Environ. Monit. and Assess.* 37:39-57.

Belnap, J., and D.A. Gillette. 1998. Vulnerability of desert biological crusts to wind erosion: the influences of crust development, soil texture and disturbance. *J. of Arid Environ.* 39:133-142.

Blackburn, W.H., and F.B. Pierson, Jr. 1994. Sources of variation in interrill erosion on rangelands. *In* Wilbert H. Blackburn, Frederick B. Pierson, Jr., Gerald E. Schuman, and R. Zartman (eds), *Variability in Rangeland Water Erosion Processes*, Soil Science Society of America, Madison, WI, pp. 1-10.

Berg, W.A., J.A. Bradford, and P.L. Sims. 1997. Long-term soil nitrogen and vegetation change on sandhill rangeland. *J. of Range Manage.* 50:482-488.

Bronson, K.F., K.G. Cassman, R. Wassmann, D.C. Olk, M. van Noordwijk, and D.P. Garrity. 1997. Soil carbon dynamics in different cropping systems in principal ecoregions of Asia. *In* R. Lal, J.M. Kimble, R.F. Follett, and B.A. Stewart (eds), *Management of Carbon Sequestration in Soil*, CRC Press, Boca Raton, FL, pp. 35-57.

Bruce, J.P., M. Frome, H. Haites, H. Janzen, R. Lal, and K. Paustian. 1999. Carbon sequestration in soils. *J. Soil and Water Cons.* 54:382-389.

Buffington, L.C., and C.H. Herbel. 1965. Vegetational changes on a semidesert grassland range from 1858 to 1963. *Ecol. Monogr.* 35:139-164.

Burke, I.C., W.K. Lauenroth, and D.G. Milchunas. 1997. Biogeochemistry of managed grasslands in central North America. *In* E.A. Paul, K. Paustian, E.T. Elliott, and C.V. Cole (eds), *Soil Organic Matter in Temperate Agroecosystems, Long-term Experiments in North America*, CRC Press, Boca Raton, FL, pp. 85-102.

Buyanovsky, G.A., and G.H. Wagner. 1998. Changing role of cultivated land in the global carbon cycle. *Biol. and Fertil. of Soils* 27:242-245.

Caldwell, M.M., J.H. Richards, D.A. Johnson, R.S. Nowak, and R.S. Dzurec. 1981. Coping with herbivory: Photosynthesis capacity and resource allocation in two semiarid *Agropyron* bunchgrasses. *Oecologia* 50:14-24.

Chadwick, O.A., E.F. Kelly, D.M. Merritts, and R.G. Amundson. 1994. Carbon dioxide consumption during soil development. *Biogeochemistry* 24:115-127.

Derner, J.D., D.D. Briske, and T.W. Boutton. 1997. Does grazing mediate soil carbon and nitrogen accumulation beneath C_4 perennial grasses along an environmental gradient? *Plant and Soil* 191:147-156.

Detling, J.K., M.I. Dyer, and D.T. Winn. 1979. Net photosysnthesis, root respiration, and regrowth of *Bouteloua gracilis* following simulated grazing. *Oecologia* 41:127-134.

Doran, J.W., E.T. Elliott, and K. Paustian. 1998. Soil microbial activity, nitrogen cycling, and long-term changes in organic carbon pools as related to fallow tillage management. *Soil & Tillage Res.* 49:3-18.

Dormaar, J.F., A. Johnston, and S. Smoliak. 1977. Seasonal variation in chemical characteristics of soil organic matter of grazed and ungrazed mixed prairie and fescue grassland. *J. of Range Manage.* 30:195-198.

Dormaar, J.F., A. Johnston, and S. Smoliak. 1984. Seasonal changes in carbon content, and dehydrogenase, phosphatase, and urease activities in mixed prairie and fescue grassland Ah horizons. *J. of Range Manage.* 37:31-35.

Dormaar, J.F., M.A. Naeth, W.D. Willms, and D.S. Chanasyk. 1995. Effect of native prairie, crested wheatgrass (*Agropyron cristatum* (L.) Gaertn.) and Russian wildrye (*Elymus junceus* Fisch.) on soil chemical properties. *J. of Range Manage.* 48:258-263.

Dormaar, J.F., and W.D. Willms. 1990. Effect of grazing and cultivation on some chemical properties of soils in the mixed prairie. *J. of Range Manage.* 43:456-460.

Dormaar, J.F., and W.D. Willms. 1998. Effect of forty-four years of grazing on fescue grassland soils. *J. of Range Manage.* 51:122-126.

Dyer, M.I., and U.G. Bokhari. 1976. Plant-animal interactions: Studies of the effects of grasshopper grazing on blue grama grass. *Ecology* 57:762-772.

Eswaran, H., E. Van den Berg, P. Reich, and J. Kimble. 1995. Global soil carbon resources. *In* R. Lal, J. Kimble, E. Levine, and B.A. Stewart (eds), *Soils and Global Change*, CRC Lewis Publishers, Boca Raton, FL, pp. 27-43.

Fan, S., M. Gloor, J. Mahlman, S. Pacaia, J. Sarmiento, T. Takahashi, and P. Tans. 1998. A large terrestrial carbon sink in North America implied by atmospheric and oceanic carbon dioxide data and models. *Science* 282:442-446.

Follett, R.F., E.A. Paul, S.W. Leavitt, A.D. Halvorson, D. Lyon, and G.A. Peterson. 1997. Carbon isotope ratios of Great Plains soils and in wheat-fallow systems. *Soil Sci. Soc. Am. J.* 61:1068-1077.

Frank, A.B., D.L. Tanaka, L. Hofmann, and R.F. Follett. 1995. Soil carbon and nitrogen of Northern Great Plains grasslands as influenced by long-term grazing. *J. of Range Manage.* 48:470-474.

Gibbens, R.P., J.M. Tromble, J.T. Hennessy, and M. Cardenas. 1983. Soil movement in mesquite dunelands and former grasslands of southern New Mexico from 1933 to 1980. *J. of Range Manage.* 36:145-148.

Gorrison, A., J.H. van Ginkel, J.J.B. Keurentjes, and J.A. van Veen. 1995. Grass root decomposition is retarded when grass has been grown under elevated CO_2. *Soil Biol. and Biochem.* 27:117-120.

Gregorich, E.G., C.F. Drury, B.H. Ellert, and B.C. Liang. 1997. Fertilization effects on physically protected light fraction organic matter. *Soil Sci. Soc. of Am. J.* 61:482-484.

Gregorich, E.G., K.J. Greer, D.W. Anderson, and B.C. Liang. 1998 Carbon distribution and losses: erosion and deposition effects. *Soil & Tillage Res.* 47:291-302.

Hassink, J., and J.J. Neeteson. 1991. Effect of grassland management on the amounts of soil organic N and C. *Netherlands J. of Agric. Sci.* 39:225-236.

Henzell, E.F., I.F. Fergas, and A.E. Martin. 1967. Accretion studies of soil organic matter. *J. Aust. Inst. of Agric. Sci.* 33:35-37.

Hobbie, S.A. 1992. Effects of plant species on nutrient cycling. *TREE* 7:336-339.

Hodgkinson, K.C., and H.G. Baas Becking. 1977. Effect of defoliation on root growth of some arid zone perennial plants. *Aust. J. Agric. Res.* 29:31-42.

Houghton, R.A. 1995. Changes in the storage of terrestrial carbon since 1850. *In* R. Lal, J. Kimble, E. Levine, and B.A. Stewart (eds), *Soils and Global Change*, CRC Lewis Publishers, Boca Raton, FL, pp. 45-65.

Houghton, J. 1997. *Global Warming: The Complete Briefing (2nd edition)*. Cambridge University Press, Cambridge. 251 pp.

Huenneke, L.F. 1995. Shrublands and grasslands of the Jornada Long-Term Ecological Research Site: desertification and plant community structure in the northern Chihuahuan Desert. *In* J.R. Barrow, E.D. McArthur, R.E. Sosebee, and R.J. Tausch (compilers), *Proceedings: Shrubland Ecosystem Dynamics in a Changing Environment*, U.S. Forest Service Intermountain Research Station, Ogden, UT, pp. 48-50.

Izaurralde, R.C., K.H. Haugen-Kozyra, D.C. Jans, W.B. McGill, R.F. Grant, and J.C. Hiley. 2000. Soil organic carbon dynamics: measurement, simulation and site to region scale-up. *In* R. Lal, R.F. Follett, and B.A. Stewart (eds), *Advances in Soil Science: Assessment Methods for Soil C Pools,* CRC Press/Lewis Publishers, Boca Raton, FL (in press).

Jameson, D.A. 1964. *Forage Plant Physiology and Soil-Range Relationships. Effect of Defoliation on Forage Plant Physiology. Special Publication 5*. American Society of Agronomy. Madison, WI. pp. 67-80.

Jeffries, R.L., and J.L. Maron. 1997. The embarassment of riches: atomospheric deposition of nitrogen and community and ecosystem processes. *TREE* 12:74-78.

Jenkinson, D.S., D.E. Adams, and A. Wild. 1991. Model estimates of CO_2 emissions from soil in response to global warming. *Nature* 351:304-306.

Johnson, M.G. 1995. The role of soil management in sequestering soil carbon. *In* Lal et al. (eds), *Soil Management and Greenhouse Effect*, Lewis Publishers, Boca Raton, FL, pp. 351-363.

Johnston, A. 1961. Comparision of lightly grazed and ungrazed range in the fescue grassland of southwestern Alberta. *Can. J. Plant Sci.* 41:615-622.

Lal, R. 1995. Global soil erosion by water and carbon dynamics. *In* R. Lal, J. Kimble, E. Levine, and B.A. Stewart (eds), *Soils and Global Change*, CRC Press, Boca Raton, FL, pp. 131-142.

Lashof, D.A., B.J. DeAngelo, S.R. Saleska, and J. Harte. 1997. Terrestrial ecosystem feedbacks to global climate change. *Annu. Rev. Energy and Environ.* 22:75-118.

LeCain, D.R., J.A. Morgan, G.E. Schuman, J.D. Reeder, and R.H. Hart. 2000. Carbon exchange of grazed and ungrazed pastures of a mixed grass prairie. *J. of Range Manage.* 53:199-206.

Manley, W.A., R.H. Hart, M.J. Samuel, M.A. Smith, J.W. Waggoner, and J.T. Manley. 1997. Vegetation, cattle, and economic responses to grazing strategies and pressures. *J. of Range Manage.* 50:638-646.

Manley, J.T., G.E. Schuman, J.D. Reeder, and R.H. Hart. 1995. Rangeland soil carbon and nitrogen responses to grazing. *J. of Soil and Water Cons.* 50:294-298.

McIvor, J.G., J. Williams, and C.J. Gardener. 1995. Pasture management influences runoff and soil movement in the semiarid tropics. *Aust. J. of Exp. Agric.* 35:55-65.

Milchunas, D.G., and W.K. Lauenroth. 1993. Quantitative effects of grazing on vegetation and soils over a global range of environments. *Ecol. Monogr.* 63(4):327-366.

Monger, H.C., D.R. Cole, J.W. Gish, and T.H. Giordano. 1998. Stable carbon and oxygen isotopes in Quaternary soil carbonates as indicators of ecogeomorphic changes in the northern Chihuahuan Desert, U.S.A. *Geoderma* 82:137-172.

Morgan, J.A., W.G. Knight, L.M. Dudley, and H.W. Hunt. 1994. Enhanced root system C-sink activity, water relations and aspects of nutrient acquisition in mycotrophic *Bouteloua gracilis* subjected to CO_2 enrichment. *Plant and Soil* 165:139-146.

Naeth, M.A., A.W. Bailey, D.J. Pluth, D.S. Chanasyk, and R.T. Hardin. 1991. Grazing impacts on litter and soil organic matter in mixed prairie and fescue grassland ecosystems of Alberta. *J. of Range Manage.* 44:7-12.

Odum, E.P. 1969. The strategy of ecosystem development. *Sci.* 164:262-270.

Owensby, C.E. 1993. Potential impacts of elevated CO_2 and above- and below-ground litter quality of a tallgrass prairie. *Water Air Soil Pollut.* 70:413-424.

Owensby, C.E., L.M. Auen, and P.I. Coyne. 1994. Biomass production in a nitrogen-fertilized, tallgrass prairie ecosystem exposed to ambient and elevated levels of CO_2. *Plant and Soil* 165:105-113.

Painter, E.L., and J.K. Detling. 1981. Effects of defoliation on net photosynthesis and regrowth of western wheatgrass. *J. of Range Manage.* 34:68-71.

Parton, W.J., J.W.B. Stewart, and C.V. Cole. 1988. Dynamics of C, N, P, and S in grassland soils: a model. *Biogeochem.* 5:109-131.

Paustian, K., O. Andren, H.H. Janzen, R. Lal, P. Smith, G. Tian, H. Tiessen, M. Van Noordwijk, and P.L. Woomer. 1997. Agricultural soils as a sink to mitigate CO_2 emissions. *Soil Use and Manage.* 13:230-244.

Rabenhorst, M.C. 1995. Carbon storage in tidal marsh soils. *In* R. Lal, J. Kimble, E. Levine, and B.A. Stewart (eds). *Soils and Global Change*, CRC-Lewis Press, Boca Raton, FL, pp. 93-103.

Raich, J.W., and W.H. Schlesinger. 1992. The global carbon dioxide flux in soil respiration and its relationship to vegetation and climate. *Tellus* 44B:81-99.

Rauzi, F., C.L. Fly, and E.J. Dyksterhuis. 1968. *Water intake of midcontinental rangelands as influenced by soil and plant cover. USDA Tech. Bull. 1390.* USDA. U.S. Government Printing Office. Washington, DC.

Reardon, P.Q., C.L. Leinweber, and L.B. Merrill. 1972. The effect of bovine saliva on grasses. *J. of Anim. Sci.* 34:897-898.

Reardon, P.Q., C.L. Leinweber, and L.B. Merrill. 1974. Response of sideoats grama to animal saliva and thiamine. *J. of Range Manage.* 27:400-401.

Reeder, J.D., G.E. Schuman, and R.A. Bowman. 1998a. Soil C and N changes on conservation reserve program lands in the Central Great Plains. *Soil and Tillage Res.* 47:339-349.

Reeder, J.D., G.E. Schuman, J.A. Morgan, D.R. LeCain, and R.H. Hart. 1998b. Impact of livestock grazing on the carbon and nitrogen balance of a shortgrass steppe. *Agronomy Abstracts.* American Society of Agronomy, Madison, WI. p. 291.

Savory, A. 1988. *Holistic Resource Management.* Island Press. Washington, D.C.

Schimel, D.S. 1995. Terrestrial ecosystems and the carbon cycle. *Global Change Biol.* 1:77-91.

Schindler, D.W., and S.E. Bayley. 1993. The biosphere as an increasing sink for atmospheric carbon: estimates from increased nitrogen deposition. *Global Biogeochem. Cycles* 7:717-733.

Schlesinger, W.H., J.F. Reynolds, G.L. Cunningham, L.F. Huenneke, W.M. Jarrell, R.A. Virginia, and W.G. Whitford. 1990. Biological feedbacks in global desertification. *Science* 247:1043-1048.

Schlesinger, W.H. 1995. An overview of the carbon cycle. *In* R. Lal, J. Kimble, E. Levine, and B.A. Stewart (eds), *Soil and Global Change*, CRC-Lewis Press, Boca Raton, FL, pp. 9-25.

Schuman, G.E., D.T. Booth, and J.W. Waggoner. 1990. Grazing reclaimed mined land seeded to native grasses in Wyoming. *J. of Soil and Water Cons.* 44:653-657.

Schuman, G.E., D.R. LeCain, J.D. Reeder, and J.A. Morgan. 2000. *Carbon dynamics and sequestration of a mixed-grass prairie as influenced by grazing. Soil Sci. Soc. America, Special Publication.* Soil Sci. Soc. of Am. Madison, WI. (In press.)

Schuman, G.E., J.D. Reeder, J.T. Manley, R.H. Hart, and W.A. Manley. 1999. Impact of grazing management on the carbon and nitrogen balance of a mixed-grass rangeland. *Ecol. Appl.* 9(1):65-71.

Shariff, A.R., M.E. Biondini, and C.E. Grygiel. 1994. Grazing intensity effects on litter decomposition and soil nitrogen mineralization. *J. of Range Manage.* 47:444-449.

Skene, J.K.M. 1966. Errors in accretion studies of soil organic matter. *J. Austr. Instit. of Agric. Sci.* 32:208-209.

Smoliak, S., J.F. Dormaar, and A. Johnston. 1972. Long-term grazing effects on *Stipa-Bouteloua* prairie soils. *J. of Range Manage.* 25:246-250.

Stevenson, T.M., and W.J. White. 1967. Root fibre production of some perennial grasses. *Sci. Agric.* 22:108-118.

Stout, J.E., and T.M. Zobeck. 1998. Earth, wind and fire: aeolian activity in burned rangeland. *In* A. Busacca, S. Lilligren, and K. Newell (eds), *Dust Aerosols, Loess Soils and Global Change*, Washington State University, College of Agriculture and Home Economics, Pullman, WA, pp. 85-88.

Tongway, D.J., and J. A. Ludwig. 1990. Vegetation and soil patterning in semi-arid mulga lands of eastern Australia. *Aust. J. of Ecol.* 15:23-34.

Trumbore, S.E., O.A. Chadwick, and R. Amundson. 1996. Rapid exchange between soil carbon and atmospheric carbon dioxide driven by temperature change. *Sci.* 272:393-396.

USDA-NRCS. 1998. *State of the Nation's Nonfederal Rangeland. 1992 NRI Summary-Nation, Region, and States.* USDA-NRCS. U.S. Government Printing Office. Washington, DC.

USDI-BLM. 1998. *National Rangeland Inventory, Monitoring and Evaluation Report, Fiscal Year 1998.* BLM. USDI. U.S. Government Printing Office. Washington, DC.

van Ginkel, J.H., and A. Gorrison. 1998. *In situ* decomposition of grass roots as affected by elevated atmospheric carbon dioxide. *Soil Sci. Soc. Am. J.* 62:951-958.

van Ginkel, J.H., A. Gorrison, and J.A. van Veen. 1996. Long-term decomposition of grass roots as affected by elevated atmospheric carbon dioxide. *J. Environ. Qual.* 25:1122-1128.

Vitousek, P.M., H.A. Mooney, J. Lubchenco, and J.M. Melillo. 1997. Human domination of earth's ecosystems. *Science* 277:494-499.

Wedin, D.A., and D. Tilman. 1996. Influence of nitrogen loading and species composition on the carbon balance of grasslands. *Science* 274:1720-1723.

Weltz, M., and M.K. Wood. 1986. Short-duration grazing in central New Mexico: effects on sediment production. *J. of Soil and Water Cons.* 41:262-266.

Wilson, D.B. 1960. *Competition among three pasture species under different levels of soil nitrogen and light intensity.* Ph.D. Thesis. Oregon State University. Corvallis, OR.

Wood, J.C., M.K. Wood, and J.M. Tromble. 1987. Important factors influencing water infiltration and sediment production on arid lands in New Mexico. *J. of Arid Environ.* 12:111-118.

CHAPTER 12

The Effects of Pasture Management Practices

R.R. Schnabel,[1] A.J. Franzluebbers,[2] W.L. Stout,[3]
M.A. Sanderson,[4] and J.A. Stuedemann[5]

Introduction

Pastureland, which includes improved, native, and naturalized pastures, accounts for 51 Mha of the 212 Mha of privately held grazing land in the U.S. (Sobecki et al., Ch. 2). This chapter focuses on improved pasture in humid regions (>625 mm mean annual precipitation).

Improved pasture is grazing land permanently producing indigenous or introduced forage species, harvested primarily by grazing, and managed to enhance forage quality and yield. Livestock raised on these lands can be grazed, confined and fed stored forages, or both.

Geographic regions and their forages

Grazing lands vary widely within and among the three major geographic regions of the eastern U.S. (Fig. 12.1). Washko (1974) divided the $5^1/_4$ Mha of pastureland in the 12 *northeastern states* (10% of the total area) into four classes: cropland pastures, improved pasture, woodland pasture, and other. The improved permanent pastures we focus on are predominantly Kentucky bluegrass (*Poa pratensis* L.)/wild white clover swards containing an indeterminate mixture of orchardgrass (*Dactylis glomerata* L.), timothy (*Phleum pratense* L.), and quackgrass (*Agropyron repens* L. Beauv.) (Rohweder and Albrecht, 1994). These pastures have been im-

[1] Soil Scientist (Deceased), Agricultural Research Service, University Park, PA.

[2] Ecologist, USDA/ARS, Watkinsville, GA, (706) 769-5631 X223, afranz@arches.uga.edu.

[3] Soil Scientist, USDA/ARS, University Park, PA, (814) 863-0947, ws1@psu.edu.

[4] Research Agronomist, USDA/ARS, University Park, PA, (814) 865-1067, mas44@psu.edu.

[5] Research Animal Scientist, USDA/ARS Watkinsville, GA, (706) 769-5631 X247, jstuedem@arches.uga.edu.

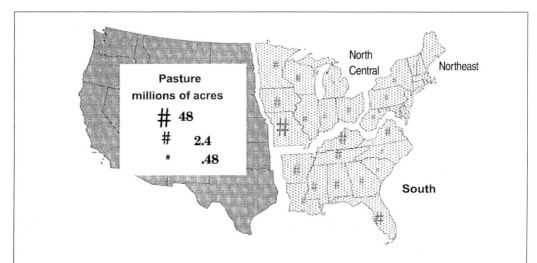

Figure 12.1. Grazing lands in the eastern U.S., 1987. Grazing land is the sum of permanent pasture, cropland, and woodland used only for grazing. Data are from Vough (1990).

tures have been improved by adding lime and fertilizer and renovating (sod/seeding legumes or grasses).

Grazing lands in the *North Central* geographic region occupy about 9 Mha. Forage species are similar to those in the northeast, except that tall fescue (*Festuca arundinacea* Schreb.) predominates in the southern reaches and smooth bromegrass is common in the northern and central portions.

In the *humid South*, pasture and forage species shift from cool season species typical of the transition zone (e.g., tall fescue, orchardgrass, red clover) to predominantly warm season species such as bermudagrass (*Cynodon dactylon*), bahiagrass (*Paspalum notatum*), and dallisgrass (*Paspalum dilatatum* Poir.) farther south. Cool season annual forages such as annual ryegrass, small grains, and several clovers (arrowleaf, crimson, subterranean, and white clover) frequently are planted as winter annual pastures on prepared seedbeds or sod/seeded into dormant warm season grass sods.

Grazed and stored forage

We consider pastures and grasslands since, even when livestock get most of their nutrients by grazing, the most efficient forage-based animal agriculture requires both grazed pastures and conserved forage. In temperate climates, pastures partially or entirely support animals for as little as 5 months to as much as the entire year. The rest of the year, pasture plants grow little or are dormant, and animals get their nutrition from conserved forages.

Stored forages are also necessary given the growth habits of temperate grasses and legumes. Common cool season forages produce less forage during the warmer months of the year (July-August) than during the spring and cool months of fall. Producers commonly stock pastures for the forage produced during July-August, which leaves more forage than grazing animals can consume during the rest of the growing season. The producer harvests excess forage and feeds it later. The periods of slow growth occur at different times and for different lengths of time in all climatic regions, but they occur, and managers must account for them.

Soil organic carbon research

Soil organic C (SOC) status and changes in response to crop management have been the subject of much research and the topic of numerous reviews and book chapters (Kononova, 1961; Tate, 1987; Jenkinson, 1988; Scow, 1997; Burke et al., 1997; Follett et al., 1995). However, relatively little of this research specifically has addressed pasture management in the U.S.

The governing principles are similar for pastures and cropland. The balance between rates of organic C (OC) input and decomposition, expressed over time, accounts for the current level of soil C in pastures. Whether pastures will sequester C in response to management or climatic perturbations depends on the relative change in these two processes. Plant, climate, animal, and soil properties control pasture productivity and OC decomposition. Most factors that change net primary productivity (NPP) also change rates of decomposition. It is reasonable to assume that conditions that encourage biomass production also maximize OC decomposition (Tate, 1987).

The majority of improved pastures in the U.S. are east of the Missouri River. They exist across a wide climatic range, from Vermont to Florida, and a broad range of precipitation amounts and distributions (Fig. 12.1). Pastures are on many soil types with varying fertility, texture, and structure. The climatic factors and soil properties have profound, generally predictable effects on SOC. Complex interactions among these variables, which are not fully understood, also may affect SOC.

Temperature, moisture, and net primary productivity

Overall, temperature and moisture are the most important climatic factors determining C flow in soils. Both factors affect NPP and microbial activity similarly. Rates of both forage production and OC decomposition are highest at temperatures of 25 to 35°C and soil moisture contents of 50 to 80% water-filled pore space, and they decline as either temperature or moisture content increase or decrease from these optimal ranges. Reductions in temperature slow decomposition more than they reduce forage production (Burke et al., 1997). Soil water status strongly influences microbial activity. If the soil is too dry, activity nearly stops, and if the soil is too wet, decomposition slows and becomes less complete.

As long as NPP is sufficient, the greatest amounts of SOC accumulation occur in cold climates or in soils that are saturated a large portion of the year (Scow, 1997). Sorption to clay, isolation in micropores, and physical protection within stable macroaggregates reduce organic matter (OM) availability (Scow, 1997). Consequently, rates of decomposition are expected to be higher in coarsely textured soils and poorly structured soils, especially those with few stable macroaggregates.

Types of management strategies

Pastureland management strategies can influence SOC. They can be strategies to manage animals, plants, and soil. Managing these variables can change relative rates of C inputs and outputs, which control steady-state levels of SOC within a climatic region on a specific soil type. SOC sequestration is generally greatest shortly after a change in management and then diminishes as rates of C input and SOC decomposition balance each other over the long term.

Effects of Animal Management

Relatively little literature deals directly with the impact of specific grazing management practices on SOC. Milchunas and Lauenroth (1993) compiled a worldwide data set of 236 grazing studies that compared one or more attributes of grazed and ungrazed sites. Nearly all of the 236 sites were rangeland that received less precipitation and fewer amendments and were far less productive than pastureland in the eastern U.S. Most studies focused on aboveground NPP. The reviewers concluded that grazing reduced aboveground NPP.

In natural ecosystems of perennial plants, annual biomass production below ground generally exceeds that above ground. Root mass was greater at grazed sites in 2/3 of the studies with measurements, and, when production was viewed at the whole plant level, grazing had no effect on plant production (Milchunas and Lauenroth, 1993). SOC or OM was measured at only 37 of the sites. No difference in SOC was found between grazed and ungrazed rangelands in which no biomass was removed from ungrazed sites. Unless stated otherwise, manure was only for grazing treatments. Ungrazed, unharvested, or mechanically harvested treatments did not receive manure.

Grazing vs. haying

In animal production systems, biomass is removed from pasture either by grazing or by mechanical harvesting, which is most commonly haying. A direct relationship exists between the level of SOC and annual additions of C to soil via crop residues (Paustian et al., 1995). Consequently, the rate of increase in SOC is higher under grazing than when hay is removed, because greater amounts of C are returned to the soil. Grazing returns 60% to 95% of ingested nutrients to the pasture as excreta (Till and Kennedy, 1981). In addition, stubble production with grazing can be up to 5% greater than with mechanical harvest (van den Pol-van Dasselaar and Lantinga, 1995; Dyer et al., 1998).

During the first 3 years of steer grazing on Coastal bermudagrass, SOC increased at a rate of 1.5 to 1.8 MT/ha/yr on Cecil-Madison-Pacolet-dominated sandy loams to sandy clay loams (clayey, kaolinitic, thermic Typic Kanhapludults) (Lovell et al., 1997). SOC under bermudagrass, that was harvested as hay or left unharvested for conservation, increased at a rate of only 0.3 to 0.4 MT/ha/yr. The similar, relatively low soil C accretion rates in unharvested and hayed management systems occurred because much of the aboveground, plant-derived C was not incorporated into the soil with either management. The higher rates of soil C accretion under cattle grazing were due to the return to the soil of much of the plant-derived C as feces that quickly became part of the SOC pool.

Following 15 to 19 years of cattle grazing on Tifton 44 or Coastal bermudagrass, SOC to a depth of 20 cm averaged 36.7 MT/ha, while three paired hayed fields contained 31.1 MT SOC/ha (Franzluebbers et al., 2000b). Most of the difference in SOC between grazed and hayed bermudagrass occurred in the surface 5 cm (Fig. 12.2). C in surface residue was also greater under grazed (1.8 MT/ha) than under hayed (1.2 MT/ha) bermudagrass.

Hassink and Neetson (1991) in a 3-year study measured 2.4 MT/ha/yr more residues returned to the system with grazing than with mowing. SOC averaged 8.9 MT/ha more in the top 25 cm in grazed pastures than in mowed grassland, with little response to fertilizer rates from 250 to 700 MT N/ha/yr. In a later study, Hassink (1994) found the effect on SOC of grazing compared to mowing to be small and inconsistent.

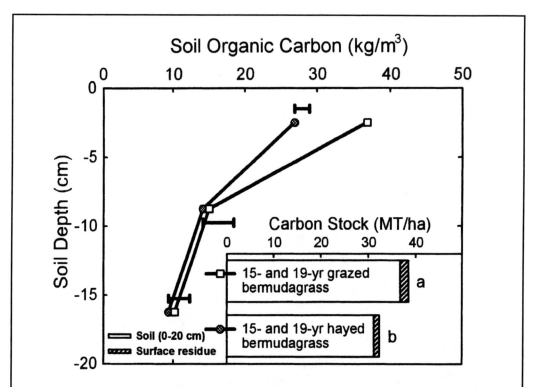

Figure 12.2. SOC depth distribution and standing stock to a depth of 20 cm under grazed and hayed bermudagrass management on a Typic Kanhapludult in Georgia (Franzluebbers et al., 2000b). LSD bars are P = 0.05 within a soil depth. Horizontal bars followed by a different letter are different at P = 0.05.

Stocking methods

The method of stocking pastures also may influence SOC. The addition of fertilizer and lime and the use of improved species make intensively managed pastures more productive. Where moisture does not constrain yield, levels of NPP are greater with intensive rotational grazing than intensive continuous grazing.

Pastures stocked continuously with few animals are least productive. For example, during the first 3 years of cattle grazing on Coastal bermudagrass, SOC increased significantly at a rate of 2.7 kg/ha for each additional grazing day within the range of 600 to 1200 grazing days/ha/yr (Stuedemann et al., 1998). The heavier foot traffic associated with the generally higher stocking densities used with more intensive vs. less intensive grazing practices may enhance breakdown of aboveground litter and its incorporation into the soil (Schuman et al., 1999). In contrast, where a moisture deficit limits production, as in western rangelands, intensive grazing may damage the stand with a concomitant loss of SOC (Hoglund, 1985; Dormaar and Willms, 1998).

Redistribution of nutrients

Spatial redistribution of plant residue and excreta from foraging areas to areas where animals congregate (camping areas) can alter SOC distribution and quantity. In six pastures on the south island of New Zealand, the difference in SOC between the main grazing areas and camping areas was significant and ranged from 3 to 14 g/kg (Haynes and Williams, 1999).

In Georgia on a Typic Kanhapludult, SOC also was concentrated most near permanent shade and water sources (i.e., camping area) in 7- to 15-year-old tall fescue pastures (Fig. 12.3). Although some literature exists on redistribution of a few different nutrients in pastures (Follett and Wilkinson, 1995; Haynes and Williams, 1999; Wilkinson et al., 1989; West et al., 1989; Peterson and Gerrish, 1996), much more research is needed to understand SOC redistribution caused by grazing and its influence on C sequestration.

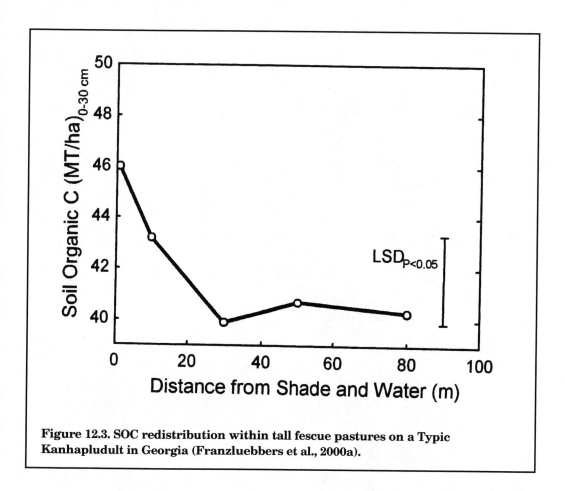

Figure 12.3. SOC redistribution within tall fescue pastures on a Typic Kanhapludult in Georgia (Franzluebbers et al., 2000a).

Effects of Forage Management

Forage and soil management obviously have interacting effects on C sequestration in grazing lands. A key to increasing C storage is increasing pasture productivity. We can either accept site conditions and select productive vegetation
more suited to the site, or we can change site conditions by fertilizing, liming, or
draining to improve pasture production.

Changes in site conditions often change the vegetative mix. Liming and adding P favors legumes. Adding N makes cool season grasses more competitive and
reduces the quantity of legumes and warm season grasses. Likewise, we may
need to alter site conditions to create an environment in which introduced species
can become established. Although sections of this chapter address forage and soil
management separately, interactions are apparent.

Animal-based agriculture in the humid eastern U.S. always has required
managing vegetation. Clearing forests and planting nonnative grasses and legumes created much of the highly productive pastureland in the humid temperate regions of the world (Tothill, 1978). The continued existence of those pastures
depends on management practices, such as grazing frequency and intensity, lime
and fertilizer inputs, and occasional replanting (Haynes and Williams, 1993).
Thus, the typical temperate, cool season grasses in these pastures (e.g., Poa, Dactylis, Festuca, Lolium) do well where rainfall and soil fertility are favorable
(Haynes, 1980).

While management is necessary to maintain productive eastern pastures, the
effect of forage management on soil C has not been an area of active research.
This section qualitatively estimates forage management's effects on soil C
through its impact on forage yield and quality — where *yield* is a surrogate for
NPP, and *quality* is a surrogate for microbial decomposition. To sequester C,
grasses should increase OC input to the soil, but, for the C to remain in the soil, it
should decompose slowly, with the overall effect that input exceeds decomposition.

Yield and carbon sequestration

Grass yields vary considerably across gradients in moisture and fertility.
Moisture deficits and low fertility generally favor warm season grasses (Stout,
1992; Stout and Jung, 1992, 1995). When no or moderate amounts of N were applied, warm season (C_4) grasses consistently produced more biomass than cool
season (C_3) grasses in Pennsylvania hill lands (Table 12.1).

Yields of all grasses increased in response to N fertilization. Yield differences
between grass types and among species within a type were smaller at the higher
fertility level. At optimum moisture and fertility, all grass types are highly productive (Table 12.2).

From these yield patterns, we infer that, at low to moderate fertility, more C is stored under warm season grasses, and the difference between warm and cool season grasses decreases as fertility increases. Wedin and Tilman (1996) reported this pattern of C storage for grasslands in Minnesota.

Much of the increase in SOC in tall grass and mixed grass prairies in recent years comes from a shift in species composition to greater proportions of warm season grasses (Schuman et al., Ch. 11). Warm season grasses generally have higher root/shoot ratios, more root biomass, and greater belowground C than cool season grasses. Also, many of the mixed and short grass warm season species are more shallowly rooted.

At low to moderate levels of N added to a Minnesota prairie soil, Wedin and Tilman (1996) reported significantly more C storage under warm season grasses than under cool season grasses. Increasing levels of N additions, however, caused a shift from grasslands dominated by highly diverse warm season grass to grasslands dominated by cool season grasses low in diversity. Therefore, given the high levels of N commonly applied, either as fertilizer, manure, or legumes, it may not be possible to shift from cool to warm season grasses to sequester C in many eastern pastures.

Table 12.1. Mean yields (MT/ha) of warm and cool season grasses in Pennsylvania.

Grass	W/O Fertilizer	75 kg N/ha/yr
Warm season grasses		
Niagara big bluestem	4.4	6.8
Blackwell switchgrass	4.2	7.9
NJ50 switchgrass	10.7	11.2
NY591 indiangrass	5.2	6.2
Cool season grasses		
KY31 tall fescue	0.8	3.8
Reed canarygrass	1.6	3.9
Pennlate orchardgrass	1.3	3.1

Table 12.2. Mean yields (1989-1991) from cool season grasses in Pennsylvania.

Grass	Yield (MT/ha)
Orchard grass	11.5
Perennial rye grass	8.9
Reed canary grass	10.1
Brome grass	9.5
Tall fescue	13.7
Timothy	10.9
Grasses were fertilized with 225-275 kg N/ha.	

Quality and carbon sequestration

Forage quality is defined in terms of the relative performance of animals when fed herbage on a nonlimiting basis. Forage quality is a function of the nutrient concentration of the herbage, its rate of intake by the animal, digestibility of the material eaten, and the efficiency with which the animal uses the metabolized products (Buxton and Mertens, 1995). Forage quality reflects the environment in which the plant was grown (including climatic, edaphic) and genetic factors.

Grass and legume species vary in forage quality, as do cultivars within species. Examples of warm season forage grasses bred for improved forage quality are 'Coastal' bermudagrass and 'Trailblazer' switchgrass, both of which were selected for improved digestibility of dry matter (Burton, 1989). A high quality forage, one that ruminants readily digest, is a forage that soil microorganisms also decompose more readily.

Many factors affect forage quality, but the overwhelming factor, and the one over which a producer has the most control, is plant maturity. All forage plants are higher in quality when young than when mature (Figs. 12.4a, 12.4b). The decrease in quality with maturity results from changes in plant morphology (Fig. 12.5) and from compositional and anatomical changes in plant tissues. As the plant matures, both cell wall concentrations and lignification typically increase, which reduces digestibility. Young plants usually contain a greater concentration

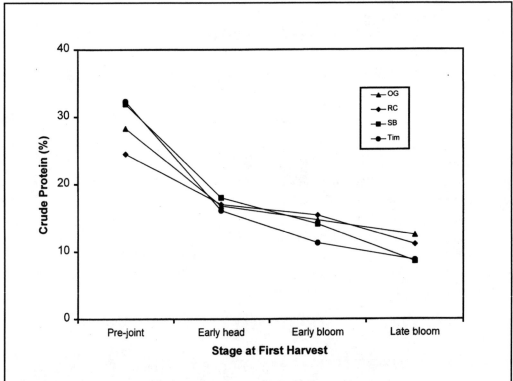

Figure 12.4a. Crude protein concentration of common pasture grasses at different developmental stages. Adapted from NE Reg. Pub. 550, 554, 557, and 570, Management and Productivity of Perennial Grasses in the Northeast, West Virginia Agricultural Experiment Station. (OG = orchard grass; RC = red clover; SB = smooth brome grass; Tim = timothy.)

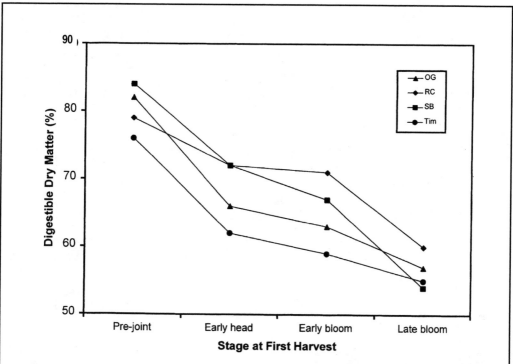

Figure 12.4b. Digestible dry matter concentration of common pasture grasses at different developmental stages. Adapted from NE Reg. Pub. 550, 554, 557, and 570, Management and Productivity of Perennial Grasses in the Northeast, West Virginia Agricultural Experiment Station. (OG = orchard grass; RC = red clover; SB = smooth brome grass; Tim = timothy.)

of N, resulting in higher protein concentration, lower C:N ratio, and greater degradability.

When grasses and legumes are harvested at the proper growth stage, legumes are usually higher in protein, and more digestible, but quality ranges widely within each group. As plants mature, they produce more biomass, but quality declines. Shorter days, higher temperatures, and lower available soil moisture cause quality losses in late summer, regardless of morphologic development. Delaying haying or grazing might increase SOC, but it does so at the expense of animal productivity.

Cool season grasses generally are of higher quality than warm season grasses, and annuals generally are of higher quality than perennials. However, forage quality also differs among cultivars of both warm and cool season grasses. For example, the switchgrass varieties *cave-in-rock* and *NJ50*, selected to provide animal feed, produce higher-quality forage than *shelter*, a variety selected more often for soil conservation purposes.

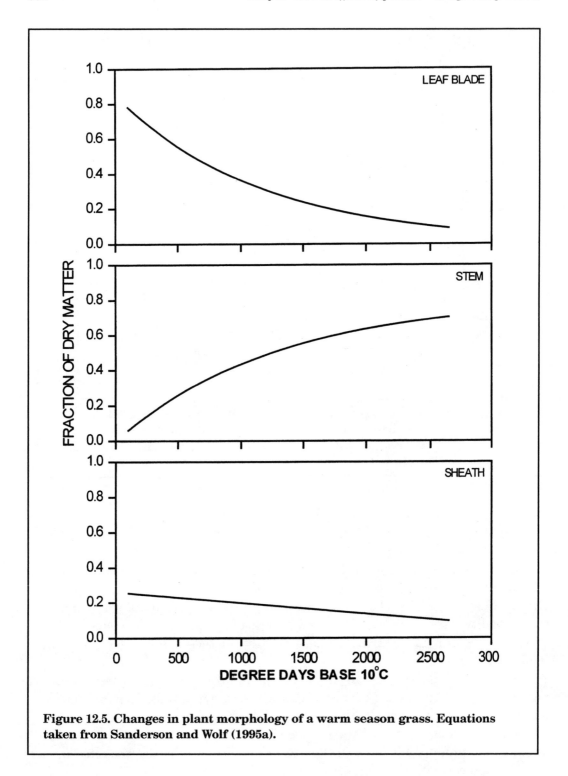

Figure 12.5. Changes in plant morphology of a warm season grass. Equations taken from Sanderson and Wolf (1995a).

Based on the forage quality patterns described above, we should be able to select grasses that decompose more slowly while maintaining relatively high levels of NPP. Managing eastern U.S. grazing lands to maximize store C storage, however, might result in suboptimum levels of animal production. Much more research is required to identify the management inputs needed to balance SOC storage with animal productivity.

The work of Franzluebbers et al. (1999b) provides an example of how grass management can affect SOC. Tall fescue is an important cool season perennial forage for many cattle producers in the humid regions of the U.S. and throughout the world. It is grown on more than 14 Mha of land in the U.S. (Buckner et al., 1979). The majority of tall fescue pastures in the U.S. is infected with a fungus, *Neotyphodium coenophialum* (Shelby and Dalrymple, 1987), which resides primarily within basal stem tissue. In Georgia, tall fescue with low endophyte infection (0% to 20%) had 29.1 MT SOC/ha in the upper 15 cm of soil, compared to 31.2 MT SOC/ha with high endophyte infection (60% to 100%) (Franzluebbers et al., 1999b). Lower potential soil microbial activity and a change in soil microbial community structure accompanied greater soil C sequestration with higher levels of endophyte infection. Biochemical alteration of the plant material caused by the endophyte may have directly or indirectly suppressed SOC decomposition.

We expect interactions among forage species and climate, forage species and grazing management, and soil management and forage quality, and ultimately all of these factors on SOC. Much of that information appears in the *NRCS National Range and Pasture Handbook* (NRCS, 1997). The end of this chapter summarizes the effects of forage management on SOC.

Effects of Soil Management

Soil management affects SOC in pastures much as it does in cropland. Management that increases NPP more than it increases decomposition results in greater SOC. When soils with low inherent fertility (i.e., low SOC) are fertilized or limed, productivity and SOC generally increase. Phosphorus availability can limit productivity where legume/grass mixtures provide forage.

Crocker and Holford (1991) surveyed 67 pasture sites in the New South Wales tablelands of Australia with SOC ranging from 20 to 29 g/kg. Adding 125 kg P/ha as superphosphate over 15 to 45 years increased SOC by 2 to 3 g/kg, with the greatest increase on soils derived from granite.

In another study in Australia, a pasture that received 4.5 MT/ha of superphosphate and 11.25 MT/ha of lime over 73 years had 67 g SOC/MT compared to 4.9 g SOC/kg in a pasture receiving neither P fertilizer nor lime (Ridley et al., 1990). In a study in New Zealand, adding 188 kg P/ha as superphosphate over 37 years substantially increased SOC (Haynes and Williams, 1992). No further increases in SOC were reported for applications of 376 kg P/ha.

In plant communities with few legumes on low-fertility sites, N amendments often increase forage production and SOC. Ladd et al. (1994) reported a linear increase in SOC with additions of N up to 80 kg N/ha. SOC in this Australian study increased for treatments that returned more crop residue to the soil.

Similar results were obtained in Kansas where smooth bromegrass yields increased with N applications up to 67 kg/ha/yr (Schwab et al., 1990). SOM paralleled yields and increased from 38 to 48 g/kg after 40 years of fertilization. When N applications were discontinued in part of the smooth brome grass experiment after 20 years, no differences in SOC appeared between the 20 and 40-year fertilized plots at the end of the 40-year period.

In Alberta, Canada, Malhi et al. (1997) reported that SOC increased by 18.5 MT/ha after 27 years of 56 kg N/ha/yr on smooth brome grass and by 23.4 MT/ha at 112 kg N/ha/yr. SOC increased most with ammonium nitrate and least with urea as the N source.

At the end of 15 years of tall fescue management with cattle grazing in Georgia, SOC was 2.6 MT/ha greater under high than under low fertilizer application rates (Table 12.3). SOC differences between fertilizer application regimes resulted from significant differences at depths of 2.5 to 7.5 and 7.5 to 15 cm. About 2/3 of the change in SOC due to fertilization was due to accumulation of an intermediately decomposed pool of particulate OC. Interestingly, the fertilizer application regime did not lead to any differences in soil microbial biomass C, basal soil respi-

Table 12.3. Effect of 15 years of fertilization of tall fescue pasture on SOC pools at Watkinsville, GA (Franzluebbers and Stuedemann, unpublished data). Low rate of fertilization was 134-15-56 kg N-P-K/ha/yr and high rate of fertilization was 336-37-139 kg N-P-K/ha /yr.

Soil Depth (cm)	Fertilization Rate	Soil Organic C (MT/ha)	Particulate Organic C (MT/ha)	Microbial Biomass C (kg/ha)	Basal Soil Respiration (kg/ha/d)
0-2.5	low	10.2	5.1	822	24.1
	high	10.9	5.9	943	28.8 *
2.5-7.5	low	11.0	4.1	585	15.0
	high	11.8 *	4.6 *	574	13.8
7.5-15	low	11.0	2.9	621	10.7
	high	11.7 *	3.1	627	10.5
15-30	low	12.8	2.7	740	7.7
	high	13.1	3.6	897	7.2
0-30	low	45.0	15.0	2769	57.5
	high	47.6 *	16.8 *	3041	60.3

Denotes significance at P = 0.05.

ration, nor their ratios with particulate OC and SOC. Higher fertilizer application rates improved plant production, which probably led to more plant root and forage residues and animal feces to supply the particulate and total OC pools.

When soils that are inherently more fertile (i.e., higher in SOC) are fertilized, the affects on productivity and SOC are less certain. Yields may increase without a response in SOC. In the long-term experiment at Rothamsted, large inputs of inorganic fertilizers had little effect on SOC (Jenkinson, 1988). After 100 years of applying 20 MT manure/ha, or either 18 or 36 kg N/ha/yr, SOC ranged from 37 to 80 g/kg, with no clear pattern. Soil microbial biomass was highest with manure but was reduced significantly by adding NPK. Without manure, there was less soil microbial biomass, but it was more active as measured by microbial respiration.

The differences in microbial activity may reflect differences in the ratio of bacteria to fungi that would be supported in the soil at the prevailing pH of treatments (Hopkins and Shiel, 1996). Withholding P from a ryegrass/clover pasture in New Zealand, with ~7.5% SOC, affected yield but not nutrient cycling within the first 2 years (Perrott et al., 1992). SOC did not respond to 10 years of low level P and S additions in another New Zealand study (Ross et al., 1995) where SOC was ~7%.

In a fertilizer rate study in Kansas with N additions up to 224 kg/ha/yr (Owensby et al., 1969), smooth bromegrass yields were greatest with the highest N application. However, SOC did not increase with increasing N. The authors speculated that increased microbial activity at higher N rates, resulting in more rapid decomposition and C oxidation, negated the effects of greater productivity on SOC.

SOC was not affected significantly by supplying N to a Coastal bermudagrass pasture, either as broiler litter or inorganic fertilizer, for 3 years (Lovell et al., 1997). The warm, moist environment in northeast Georgia probably led to rapid decomposition of the ~2.2 MT/ha/yr of added C in broiler litter.

Fertilizer additions also may result in less SOC by reducing the amount of photosynthate directed to roots or by producing more biodegradable roots and shoots. Litter decomposition links with the C/N ratio of the biomass, and the added N may produce a more N-rich and hence more degradable forage. Poorly drained pastures at North Wyke, southern England, receiving 200 kg N/ha/yr for 50 years, had 110 g/kg less SOC in the upper 10 cm than pastures receiving no N (Lovell et al., 1995).

Reduced total, microbial, and mineralizable C pools in fertilized compared with unfertilized pastures were attributed to less root production at the expense of increased herbage production. Application of high rates of N leads to reductions in root mass and root C input to the soil (Ennick et al., 1980). The amount of root C contribution to the soil is often less than the increase in herbage returned to the soil following fertilization (Hassink, 1994).

Finally, SOC in highly fertile soil may increase further in response to fertilizer application. For example, Hatch et al. (1991) reported greater accumulation of SOC under perennial ryegrass fertilized with 420 kg N/ha/yr than under perennial ryegrass/white clover.

The CCGRASS model (van den Pol-van Dasselaar and Lantinga, 1995) predicted that the rate of increase for SOC would be greatest at low to moderate rates of N (100 to 250 kg N/ha/yr). Much of the available literature supports those predictions.

Carbon Sequestered in Pasture Soils

Pastureland soils generally have high levels of SOC and high structural stability. SOC accumulates when cultivation of arable soils ceases. For example, SOC concentration of the 0- to15-cm depth increased from ~33 to 38 g/kg during the first 5 years of pasture growth following crop cultivation in Argentina (Studdert et al., 1997). Concomitant with the increase in total C were increases in light-fraction C, soil microbial biomass, and aggregate stability. Similar improvements in aggregation and SOC concentration occurred with the conversion from annual cropping to alfalfa production during the first 5 years in Quebec, Canada (Angers, 1992).

In a longer-term study, bermudagrass, maintained for at least 35 years in south-central Texas, contained 49 MT SOC/ha in the upper 20 cm of soil. Similar soils under conventionally tilled cropping systems had only 20 to 25 MT SOC/ha (Franzluebbers et al., 1998). In another long-term study, conventionally tilled cropland was converted to either tall fescue/common bermudagrass pasture or minimum-till cropland on a Typic Kanhapludult in Georgia. After 25 years, 7.6 MT/ha more SOC had accumulated under the pasture to a depth of 20 cm than under the minimum-till cropland, which produced a summer crop and a winter cover crop with minimal soil disturbance (Fig. 12.6).

Lewis et al. (1987) reported a linear increase in SOC of 0.2 to 0.3 g/kg/yr with age of pasture by applying superphosphate and seeding legumes. In a permanent rotation trial started in 1925 and fertilized with P, Grace et al. (1995) reported a linear decline in SOC with years in fallow and years of cropping after pasture. SOC ranged from 25 g/kg under permanent pasture to 10 g/kg for the wheat–fallow rotation.

SOC generally increases following conversion of cropland to pasture, although it could take many years to reach a new steady state. Dormaar and Smoliak (1985) and McConnell and Quinn (1988) reported that it took 50 plus years for

SOC of abandoned cropland to approach the level of native rangeland. The difference in SOC to a depth of 20 cm on a Typic Kanhapludult in Georgia between tall fescue pastures established for 10 vs. ~50 years was 8.7 MT/ha (Fig. 12.7) and between Coastal bermudagrass hay land established for 10 vs. ~40 years was 2.8 MT/ha (Fig. 12.8). SOC reached a new steady state only after 200 years when arable land was allowed to revert to grass at Rothamsted (Jenkinson, 1988). Fifty percent of the change in SOC occurred in the first 25 years.

Others have reported relatively rapid SOC changes in response to land use changes. In the Great Plains, 21% (0.8 MT/ha/yr) of the C lost by decades of intensive tillage was recovered within 5 years under the Conservation Reserve Program, where cropland was converted to unharvested grasslands (Gebhart et al., 1994). In another study, 5 years of grass pasture increased SOC from 46 to 77 g/kg

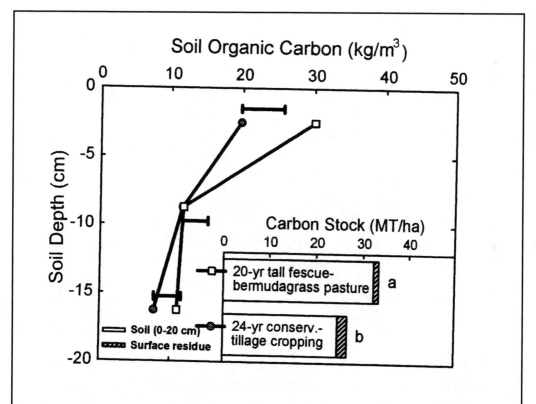

Figure 12.6. SOC depth distribution and standing stock to a depth of 20 cm at the end of 25 years of conversion from conventionally tilled cropland to tall fescue-bermudagrass pasture or conservation tillage cropping on a Typic Kanhapludult in Georgia (Franzluebbers et al., 2000b). LSD bars are P = 0.05 within a soil depth. Horizontal bars followed by a different letter are different at P = 0.05.

(Douglas and Goss, 1982). Richter et al. (1990) reported that 24% of SOC stored in grassland was lost in 6 years of annual tillage, mostly from a loss of root mass.

Much of the pastureland in the eastern U.S. was or could potentially support forest, because of abundant precipitation. Land use before pasture establishment could affect short- and mid-term status of SOC by altering its initial depth distribution and total quantity.

In two paired comparisons of 15-year-old bermudagrass (Tifton 44 pasture and Coastal hayland) in Georgia, SOC to a depth of 20 cm averaged 34.9 MT/ha after clearing mixed hardwood-pine forest and 29.0 MT/ha after several years of cropping (Fig. 12.9). Increases in SOC after forest clearing occurred at the soil surface, but also at a depth of 12.5 to 20 cm, perhaps due to incorporation, during the planting process, of a large pool of C-rich surface residue, which typically oc-

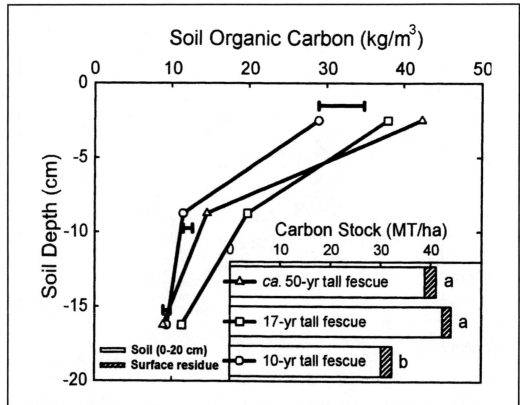

Figure 12.7. SOC depth distribution and standing stock to a depth of 20 cm as affected by time under 'Kentucky-31' tall fescue pasture on a Typic Kanhapludult in Georgia (Franzluebbers et al., 2000b). LSD bars are P = 0.05 within a soil depth. Horizontal bars followed by a different letter are different at P = 0.05.

curs in forest ecosystems (Fig. 12.10). This pool of surface residue in forests may be fairly recalcitrant when mixed into the soil after clearing.

Importance of Soil Organic Carbon to Soil Quality

Pasture management, compared with crop management, typically increases SOC concentration nearest the surface (Fig. 12.11). This increase affects soil physical properties such as water infiltration. Greater rates of infiltration in turn can improve efficiency of water use and deter erosion.

Surface residues can buffer rainfall energy, thus protecting surface structure and improving infiltration. Surface residue in pastures also can be an important physical barrier to water movement across the landscape. However, removing the

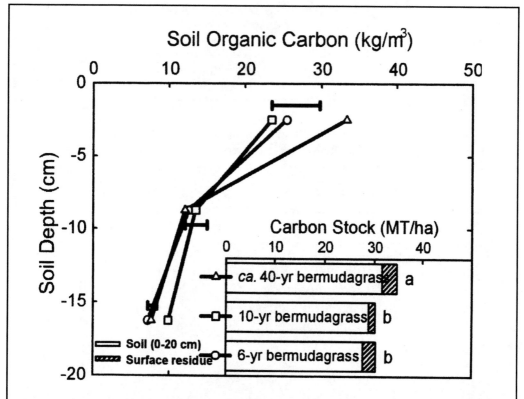

Figure 12.8. SOC depth distribution and standing stock to a depth of 20 cm as affected by time under Coastal bermudagrass hayland on a Typic Kanhapludult in Georgia (Franzluebbers et al., 2000b). LSD bars are P = 0.05 within a soil depth. Horizontal bars followed by a different letter are different at P = 0.05.

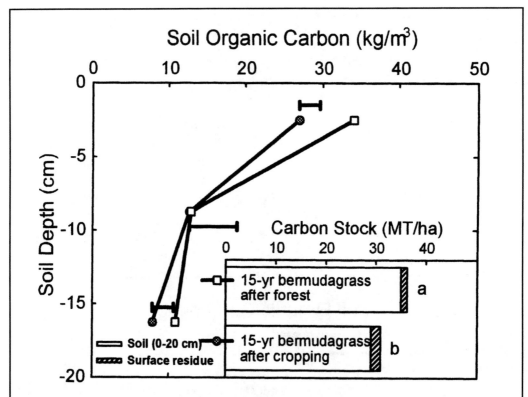

Figure 12.9. SOC depth distribution and standing stock to a depth of 20 cm as affected by previous land use before establishment of either Coastal (hayed) or Tifton 44 (grazed) bermudagrass on a Typic Kanhapludult in Georgia (Franzluebbers et al., 2000b). LSD bars are P = 0.05 within a soil depth. Horizontal bars followed by a different letter are different at P = 0.05.

surface residue before single-ring infiltration measurements (30-cm diameter) did not reduce steady-state water infiltration, which averaged 17 cm/hr during September in 8-year tall fescue pastures (Franzluebbers, unpublished data). These results suggest that greater infiltration rates in well managed pasture may link more closely to improvements in SOC than to accumulation of surface residue.

Despite a twofold difference in SOC in a 30-year Bavarian study, NMR & FTIR spectra of fulvic and humic acids reflected no differences in chemical stability (Capriel et al., 1992). The authors attributed differences in SOC to physical stabilization of the OM.

Soil's physical improvements with increasing SOC are known. Douglas and Goss (1982) found a linear relationship between a wet stability index and OM con-

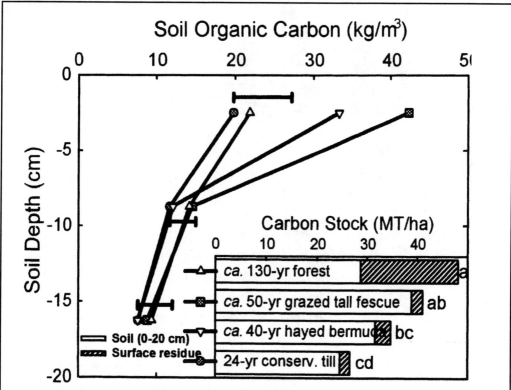

Figure 12.10. SOC depth distribution and standing stock to a depth of 20 cm as affected by long-term management systems on a Typic Kanhapludult in Georgia (Franzluebbers et al., 2000b). LSD bars are P = 0.05 within a soil depth. Horizontal bars followed by a different letter are different at P = 0.05.

tent. Haynes and Swift (1990) reported nearly twice as much SOC (36 vs. 20 g/kg) and more stable aggregates under 25-year-old pasture than under cropland.

The large increase in SOC near the soil surface in pasture typically leads to lower bulk density and greater porosity (Fig. 12.12), which allows water to pass through the soil surface more rapidly. The data of Carreker et al. (1977) illustrate the dependence of water infiltration on SOC concentration in the Southern Piedmont region of the U.S. Various crop/pasture rotations, compared with continuous cropping, led to increases in SOC concentration which, in turn, led to increases in the rate of water infiltration and the amount of time elapsed before runoff occurred (Fig. 12.13).

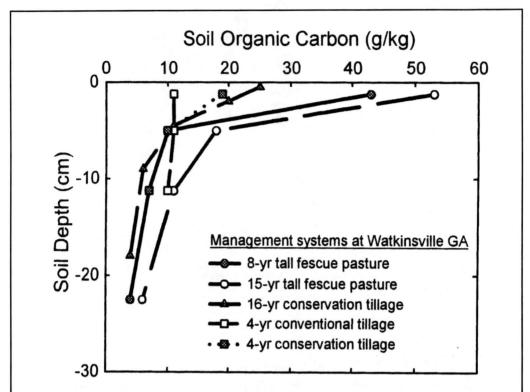

Figure 12.11. SOC concentration with soil depth under various crop and pasture management systems on Typic Kanhapludults. Data from tall fescue pastures are from Franzluebbers et al. (1999b). Data from 16-year conservation tillage are from Bruce and Langdale (1997). Data from 4-year conservation tillage are from Franzluebbers et al. (1999a).

Potential Negative Environmental Consequences of Pasture Development

Sequestration of soil C can increase with increased pasture productivity and conversion of marginal cropland to pasture. In the long term, as a steady state approaches, the rate of increase will diminish, approaching zero. However, while major, positive environmental benefits will result from an increase in pasture development, negative environmental consequences also may occur. Some potential environmental costs of such a shift include water quality problems, potential C costs related to fertilization, effects of various nutrients' buildup in soils, etc.

Dairy and cow-calf industries dominate ruminant animal agriculture in the eastern U.S. Use of grasslands through grazing has always been the major source of energy and protein for cow-calf herds. Recently, a growing number of small and medium sized dairy producers have adopted grazing to provide a significant part

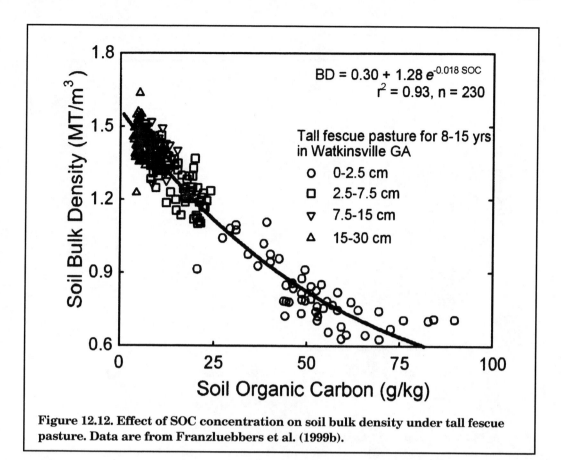

Figure 12.12. Effect of SOC concentration on soil bulk density under tall fescue pasture. Data are from Franzluebbers et al. (1999b).

of the energy and protein for both their milking and dry cow herds (Fales et al., 1995).

This shift has been primarily for economic reasons and has necessitated a conversion of croplands to grazing lands. The increase in soil C from this conversion also may incur environmental and economic costs. For example, water quality concerns could change from those dominated by sediment and pesticides in cropland to those dominated by fecal organisms and nutrients in pastures.

In intensively managed grazing systems adopted by dairy farmers in the northeastern U.S., animals are rotated frequently through a series of pastures during the grazing season, causing frequent defoliation of the pasture swards. The main adaptive response of plants to these episodes of defoliation is the priority allocation of C from roots to shoots and increased respiration in order to restore light interception and C assimilation through increased shoot growth. The rate of shoot growth and new leaf expansion depends on the quantity of reserve protein in the plant and the rate at which this protein is recycled within the plant.

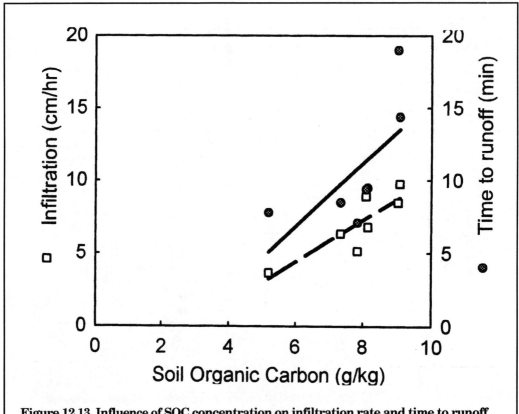

Figure 12.13. Influence of SOC concentration on infiltration rate and time to runoff as a result of various crop/pasture rotations. Data are from Carreker et al. (1977).

In other words, the richer a grassland system is in N, the faster it can recover from defoliation and resume sequestering C in the root system. However, the richer a grassland system is in N, the greater the threat to water quality from the leaching of urine N or applied fertilizer N (Ryden et al., 1984; Jarvis et al., 1989; Whitehead, 1995; Stout et al., 1997). Thus, there can be a trade-off between capturing CO_2 from the atmosphere by increasing C sequestered in the soil and improving water quality.

Legumes or fertilizer N can maintain N fertility. Legumes not only supply N to increase C sequestration and primary biomass production but also increase the quality of forage, thereby increasing production per animal. However, pasture systems driven by legume N are not as productive as those driven by fertilizer N. This means that legume systems are more profitable on a per animal basis but less profitable on a per hectare basis. If land prices were low, profits could be maximized on a per animal basis and use of legumes would be desirable. However, if land prices were high, profits would need to be maximized on a per hectare basis, and fertilizer N systems would be needed.

If fertilizer N is required to maximize profits, additional C costs are associated with manufacturing, transporting, and spreading it. Since N fertilization rates on intensively managed pastures can be higher than those on corn (Whitehead, 1995; Fales et al., 1995), the C costs of this additional N must be charged against any additional C sequestered by N-fertilized pastures. The C costs associated with mining, hauling, and grinding the lime used to neutralize the soil acidity, which N fertilization causes, also must be charged against increased soil C. There is also a direct C cost in neutralizing acid from the nitrification of ammonium nitrate or urea. As an example, 0.43 kg C could be released to the atmosphere as CO_2 if calcium carbonate is used to neutralize the acidity generated per kg N.

Another hidden problem associated with C sequestration by intensively managed pasturelands arises when energy and nutrient-dense feedstuffs are imported into grazing areas. Intensively grazed pastures are low in energy and high in protein relative to the needs of the grazing animal. This is especially true for the grazing dairy cow, where low intake of energy from pasture herbage can result in low milk production and high levels of urea N in milk.

Producers balance the animal's ration by supplementing the herbage with off-farm, energy-dense feed grains, much of it imported from outside the geographic region. The manure generated by feeding this grain cannot economically be returned to the fields where the grain was grown but typically remains at its destination (Lanyon, 1995). This results in a buildup of nutrients in soil (particularly P and K) on grazing dairy farms. Increased levels of P can contaminate fresh surface waters and have been associated with outbreaks of pfisteria. High levels of soil K cause concerns because high levels of K in pasture herbage have a deleterious effect on the health of grazing animals, especially dry cows.

Conclusion

Our review of SOC status and dynamics suggests we can store more C in the grazing land soils of the eastern U.S. (Table 12.4). The magnitude and duration of this storage is difficult to estimate for the region. Converting marginal cropland to pasturelands will increase SOC. Changes in how animals, plants, and soils are managed also can affect the balance between C inputs to the soil via plant fixation and losses of SOC to the atmosphere via decomposition.

Where pasturelands are highly productive and SOC is already high, small or no increases in C storage can be expected. Larger increases may be made on marginally productive pasturelands by improving soil fertility or managing animals and plants better. Making pastures more productive, however, could compromise efforts to improve water quality and could decrease farm profitability.

Table 12.4. The effects of pasture management methods on C storage in soil.

Factor	Measured Effect on NPP or Forage Quality	Measured or Inferred Effect on C Storage in Soil
ANIMAL MANAGEMENT		
Grazing grasslands	More C returned to soil for rapid incorporation.	Increase SOC.
Intensive grazing	With adequate moisture, intensive management increases NPP. Increased foot traffic breaks down residue.	Increase SOC.
	With limited moisture, increased stocking can damage stands.	Decrease SOC.
FORAGE MANAGEMENT		
Replacing C_3 grasses with C_4 grasses	At low to moderate fertility, increase NPP and reduce forage quality.	Increase SOC.
	At high fertility, little change in NPP.	Little change in SOC. May not be sustainable.
Replace endophyte infected fescue with uninfected fescue	Increase forage quality.	Decrease SOC.
Increase harvest frequency	Reduce NPP, increase forage quality.	Decrease SOC.
Delay harvest or grazing	Reduce forage quality.	Increase SOC.
SOIL MANAGEMENT		
Liming	Increases P availability and NPP.	Increases SOC.
P fertilization	If P deficient, increase NPP.	Increase SOC.
	If P is adequate or in excess, no change.	No change.
N fertilization	Low inherent fertility, increase NPP and forage quality.	Increase SOC.
	High inherent fertility; NPP, and decomposition of SOC, no change or increase.	No change, decrease, or increase in SOC, depending on relative change in NPP and decomposition.
Manuring	Increases NPP if fertility limits growth.	Increases SOC.
Drainage	Increases NPP, increases SOC decomposition.	Decreases SOC.

Follett, Kimble, and Lal, editors

In Memory of Ron Schnabel

Dr. Ronald R. Schnabel died on February 21, 2000, in State College, PA. Ron was born on March 19, 1951, in Eureka, SD. He was the son of Ruth Ketterling Schnabel and the late Gottfried Schnabel.

Ron graduated from South Dakota State University in March 1974 with a B.S. in Environmental Management and in 1977 with a M.S. in Soil Science. He received his Ph.D. in Soil Science from Washington State University in 1981. His Ph.D. research was innovative and forward-looking. He successfully applied soil physics and soil chemistry to the modeling of nitrate leaching and denitrification losses from furrow irrigated soils.

Ron joined the Agricultural Research Service at University Park, Pa., in 1980 as a Soil Chemist. He was a research scientist and project leader at the Northeast Watershed Research Center and then at the Pasture Systems and Watershed Management Research Laboratory. He was extremely productive and was best known for developing innovative field-based methodologies to measure inorganic N losses, and for his stream riparian zone work. He developed methods for sampling and estimating nitrate and ammonium leaching in soil, using ion exchange resins, and for determining gas diffusivities in soils, especially nitrous oxides. He showed that stream riparian zones and floodplains in northeastern hill land watersheds could intercept and denitrify excess nitrate en route to streams, but he also cautioned us with work that showed this beneficial site-specific effect could be much reduced at the watershed scale. Prior to his untimely death, Ron had expanded his research to look at carbon sources and carbon-N dynamics affecting N availability and losses from stream riparian zones and heavily manured soils. He also had expanded his riparian zone management work to include the impact on benthic organisms.

Ron was a gifted and giving scientist who made a difference. He worked very effectively with others and very much enjoyed working with undergraduate and graduate students. In addition to his professional activities, Ron served his community as a volunteer instructor at the Mid-State Literacy Council. In this capacity he taught reading, writing, math, and life skills. His fellow scientists, collaborators, co-workers, and friends will sorely miss him.

References

Angers, D.A. 1992. Changes in soil aggregation and organic carbon under corn and alfalfa. *Soil Sci. Soc. of Am. J.* 56:1244-1249.

Bruce, R.R., and G.W. Langdale. 1997. Soil carbon level dependence upon crop culture variables in a thermic-udic region. *In* E.A. Paul, K. Paustian, E.T. Elliott, and C.V. Cole (eds), *Soil Organic Matter in Temperate Agroecosystems: Long-Term Experiments in North America*, CRC Press, Boca Raton, FL, pp. 247-261.

Buckner, R.C., J.B. Powell, and R.V. Frakes. 1979. Historical development. *In* R.C. Buckner and L.P. Bush (eds), *Tall Fescue. Agron. Monogr. 20*, ASA, Madison, WI, pp. 1-8.

Burke, I.C., W.K. Lauenroth, and D.G. Milchunas. 1997. Biogeochemistry of managed grasslands in central North America. *In* E.A. Paul, K. Paustian, E.T. Elliott, and C.V. Cole (eds), *Soil Organic Matter in Temperate Agroecosystems: Long-Term Experiments in North America*, CRC Press, Boca Raton, FL, pp. 85-102.

Burton, G.W. 1989. Progress and benefits to humanity from breeding warm season forage grasses. *Contributions from Breeding Forage and Turf Grasses. Special Publication No. 15.* Crop Science Society of America, Madison, WI. p. 21-29.

Buxton, D.R., and D.R. Mertens. 1995. Quality-related characteristics of forages. *In* R.F. Barnes, D.A. Miller, and C.J. Nelson (eds), *Forages, The Science of Grassland Agriculture.*, Vol. 11., 5th ed., Iowa State Univ. Press, Ames, IA, pp. 83-96.

Capriel, P., P. Harter, and D. Stephenson. 1992. Influence of management on the organic matter of a mineral soil. *Soil Sci.* 153:122-128.

Carreker, J.R., S.R. Wilkinson, A.P. Barnett, and J.E. Box. 1977. *Soil and Water Management Systems for Sloping Land. Agricultural Research Service, USDA, ARS-S-160.* U.S. Government Printing Office. Washington, DC.

Crocker, G.K., and I.C.R. Holford. 1991. Effects of pasture improvement with superphosphate on soil pH, nitrogen and carbon in a summer rainfall environment. *Aust. J. of Exp. Agric.* 31:221-224.

Dormaar, J.F., and S. Smoliak. 1985. Recovery of vegetative cover and soil organic matter during revegetation of abandoned farmland in a semiarid climate. *J. of Range Manage.* 38:487-491.

Dormaar, J.F., and W.D. Willms. 1998. Effect of forty-four years of grazing on fescue grassland soils. *J. of Range Manage.* 51:122-126.

Douglas, J.T., and M.J. Goss. 1982. Stability and organic matter content of surface soil aggregates under different methods of cultivation and in grassland. *Soil & Tillage Res.* 2:155-175.

Dyer, M.I., C.L. Turner, and T.R. Seastedt. 1998. Biotic interactivity between grazers and plants: Relationships contributing to atmospheric boundary layer dynamics. *J. Atm. Sci.* 55:1247-1259.

Ennick,G.C., M. Gillet, and L. Sibma. 1980. Effect of high nitrogen supply on sward deterioration and root mass. *In* W.H. Prins and G.H. Arnold (eds), *The Role of N in Intensive Grassland Production*, Pudoc, Wageningen, pp. 67-76.

Fales, S.L., L.D. Muller, S.A. Ford, M. O'Sullivan, R.J. Hoover, L.A. Holden, L.E. Lanyon, and D.R. Buckmaster. 1995. Stocking rate affects production and profitability in a rotationally grazed pasture system. *J. of Prod. Agric.* 8:88-96

Follett, R.F., and S.R. Wilkinson. 1995. Nutrient management in forages. *In* R.F. Barnes, D.A. Miller, and C.J. Nelson (eds), *Forages: Vol. II*, Iowa State University Press, Ames, IA.

Franzluebbers, A.J., F.M. Hons, and D.A. Zuberer. 1998. *In situ* and potential CO_2 evolution from a Fluventic Ustochrept in south-central Texas as affected by tillage and cropping intensity. *Soil and Tillage Res.* 47:303-308.

Franzluebbers, A.J., J.A. Stuedemann, H.H. Schomberg, and S.R. Wilkinson. 2000a. Spatial distribution of soil carbon and nitrogen pools under grazed tall fescue. *Soil Sci. Soc. Am. J.* 64:635-639.

Franzluebbers, A.J., J.A. Stuedemann, H.H. Schomberg, and S.R. Wilkinson. 2000b. Soil organic C and N pools under long-term pasture management in the Southern Piedmont, USA. *Soil Biol. Biochem.* 32:469-478.

Franzluebbers, A.J., G.W. Langdale, and H.H. Schomberg. 1999a. Soil carbon, nitrogen, and aggregation in response to type and frequency of tillage. *Soil Sci. Soc. of Am. J.* 63:349-355.

Franzluebbers, A.J., N. Nazih, J.A. Stuedemann, J.A. Fuhrmann, H.H. Schomberg, and P.G. Hartel. 1999b. Soil carbon and nitrogen pools under low- and high-endophyte-infected tall fescue. *Soil Sci. Soc. of Am. J.* 63:1687-1694

Grace, P.R., J.M. Oades, H. Kieth, and T.W. Hancock. 1995. Trends in wheat yield and organic carbon in the permanent rotation trial at the Waite Agricultural Research Institute, South Australia. *Aust. J. of Exp. Agric.* 35:857-864.

Gebhart, D.L., H.B. Johnson, H.S. Mayeux, and H.W. Polley. 1994. The CRP increases soil organic carbon. *J. of Soil and Water Cons.* 49:488-492.

Hassink, J. 1994. Effects of soil texture and grassland management on soil organic C and N and rates of C and N mineralization. *Soil Biol. and Biochem.* 26:1221-1231.

Hassink, J., and J.J. Neetson. 1991. Effects of grassland management on the amounts of soil organic N and C. *Netherlands J. of Agric. Sci.* 39:225-236.

Hatch, D.J., S.C. Jarvis, and S.E. Reynolds. 1991. An assessment of the contribution of net mineralization to N cycling in grass swards using a field incubation method. *Plant and Soil* 138:23.

Haynes, R.J. 1980. Competitive aspects of the grass-legume association. *Adv. Agron.* 33:227-261.

Haynes, R.J., and R.S. Swift. 1990. Stability of soil aggregates in relation to organic constituents and soil water content. *J. of Soil Sci.* 41:73-83.

Haynes, R.J., and P.H. Williams. 1992. Accumulation of soil organic matter and the forms, mineralization potential and plant-availability of accumulated organic sulphur: effects of pasture improvement and intensive cultivation. *Soil Biol. and Biochem.* 24:209-217.

Haynes, R.J., and P.H. Williams. 1993. Nutrient cycling and soil fertility in the grazed pasture ecosystem. *Adv. Agron.* 49:119-199.

Haynes, R.J., and P.H. Williams. 1999. Influence of stock camping behaviour on the soil microbiological and biochemical properties of grazed pastoral soils. *Biol. and Fertil. of Soils* 28:253-258.

Hopkins, D.W., and R.S. Shiel. 1996. Size and activity of soil microbial communities in long-term experimental grassland plots treated with manure and inorganic fertilizers. *Biol. and Fertil. of Soils* 22:66-70.

Hoglund, J.H. 1985. Grazing intensity and soil nitrogen accumulation. *Proceedings of the New Zealand Grassland Association* 46:65-69.

Jarvis, S.C., D.J. Hatch, and D.H. Roberts. 1989. The effects of grassland management on nitrogen loss from grazed swards through ammonia volatilization, the relationship to excretal N returns from cattle. *J. of Agric. Sci. Cambridge* 112:205-216.

Jenkinson, D.S. 1988. Soil organic matter and its dynamics. *In* A. Wild (ed), *Russell's Soil Conditions and Plant Growth*, John Wiley & Sons, New York.

Kononova, M.M. 1961 *Soil Organic Matter: Its Nature, Its Role in Soil Formation and in Soil Fertility*. Pergamon Press. New York.

Ladd, J.N., M. Amato, Z. Li-Kai, and J.E. Schultz. 1994. Differential effects of rotation, plant residue and nitrogen fertilizer on microbial biomass and organic matter in an Aust. alfisol. *Soil Biol. and Biochem.* 26:821-831.

Lal, R., J.M. Kimble, R.F. Follett, and C.V. Cole. 1998. *The Potential of U.S. Cropland to Sequester Carbon and Mitigate the Greenhouse Effect*. Ann Arbor Press. Chelsea, MI.

Lanyon, L.E. 1995. Does nitrogen cycle? Changes in the spatial dynamics of nitrogen with industrial nitrogen fixation. *J. of Prod. Agric.* 8:70-78.

Lewis, D.C., A.L. Clarke, and W.B. Hall. 1987. Accumulation of plant nutrients and changes in soil properties of sandy soils under fertilized pasture in Southeastern South Australia. II. Total sulfur and nitrogen, organic carbon and pH. *Aust. J. of Soil Res.* 25:203-210.

Lovell, R.D., S.C. Jarvis, and R.D. Bardgett. 1995. Soil microbial biomass and activity in long-term grassland: Effects of management changes. *Soil Biol. and Biochem.* 27:969-975.

Lovell, A.D., S.R. Wilkinson, J.A. Stuedemann, D.H. Seman, and A.J. Franzluebbers. 1997. Broiler litter and grazing pressure impacts on soil organic C and N pools. *Agronomy Abstracts*. American Society of Agronomy, Madison, WI. p. 217.

Malhi, S.S., M. Nyborg, J.T. Harapiak, K. Heier, and N.A. Flore. 1997. Increasing organic C and N in soil under bromegrass with long-term N fertilization. *Nutri. Cycling in Agroecosys.* 49:255-260.

McConnell, S.G., and M.L. Quinn. 1988. Soil productivity of four land use systems in southeastern Montana. *Soil Sci. Soc. of Am. J.* 52:500-506

Milchunas, D.G., and W.K. Lauenroth. 1993. Quantitative effects of grazing on vegetation and soils over a global range of environments. *Ecological Monographs* 63:327-366.

NRCS. 1997. *National Range and Pasture Handbook*. NRCS, Grazing Lands Technology Institute. U.S. Government Printing Office. Washington, DC.

Owensby, C.E., K.L. Anderson, and D.A. Whitney. 1969. Some chemical properties of a silt loam soil after 20 years nitrogen and phosphorus fertilization of smooth brome-grass (*Bromus inermus* Leyss). *Soil Sci.* 108:24-29.

Paustian, K., G.P. Robertson, and E.T. Elliot. 1995. Management impacts on carbon storage and gas fluxes (CO_2, CH_4) in mid-latitude cropland. *In* R. Lal, J. Kimble, E. Levine, and B.A. Srewart (eds), *Soil Management and Greenhouse Effect, Advances in Soil Science*, Lewis, Boca Raton, FL, pp. 69-83.

Perrot, K.W., S.U Sarathchandra, and B.W. Dow. 1992. Seasonal and fertilizer effects on the organic cycle and microbial biomass in a hill country soil under pasture. *Aust. J. of Soil Res.* 30:383-394.

Peterson, P.R., and J.R. Gerrish. 1996. Grazing systems and spatial distribution of nutrients in pastures: Livestock management considerations. *In* R.E. Joost, and C.A. Roberts (eds), *Nutrient Cycling in Forage Systems*, Potash Phosphate Inst. and Foundation Agron. Res., Manhattan, KS, pp. 203-212.

Richter, D.D., L.I. Babbar, M.A. Huston, and M. Jaeger. 1990. Effects of annual tillage on organic C in a fine-textured udalf: the importance of root dynamics to soil carbon storage. *Soil Sci.* 149:78-83.

Ridley, A.M., W.J. Slattery, K.R. Helyar, and A. Cowling. 1990. The importance of the carbon cycle to acidification of a grazed annual pasture. *Aust. J. of Exp. Agric.* 30:529-537.

Rohweder, D.A., and K.A. Albrecht. 1994. Permanent pasture ecosystems. *In* R.F. Barnes, D.A. Miller, and C.J. Nelson (eds), *Forages: Vol. 11, The Science of Grassland Agriculture.*, 5th ed., Iowa State University Press, Ames, IA, pp. 207-223.

Ross, D.J., T.W. Speir, H.A. Kettles, and A.D. Mackay. 1995. Soil microbial biomass, C and N mineralization and enzyme activities in a hill pasture: influence of season and slow-release P and S fertilizer. *Soil Biol. and Biochem.* 27:1431-1443.

Ryden, J.C., P.R. Ball, and E.A. Garwood. 1984. Nitrate leaching from grassland. *Nature (London)* 311:50-53.

Sanderson, M.A., and D.D. Wolf. 1995a. Switchgrass biomass composition and morphological development in diverse environments. *Crop Sci.* 35:1433-1438.

Schuman, G.E., J.D. Reeder, and W.A. Manley. 1999. Impact of grazing management on the carbon and nitrogen balance of a mixed-grass rangeland. *Ecol. Appls.* 9:65.

Schwab, A.P., C.E. Owensby, and S. Kulyingyong. 1990. Changes in soil chemical properties due to 40 years of fertilization. *Soil Sci.* 149:35-43.

Scow, K.M. 1997. Soil microbial communities and carbon flow in agroecosystems. *In* L.E. Jackson (ed), *Ecology in Agriculture*, Academic Press, San Diego, CA, pp. 367-413.

Shelby, R.A., and L.W. Dalrymple.1987. Incidence and distribution of the tall fescue endophyte in the United States. *Plant Dis.* 71:783-786.

Stout, W.L. 1992. Water-use efficiency of grasses as affected by soil, nitrogen and temperature. *Soil Sci. Soc. of Am. J.* 56:897-902.

Stout, W.L., and G.A. Jung. 1992. Influences of soil environment on biomass and nitrogen accumulation rates of orchardgrass. *Agron. J.* 84:1011-1019.

Stout, W.L., and G.A. Jung. 1995. Effects of soil and environment on biomass accumulation of switchgrass. *Agron. J.* 87:663-669.

Stout, W.L., S.L. Fales, L.D. Muller, R.R. Schnabel, W.E. Priddy, and G.F. Elwinger. 1997. Nitrate leaching from cattle urine and feces in northeast U.S. *Soil Sci. Soc. of Am. J.* 61:1787-1794.

Studdert, G.A., H.E. Echeveria, and E.M. Cassanovas,. 1997. Crop-pasture rotation for sustaining the quality and productivity of a Typic Argiudoll. *Soil Sci. Soc. of Am. J.* 61:1466-1472.

Stuedemann, J.A.., A.J. Franzluebbers, D.H. Seman, S.R. Wilkinson, R.R. Bruce, A.D. Lovell, and S.W. Knapp. 1998. Role of the grazing animal in soil carbon restoration in the Southern Piedmont. *Agron. Abstr.* p. 57.

Tate, R.L. III. 1987. *Soil Organic Matter: Biological and Ecological Effects.* John Wiley & Sons. New York.

Till, A.R., and A.P. Kennedy. 1981. The distribution in soil and plant of 35S sulfur isotope from sheep excreta. *Aust. J. of Agric. Res.* 32:339-351.

Tothill, J.C. 1978. Comparative aspects of the ecology of pastures. *In* J.R. Wilson (ed), *Plant Relations in Pastures*, CSIRO, Australia, pp. 385-420.

Van den Pol-Van Dasselaar, A. and E.A. Lantinga. 1995. Modelling the carbon cycle of grassland in the Netherlands under various management strategies and environmental conditions. *Netherlands J. of Agric. Sci.* 43:183-194.

Vough, L.R. 1990. Grazing lands in the east. *Rangelands* 12:251-255.

Washko, J.B. 1974. Forages and grassland in the northeast. *In* H.B. Sprague (ed), *Grasslands of the United States: Their Economic and Ecologic Importance*, Iowa State University Press, Ames, IA, pp. 98-112.

Wedin, D.A., and D. Tilman. 1996. Influence of nitrogen loading and species composition on the carbon balance of grasslands. *Sci.* 274:1720-1723.

West, C.P., A.P. Mallarino, W.F. Wedin, and D.B. Marx. 1989. Spatial variability of soil chemical properties in grazed pastures. *Soil Sci. Soc. Am. J.* 53:784-789.

Whitehead, D.C. 1995. *Grassland Nitrogen.* CAB International. Oxon, UK.

Wilkinson, S.R., J.A. Stuedemann, and D.P. Belesky. 1989. Soil potassium distribution in grazed K-31 tall fescue pastures as affected by fertilization and endophytic fungus infection level. *Agron. J.* 81:508-512.

CHAPTER 13

The Effects of Fire and Grazing on Soil Carbon in Rangelands

C.W. Rice[1] and C.E. Owensby[2]

Introduction

Rangelands occupy 47% of the earth's land area (Williams et al., 1968) and 161 Mha of the U.S. (Sobecki et al., Ch. 2). A majority of the rangeland in the U.S. is in the central prairie region, with a gradient from short grass to tall grass prairie along a moisture gradient. A large percentage of total net primary productivity in grassland ecosystems occurs below ground, and much of the fixed C is processed by heterotrophic organisms (Elliot et al., 1988). The total soil organic C (SOC) results from the interaction of vegetation and soil organisms, climate, parent material, time, and disturbances (Jenny, 1941). Thus, the response of rangelands to land management and environmental change is potentially significant to the global C budget.

Various modeling studies (e.g., Hunt et al., 1991; Cole et al., 1993; Ojima et al., 1990) indicate that grasslands could function as either sinks or sources of C, depending on land management regimes. Two major rangeland management practices are fire and grazing. This chapter assesses their impacts on soil C.

Effects of Fire

Plant community structure

Most range areas developed with fire as a major determinant in plant community structure. Fire was probably the most important controlling factor in devel-

---

[1] Professor of Soil Microbiology and Director, KS EPA-EPSCoR Program, 2004 Throckmorton Plant Sciences Center, Dept. of Agronomy, Kansas State Univ., Manhattan, KS 66505-5501, phone (785) 532-7217, fax (785) 532-6094, e-mail cwrice@ksu.edu.

[2] Professor of Range Management, 2004 Throckmorton Plant Sciences Center, Dept. of Agronomy, Kansas State Univ., Manhattan, KS 66505-5501, phone (785) 532-7217, fax (785) 532-6094, e-mail owensby@ksu.edu.

oping the grassland climax (Sauer, 1950; Stewart, 1951). Fires favor grasslands and grasslands favor fire. Probably none of the major grasslands escaped periodic natural fires; these fires, along with grazing by ungulate herbivores, shaped the grassland community structure greatly. Climatic influences were probably the dominant forces during development, but fire was likely to be a major sustaining force in the stability of most range plant communities.

Natural fires primarily were caused by lightning in the initial development of the prairie (Komarek, 1966), but later primitive humans set many fires for different purposes. Catlin (1848) and Lewis (1969) indicated natural fires were frequent during pre-settlement times, and aboriginal humans burned areas to attract buffalo during their tenure on the prairie (Gleason, 1913; Lewis, 1969). Bragg (1974) indicated the presence of humans increased the likelihood of accidental fires, as well.

Frequency of fire on the prairie was regular, occurring almost annually (Catlin, 1848; Newberry, 1873; Shimek, 1911) and at least two to three times in a given area in a 5-year period. Other range types, such as open forest, woodland, savanna, and desert shrub, had natural fires as well, their frequency varying from 2- to 30-year intervals (Mutch, 1970; Kozlowski and Ahlgren, 1974).

Mutch (1970) hypothesized that, if plant species have developed numerous fire-resistant mechanisms, they also could have developed flammability characteristics which contribute to maintaining fire-dependent plant communities. Grasslands with large amounts of highly flammable fire fuels are particularly susceptible to frequent burning and usually are made of fire-resistant species. Borchert (1950) stated that the climate of grassland regions also favors fire.

Tall grass prairie

Fire is particularly important in maintaining tall grass prairie by limiting woody plant invasion. Fire can have a direct effect on soil organic matter (SOM), since aboveground plant residues are lost by ignition, as much as 89% per burn (Ojima et al., 1990). In tall grass prairie, the increased plant production following the fire compensates for the loss of plant C by ignition.

Knapp (1985) reported reduced photosynthetic capacity for tall grass prairie dominants when large amounts of surface litter were allowed to accumulate due to lack of fire or grazing. Anderson et al. (1970) and Owensby and Anderson (1967) indicated that, under grazing, the 17-year average yearly production was equal on areas burned and unburned in late spring. On areas with no surface litter removal, annual net plant productivity is greater on annually burned tall grass prairie (Knapp et al., 1998; Towne and Owensby, 1984). The positive response to fire is due to removal of plant litter, which increases light and soil temperatures (Knapp and Seastedt, 1986).

The studies in tall grass prairie generally were planned around a specific time of burning. Time of burning is critical to the plant's response to fire. The longer the surface remains barren following fire, the greater is the loss of soil water through surface evaporation or through increased runoff.

Short grass and mixed prairie

The positive response to burning is not universal; the use and effects of fire vary greatly with vegetation type. On mixed grass prairies, fire results in lower plant productivity (Redman, 1978). Apparently the loss of plant residues decreases soil water, which is limiting in mixed and short grass prairies. In arid areas, fire frequency may be low because of detrimental side effects not related to its intended purpose. On the other hand, certain areas, particularly in humid regions, may require fire to maintain the ecosystem's integrity.

Burning in the central Great Plains short grass region generally has reduced herbage yields (Hopkins et al., 1948; Launchbaugh, 1964; Harris, 1973). Most studies have reported on effects of accidental fires that came at inopportune times. Launchbaugh (1972) indicated that proper timing may reduce the detrimental effects of fire in short grass prairie (Table 13.1). They showed that herbage yields were reduced only during the last year, after burning the same plots for 3 years.

Improperly timed fire in the short grass prairie of the southern Great Plains generally has reduced herbage yields. Dwyer and Peiper (1967) studied the effects of an accidental fire in pinyon-juniper/blue grama vegetation of New Mexico.

Table 13.1. Grass and forb dry weight production on clay upland range at Hays, KS, burned at different dates (adapted from Launchbaugh, 1972).

Date of Burn	Component	Herbage Yield (MT/ha) Unburned	Burned	Yield Change due to Burning (MT/ha)
November 22, 1944	grass	3.02	2.15	−0.87
	forb	.15	.50	+.36
	Total	3.17	2.65	−0.52
March 27, 1945	grass	3.02	.71	−2.32
	forb	.15	1.36	+1.21
	Total	3.17	2.06	−1.11
March 18, 1959	grass	3.83	1.33	−2.50
	forb	.43	.21	−0.21
	Total	4.26	1.55	−2.71
April 26, 1975[1]	grass	1.75	2.46	+.72
	forb	.11	.24	+.12
	Total	1.86	2.70	+.84

[1]Prescribed burn; others were wildfires.

The Potential of U.S. Grazing Lands to Sequester Carbon and Mitigate the Greenhouse Effect

Forage production in the burned area was reduced in the first season but was not reduced by the end of the second season, compared with that of an adjacent unburned area. Properly timed fires generally do not reduce herbage yields (Trlica and Schuster, 1969; Heirman and Wright, 1973). However, when Trlica and Schuster (1969) burned Texas High Plains range in the fall, summer, and spring, all times of burning reduced herbage yields.

Fire in the short grass prairie of the southern Great Plains has its primary value in woody plant control. Junipers and cacti appear the most likely candidates for control. The reduction in heavy infestation of those species may overcome any temporary yield reduction of desirable species.

As with most regions, research reports on fire in the semiarid mixed prairie of the northern Great Plains are based largely on one fire at one time, with the result being data insufficient to indicate whether prescribed burning is detrimental. Research reports from the region indicate reduced herbage yields associated with range burning (Clarke et al., 1943; Coupland, 1973; Dejong and McDonald, 1975). Redmann (1978) and Dejong and McDonald (1975) reported that reductions in yields of herbage most likely resulted from lower soil moisture on burned areas than on unburned ones.

Proper timing of prescribed burns in the mesic mixed prairie area did not result in reduced herbage yields when Gartner and Thompson (1972) burned in late April in the Black Hills forest-grass ecozone to reduce pine invasion into natural grassland areas. Dix (1960) found that herbage yields were reduced on two of three areas that were burned in western North Dakota, but these fires occurred on August 14, September 30, and May 29, none of which is a desirable burning dates. Fire in the region generally favors warm season grasses over cool season ones. Since properly timed prescribed burns usually are not detrimental to herbage yields or vegetative composition, range burning in the mesic mixed prairie of the northern Great Plains provides an excellent management tool to control vegetative composition, reduce woody species invasion, and promote grazing distribution.

The fescue prairie of the northern Great Plains has suffered increases in tree and shrub invasion due to lack of fire (Johnston and Smoliak, 1968; Bailey and Wroe, 1974; Bailey and Anderson, 1978). Bailey and Anderson (1978) reported that annual herbage production was not reduced by either a fall or spring burn if the desirable species were dormant.

Concerns

Loss of organic matter

The major concern about fire in prairies is that the destruction of organic matter (OM) and subsequent nutrient loss and soil surface exposure will do irreparable damage to the soil. However, since prairies have been subjected to repeated fires during their evolution, compensating mechanisms within the ecosystem probably prevent catastrophic changes in the soil's physical and chemical properties. The nature of the root system of grasses, with large numbers of small roots near the soil surface, precludes significant soil erosion when plant densities are relatively high. The roots hold the soil mass against erosive wind and water. Root biomass turnover rates insure a continual, renewed supply of OM to the soil system.

OM is essential in soil aggregation, which improves infiltration capacity and resists erosion. SOM in grassland soils is almost exclusively a product of the short-lived root system of grasses, and as much as 95% of the OM comes from the belowground plant parts. Generally, an amount equal to the entire belowground biomass dies and is replaced within a 2- to 4-year period (Dahlman and Kucera, 1965).

Loss of nutrients

Among the many elements in the fuels which range fires consume, only C and N are lost appreciably due to volatilization. Other nutrients in the particulates of smoke also are lost. Combustion losses from burned Kansas Flint Hills range were 80% for C, 73% for N, and 33% for phosphorus (Ojima et al., 1990). The extent of the loss largely depends on fire intensity, with extremely hot fires which burn to white ash completely volatilizing the N (White et al., 1973).

Nitrogen

Since some or most of the N in the aboveground biomass is lost during range burning, the common assumption has been that the burning will reduce soil N (Sharrow and Wright, 1977a). However, long-term burning studies in the Kansas Flint Hills have failed to substantiate that conclusion. Owensby and Wyrill (1973) reported no reductions in total soil N after 48 years of annual burning. Aldous (1934) reported similar results on areas burned for a shorter period.

Sharrow and Wright (1977b) hypothesized that, on tobosagrass sites in Texas, it would take 5 to 8 years to restore N to preburn levels on burned sites. That hypothesis proved wrong.

They sampled during July, when a high proportion of the N was in the aboveground biomass. Aboveground biomass was greater and higher in N content on burned areas than on unburned ones. Subsequent translocation of N to storage areas below ground might reduce or eliminate differences between burned and unburned areas in total soil N. Total soil N values should be determined just before spring growth in order to account for differences due to aboveground biomass N. Since N is lost from vegetation consumed in the fire, why do soils on burned areas have similar N content to that of unburned areas?

Apparently, denitrification losses from soil on burned areas in tall grass prairie are extremely low compared to those on unburned areas (Ojima et al., 1990; Groffman et al., 1993). Denitrification occurs under anaerobic conditions, and moist conditions prevail under unburned conditions for a greater period than under burned conditions.

Eisele et al. (1990), working in the same region, also showed that nonsymbiotic N fixation by *Nostoc* sp. was enhanced greatly by the addition of the ash to the soil's surface. They concluded that the elemental phosphorus in the ash stimulated N-fixation. Therefore, the reduced denitrification losses and increased N-fixation on burned areas offset the combustion losses.

On more arid areas, the effect of combustion losses may not be overcome by reduced denitrification and increased fixation. Sharrow and Wright (1977a) showed lower nitrate levels in soils on burned areas than on unburned ones.

Other elements

Since N is the only element lost in any great amount to volatilization, other minerals should not decrease in burned more than in unburned areas. Owensby and Wyrill (1973) showed generally higher amounts of available soil calcium, magnesium, and potassium on burned areas than on unburned areas, with essentially no difference in P content of soils on burned and unburned areas. Ueckert et al. (1978) reported that burning may increase salinity, sodium, and potassium of soils on tobosa-mesquite range during wet years.

Apparently little change can be expected in soil chemistry as a result of range burning. Changes in the rate and time of mineralization of nutrients may occur as a result of differing soil temperatures between burned and unburned areas within a given season, which alters microbial degradation of SOM.

Soil temperature

The temperatures at the soil surface during a burn are high only for a short period and do not warm the soil appreciably, usually less than 1°C. The soil temperature differences between burned and unburned ranges come later. Removing the insulating mulch and exposing the mineral soil surface increases soil temperature (Kelting, 1957; Kucera and Ehrenreich, 1962).

Aldous (1934) reported higher soil temperatures at 2.5, 7.6, and 17.8 cm, from early spring to midsummer, on areas burned in early spring than on unburned areas (Fig. 13.1). With a later spring burn, Hulbert (1969) reported similar effects, to a depth of 40 cm, which lasted until late August. Soil temperature affects microbial activity, evaporation, and plant growth and development and probably is basic to many of the differences that result from range burning.

Sharrow and Wright (1977a) reported that higher soil temperatures on burned than on unburned tobosagrass range (Fig. 13.2) resulted in increased nitrate production early in the season, but increased plant growth caused increased

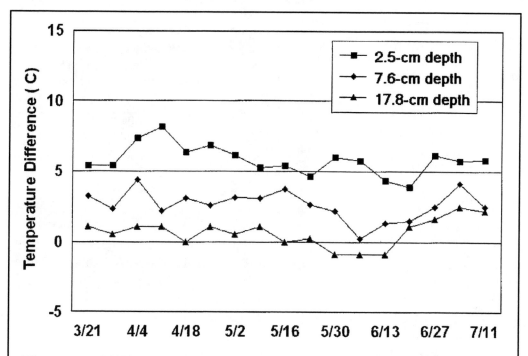

Figure 13.1. Difference in maximum soil temperatures at indicated depths on burned vs. unburned Kansas bluestem range (adapted from Aldous, 1934).

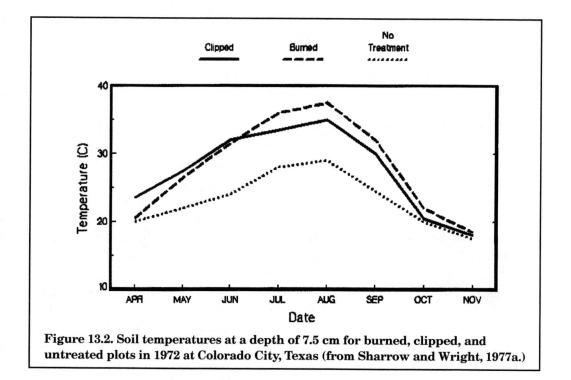

Figure 13.2. Soil temperatures at a depth of 7.5 cm for burned, clipped, and untreated plots in 1972 at Colorado City, Texas (from Sharrow and Wright, 1977a.)

The Potential of U.S. Grazing Lands to Sequester Carbon and Mitigate the Greenhouse Effect

uptake so that soil nitrate levels were lower on burned than unburned areas (Table 13.2). Soil microbial activity is stimulated by warmer soils on burned areas. The net result is an increased rate of mineralization of SOM, making soil nutrients available earlier in the growing season during the most active growth period.

Table 13.2. Average soil nitrate N (mg/g) at three depths during the spring growth period at Colorado City, TX (adapted from Sharrow and Wright, 1977a).

	Soil Depth		
Treatment	0-2.5 cm	2.5-5.0 cm	5.0-12.5 cm
No treatment	94	37	25
Burned	47	25	17

Net effect

Even though most aboveground OM oxidizes during range burning, the net effect on the OM content of the mineral soil is negligible. In those rangelands that are fire-positive, burning causes greater C inputs into the soil. The increase in aboveground plant production also shows in root biomass (Ojima et al., 1990; Seastedt and Ramundo, 1990; Rice et al., 1998). The average increase on ungrazed areas is 23%.

After 10 years of annual burning, SOC was not significantly less for burned than for unburned tall grass prairie (Seastedt and Ramundo, 1990; Dell, 1998). Owensby and Wyrill (1973), working in the Kansas Flint Hills, reported no reduction in SOM following 48 years of annual burning. Indeed, plots burned in early spring had higher SOM than plots not burned or burned in late spring.

Aldous (1934) reported similar results from earlier research on the same plots. Ueckert et al. (1978) concluded that burning in a tobosa-mesquite area in Texas increased SOC above that of unburned areas when the clayey soils were cracked, but that when soils were not cracked, burning did not affect SOC content. Other authors, reporting on the effects of one year of burning, concluded that burning does not lower SOM content below that of unburned areas (Reynolds and Bohning, 1956; Wolfe, 1972).

Simulation modeling of fire's effects in tall grass prairie shows a depletion of SOM over a period of 20 to 100 years (Risser and Parton, 1982; Ojima, 1987; Ojima et al., 1990, 1994) but validation of these results is limited because of the decade to century required to measure these changes.

The lower predicted SOM levels with fire, in spite of higher ANPP than on unburned prairie, apparently result from enhanced SOM turnover because of increased soil temperatures after fire (Ojima et al., 1990). Fire has increased the proportion of SOC in the labile fraction after 5 years (Rice and Garcia, 1994) (Fig. 13.3). This is probably the result of increased plant production returning a greater mass of C to the soil. This higher ratio of active C to total OC may indicate long-term changes in total SOC (Anderson and Domsch, 1989; Insam et al., 1989; Insam, 1990). This is in contrast to the model discussed earlier, where OM decreased with long-term annual burning. Thus, at this time, it appears that fire's effects on SOC are negligible.

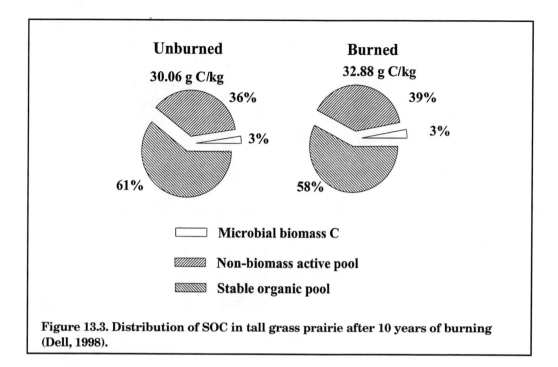

Figure 13.3. Distribution of SOC in tall grass prairie after 10 years of burning (Dell, 1998).

Effect of Grazing

The impact of grazing on ecosystem processes depends largely on the extent of the removal of plant parts used for photosynthesis or for storage of energy reserves. The amount of photosynthetic tissue available to capture energy, determined in part by grazing intensity, regulates productivity. Up to a certain point, the consequences of increasing grazing's intensity are minimal, but the range ecosystem changes with further increases.

General productivity

Productivity depends on the rate at which matter cycles through the ecosystem, and energy is needed for that cycling. If the capture of energy is reduced (reduced photosynthetic area), then productivity declines. Therefore, energy flow through the ecosystem slows (reduced productivity).

Other ecosystem attributes also are changed. Biogeochemical cycles are disrupted. Aerial and subterranean layering are altered. In short, the output from the system is reduced, and the plant community is altered. Those species which are near their limits of tolerance, on the more mesic side, decline, and those on the more xeric side increase.

Intensity, season, duration, and frequency of grazing determine the extent of community change. In most instances, total biomass production declines as grazing intensity increases. The effect usually is not linear on aboveground biomass production, however, which remains essentially unchanged with light and moderate grazing intensities and then declines with heavy use. Some range areas appear to sustain high aboveground biomass production even with heavy grazing, particularly in arid areas where short grasses dominate.

Top growth

Range plants evolved with grazing influence and developed a certain amount of tolerance to top removal without reducing C reserves or drastically slowing growth. Many research studies have shown that partial removal of the top growth of grasses does not affect production within the season or in subsequent seasons (Sampson and McCarty, 1930; Cooper, 1956; Driscoll, 1957). Other studies showed increased yield with only partial removal (Canfield, 1939; Merrill and Young, 1959; Eck et al., 1975).

More severe removal results in reduced biomass production. The energy reserves become deficient, and regrowth following dormancy or severe top removal is reduced. Only when defoliation occurs to the extent that the plant cannot sustain shoot growth on the remaining leaf area does the plant use C reserves to initiate new growth from a new apical meristem. In order to maintain accelerated growth, it is necessary that shoot removal not be so severe as to move the plant's growth rate back into the initial slow growth period.

The amount of aerial growth depends on the amount of leaf area. Humphreys and Robinson (1966), working in Australia, concluded that, when moisture and nutrients were in adequate supply, subtropical grass growth depends more on leaf area than on the status of reserve energy during the growing season.

Belowground growth

The impact of grazing is not the same for aboveground and belowground growth. Branson (1956) reported that simulated grazing (frequent clipping to 1- and 2.5-cm stubble heights) reduced both aboveground and belowground biomass (Table 13.3). Note that shoot growth on severely clipped plants was reduced, but not so much as root growth, which indicates that the aboveground sink has priority over the belowground when there is not sufficient photosynthetic capacity to sustain sufficient capture of energy to continue normal growth.

The grazing impact on an ecosystem's C stores is primarily through reduced belowground biomass production. Belowground plant parts depend on aboveground growth for energy, and severe defoliation can affect not only their quality (energy content of reserve storage organs) but also their quantity. Immediately

Table 13.3. Total yields of tops and roots of five grass species subjected to indicated clipping treatments (from Branson, 1956).

	Western Wheatgrass	Bluebunch Wheatgrass	Needle-and-thread	Kentucky Bluegrass	Blue Grama
Clipping Frequency			Top Weights (g)		
2-week intervals	15.8	31.1	19.7	21.7	12.2
4-week intervals	30.5	38.3	ND	ND	ND
At end of 14 weeks	73.5	84.3	163.1	91.6	33.0
			Root Weights (g)		
2-week intervals	1.1	0.7	21.5	0.5	0.8
4-week intervals	3.4	0.7	ND	ND	ND
*ND = no data					

Table 13.4. Root weights of different species clipped frequently and unclipped (from Biswell and Weaver, 1933).

	Root Dry Weight (g/m^2)		
Species	Clipped	Unclipped	% of Unclipped
Big bluestem (lowland)	4.44	82.30	5.3
Big bluestem (upland)	3.90	100.00	3.9
Buffalograss	6.22	22.74	27.3
Switchgrass (lowland)	3.10	116.22	2.6
Switchgrass (upland)	1.84	51.12	3.6
Kentucky bluegrass	0.82	3.98	20.6
Little bluestem	1.58	19.88	7.9
Blue grama	0.82	19.50	4.2
Sideoats grama	2.92	18.64	25.6
		Average	**10.1**

after severe defoliation, root growth stops for periods proportional to the severity of removal (Crider, 1955). After frequent, severe defoliation, plant root systems are reduced (Biswell and Weaver, 1933) (Table 13.4).

Dwyer et al. (1963) severely clipped warm season grasses for 6 years, finding that reductions in amounts of both shoots and roots directly relate to the severity of removal of aboveground biomass. Launchbaugh (1957) extracted roots from a short grass range near Hays, Kansas, that had been grazed at heavy, moderate, and light intensities from 1946 to 1956. Following the drought of the early 1950s, roots in the heavily grazed pasture penetrated to only 1.4 m, while roots in moderately and lightly grazed pastures extended to 2.1 m. Reduced root growth also results in reduced moisture and nutrient uptake, rendering the plant less competitive and less productive.

Incomplete growth response

Obviously, the direct effect of grazing is the removal of aboveground plant matter and its associated C. However, some of the plant C removed by the grazer is returned to the soil as feces, and neither the removal of plant matter nor the fertilization has a consistent impact on net primary production across grasslands.

Milchunas and Lauenroth (1993) reviewed the literature related to grazing's effects on aboveground plant production. They found that aboveground production generally decreased with grazing. Burke et al. (1997) suggested that grasslands with a long evolutionary history of grazing either show no aboveground response or a positive response to grazing. Grasslands with a short evolutionary history of grazing often show a significant reduction of aboveground productivity. Results for mixed grass prairie (Holland et al., 1992) and tall grass prairie (Turner et al., 1993) suggest a positive response in aboveground net primary production. However, Biondini et al. (1998) found no effect in northern mixed grass prairie.

The belowground plant response to grazing is as variable as the aboveground plant response. The review by Milchunas and Lauenroth (1993) showed that root biomass responded positively to grazing at more sites than negatively. The average response was a 20% increase in root biomass. In short grass systems, root biomass was reduced or not different between grazed areas and ungrazed rangelands (Schuster, 1964; Leetham and Milchunas, 1985; Milchunas and Lauenroth, 1989). In mixed grass prairie, root biomass significantly increased and decreased with grazing (Brand and Goetz, 1986; Holland and Detling, 1990).

Inconsistent carbon response

Like plants, SOC responds to grazing inconsistently. Milchunas and Lauenroth (1993) reported a nearly equal number of positive and negative responses to grazing in their comprehensive review. Soil C in short grass prairie decreases with grazing, due to lower C inputs and increased turnover (Holland and Detling, 1990; Holland et al., 1992). In northern mixed grass prairie, grazing had no significant effect on soil C (Kieft, 1994; Matthews et al., 1994; Frank and Groffman, 1998; Biondini et al., 1998).

Others have reported an increase in soil C with moderate grazing (Dormaar et al., 1990, Dormaar and Willms, 1990; Frank et al., 1995; Manley et al., 1995). Derner et al. (1997) examined a gradient from tall grass to short grass prairie in the central U.S. grasslands. The tall and mixed grass prairie had significantly reduced soil C with long-term (>25 yr) grazing, while the short grass prairie had increased levels of soil C. In the grasslands of the flooding Pampa (Argentina), grazing had no effect on SOC (Lavado et al., 1996).

Effect of Increased Atmospheric Carbon Dioxide

Although most CO_2 enrichment experiments in undisturbed systems show increased net ecosystem CO_2 uptake (Koch and Mooney, 1996), no one has explicitly demonstrated increased soil C stocks after extended exposure to elevated CO_2 under field conditions. The observed accumulations of soil C under elevated CO_2 are supported by what we know about belowground inputs in this system. Although the observed plants were dominantly C_4 plants, both above- and belowground production were stimulated under elevated CO_2 when moisture was limiting (Owensby et al., 1996).

Information from root ingrowth bags suggests that root production increased by an annual average of 41% under elevated CO_2 (Owensby et al., 1999). In a study with elevated CO_2, we have measured a nonsignificant increase in SOM after 3 years (Rice et al., 1994); however, both above- and belowground plant production increased in this environment.

After 8 years under twice-ambient CO_2, SOC in the surface 30 cm significantly increased by an average of 59 g/m^2/yr relative to the two ambient CO_2 controls. This has resulted in a significantly greater concentration of SOC under elevated CO_2 (Williams et al., submitted; Fig. 13.4).

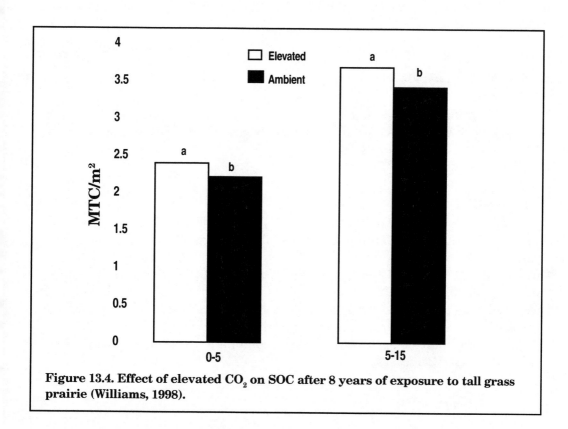

Figure 13.4. Effect of elevated CO_2 on SOC after 8 years of exposure to tall grass prairie (Williams, 1998).

Conclusion

Most rangelands have coevolved with fire and grazing. Therefore, fire, when properly managed, can promote growth of many rangeland grasses. In tall grass prairie, increased plant growth compensates for the loss of aboveground C from burning. The increase in plant production results in little change in SOC. In mixed and short grass prairie, burning can reduce plant production; however, the impact on SOC is not always evident.

The impact of grazing on rangelands depends largely on grazing intensity. Low or moderate levels of grazing have minimal impact on production and soil C. In some situations, grazing can increase soil C. High grazing intensities often have negative effects on plant production and soil C.

How rangelands respond to increasing levels of atmospheric CO_2 will greatly determine their sequestering of soil C. The extent to which ranglands sequester more C in response to increased CO_2 depends on how their plant communities respond to the CO_2 increase. Few CO_2 enrichment experiments have been conducted in the field for relatively long times of exposure. In an 8-year CO_2 enrichment study on intact tall grass prairie, both above- and belowground plant production increased. The additional plant C in the soil system resulted in an average increase of 59 g $C/m^2/yr$ over the 8-year period. After 8 years, SOC was significantly higher under elevated CO_2.

Acknowledgments

This research was supported by the U.S. Department of Energy, Carbon Dioxide Research Division, NSF (LTER) Program, Konza Prairie LTER, and the Kansas Agricultural Experiment Station.

References

Aldous, A.E. 1934. Effect of burning on Kansas bluestem pastures. *Kansas Agric. Exp. Sta. Bull.* 38.

Anderson, K.L., Clenton E. Owensby, and Ed F. Smith. 1970. Burning bluestem range. *J. Range Manage.* 23:81-92.

Anderson, J.P.E., and K.H. Domsch. 1989. Ratios of microbial biomass carbon to total carbon in arable soils. *Soil Biol. Biochem.* 21:471-479.

Bailey, A.W., and R.A. Wroe. 1974. Aspen invasion in a portion of the Alberta parklands. *J. Range Manage.* 27:263-266.

Bailey, A.W., and M.L. Anderson. 1978. Prescribed burning of a fescue-stipa grassland. *J. Range Manage.* 31:446-449.

Biondini, M.E., B.D. Patton, and P.E. Nyren. 1998. Grazing intensity and ecosystem processes in a northern mixed-grass prairie, USA. *Ecol. Appl.* 8:469-479.

Biswell, H.H., and J.E. Weaver. 1933. Effect of frequent clipping on the development of roots and tops of grasses in prairie sod. *Ecol.* 14(4):368-390.

Borchert, J.R. 1950. The climate of the central North American grassland. *Ann. Assoc. Am. Georg.* 40:39.

Bragg, T.B. 1974. *Woody plant succession on various soils of unburned bluestem prairie in Kansas.* Ph.D. Thesis, Kansas State University. Manhattan, KS.

Brand, M.D., and H. Goetz. 1986. Vegetation of exclosures in southwestern South Dakota. *J. Range Manage.* 39:434-437.

Branson, F.A. 1956. Quantitative effects of clipping treatments on five range grasses. *J. Range Manage.* 9:86-88.

Burke, I.C., W.K. Lauenroth, and D.G. Milchunas. 1997. Biogeochemistry of managed grasslands in Central North America. *In* E.A. Paul et al. (eds), *Soil Organic Matter in Temperate Agroecosystems: Long-Term Experiments in North America*, CRC Press, Boca Raton, FL, pp. 85-102.

Canfield, R.H. 1939. The effect of intensity and frequency of clipping on density and yield of black grama and tobosagrass. *USDA Tech. Bul.* 681.

Catlin, G. 1848. *Illustrations of the Manners, Customs, and Conditions of the North American Indians.* Vol. 2, 7th Ed. Henry Bohn. Yorks St., London.

Clarke, S.E., E.W. Tisdale, and N.A. Skoglund. 1943. The effects of climate and grazing on short grass prairie vegetation. *Can. Dept. Agric. Tech. Bull.* 46.

Cole, C.V., K. Paustian, E.T. Elliott, A.K. Metherell, D.S. Ojima, and W.J. Parton. 1993. Analysis of agroecosystem carbon pools. *Water Air Soil Pollut.* 70:357-371.

Cooper, C.S. 1956. The effect of time and height of cutting on yield, crude protein, content, and vegetative composition of a native flood meadow in eastern Oregon. *Agron. J.* 48:257-258.

Coupland, R.T. 1973. *Producers: 1. Dynamics of aboveground standing crop. Matador Project, Tech. Rep. No. 27.* Canadian Int. Biol. Prog. Saskatoon, SK.

Crider, F.J. 1955. Root-growth stoppage resulting from defoliation of grass. *USDA Tech. Bull.* 1102.

Dahlman, R.C., and C.L. Kucera. 1965. Root productivity and turnover in native prairie. *Ecol.* 46:84-89.

Dejong, E., and K.B. McDonald. 1975. The soil moisture regime under native grassland. *Geoderma* 14:20-221.

Dell, C.J. 1998. *The impact of fire on nitrogen cycling in tall grass prairie.* Ph.D. dissertation, Kansas State University. Manhattan, KS.

Derner, J.D., D.D. Briske, and T.W. Boutton. 1997. Does grazing mediate soil carbon and nitrogen accumulation beneath C4, perennial grasses along an environmental gradient? *Plant Soil* 191:147-156.

Dix, R.L. 1960. The effects of burning on the mulch structure and species composition of grasslands in western South Dakota. *Ecol.*41:49-55.

Dormaar, J.F., S. Smoliak, and W.D. Willms. 1990. Distribution of nitrogen fractions in grazed and ungrazed fescue grassland Ah horizon. *J. Range Manage.* 43:6-9.

Dormaar, J.F., and W.D. Willms. 1990. Effect of grazing and cultivation on some chemical properties of soils in the mixed prairie. *J. Range Manage.* 43:456-460.

Driscoll, R.S. 1957. Effects of intensity and date of herbage removal on herbage production of elk sedge. *J. Range Manage.* 10:212.

Dwyer, D.D., W.C. Elder, and G. Singh. 1963. Effects of height and frequency of clipping on pure stands of range grasses in north central Oklahoma. *Okla. Agric. Expt. Sta. Bull.* B-614.

Dwyer, D.D., and R.D. Peiper. 1967. Fire effects on bluegrama-pinyon- juniper rangeland in New Mexico. *J. Range Manage.* 20:359-362.

Eck, H.V., W.G. McCully, and J. Stubbendieck. 1975. Response of short grass plains vegetation to clipping, precipitation, and soil water. *J. Range Manage.* 28:194-197.

Eisele, K.A., D.S. Schimel, L.A. Kapuska, and W.J. Parton. 1990. Effects of available P and N:P ratios on non-symbiotic dinitrogen fixation in tall grass prairie soils. *Oecologia* 79:471-474.

Elliott, E.T., H.W. Hunt, and D.E. Walter. 1988. Detrital foodweb interactions in North American grassland ecosystems. *Agric., Ecosys. and Environ.* 24:41-56.

Frank, A.B., D.L. Tanaka, L. Hofmann, and R.F. Follet. 1995. Soil carbon and nitrogen of northern Great Plains grasslands as influenced by long-term grazing. *J. Range Manage.* 48:470-474.

Frank, D.A., and P.M. Groffman. 1998. Ungulate vs. landscape control of soil C and N processes in grasslands of Yellowstone National Park. *Ecol.*79:2229-2241.

Gartner, F.R. and W.W. Thompson. 1972. Fire in the Black Hills forest-grass ecotone. *Proc. Tall Timbers. Fire Ecol. Conf.* 12:37-68.

Gleason, H.A. 1913. The relation of forest distribution and prairie fires in the midwest. *Torreya* 13:173-181.

Groffman, P.M., C.W. Rice, and J.M. Tiedje. 1993. Denitrification in a tall grass prairie landscape. *Ecol.*74:855-862.

Harris, W.D. 1973. *Some effects of wildfire on adjacent grazed and ungrazed western Kansas mixed prairie.* Masters Thesis, Ft. Hays State University. Hays, KS.

Heirman, A.L., and H.A. Wright. 1973. Fire in medium fuels of West Texas. *J. Range Manage.* 26:331-335.

Holland, E.A., and J.K. Detling. 1990. Plant response to herbivory and belowground nitrogen cycling. *Ecol.*71:1040-1049.

Holland, E.A., W.J. Parton, J.K. Detling, and L. Coppock. 1992. Physiological responses of plant populations to herbivory and their consequences for ecosystem nutrient flow. *Am. Naturalist* 140:685-706.

Hopkins, H., F.W. Albertson, and A. Riegel. 1948. Some effects of burning on a prairie in West-Central Kansas. *Trans. Kansas Acad. Sci.* 51:131-141.

Hulbert, L.C. 1969. Fire and litter effects in undisturbed bluestem prairie in Kansas. *Ecol.*50:874-877.

Humphreys, L.R., and A.R. Robinson. 1966. Interrelations of leaf area and non-structural carbohydrate status as determinants of the growth of subtropical grasses. *Proc. Tenth International Grassland Congress* 10:113-116.

Hunt, H.W., M.J. Trlica, E.F. Redente, J.C. Moore, J.K. Detling, T.G.F. Kittel, D.E. Walter, M.C. Fowler, D.A. Klein, and E.T. Elliott. 1991. Simulation model for the effects of climate change on temperate grassland ecosystems. *Ecol. Modeling* 53:205-246.

Insam, H. 1990. Are the soil microbial biomass and basal respiration governed by the climatic regime? *Soil Biol. Biochem.* 22:525-532.

Insam, H., D. Parkinson, and K.H. Domsch. 1989. Influence of macroclimate on soil microbial biomass. *Soil Biol. Biochem.* 21:211-221.

Jenny, H. 1941. *Factors of Soil Formation.* McGraw-Hill. New York.

Johnston, A., and S. Smoliak. 1968. Reclaiming brushland in Southwestern Alberta. *J. Range Manage.* 21:404-406.

Kelting, R.W. 1957. Winter burning in central Oklahoma grassland. *Ecol.* 38:520-522.

Kieft, T.L. 1994. Grazing and plant canopy effects on semi-arid soil microbial biomass and respiration. *Biol. Fertil. Soils.* 18:155-162.

Knapp, A.K. 1985. Effect of fire and drought on the ecophysiology of *Andropogon gerardii* and *Panicum virgatum* in a tall grass prairie. *Ecol.* 66:1309-1320.

Knapp, A.K., and T.R. Seastedt. 1986. Detritus accumulation limits productivity in tall grass prairie. *BioSci.* 36:662-668.

Knapp, A.K., J.M. Briggs, J.M. Blair, and C.L. Turner. 1998. Patterns and controls of aboveground net primary production in tall grass prairie. *In* A.K. Knapp, J.M. Briggs, D.C. Hartnett, S.L. Collins (eds), *Grassland Dynamics: Long-Term Ecological Research in Tall Grass Prairie,* Oxford University Press, New York, pp. 193-221.

Koch, G., and H. Mooney. 1996. *Carbon Dioxide and Terrestrial Ecosystems. Physiological Ecology Series.* Academic Press. New York.

Komarek, E.V., Sr. 1966. The meteorological basis for fire ecology. *Proc. 5th Tall Timbers Fire Ecol. Conf.* 5:85-126.

Kozlowski, T.T., and C.E. Ahlgren. 1974. *Fire and Ecosystems.* Academic Press. New York.

Kucera, C.L., and J.H. Ehrenreich. 1962. Some effects of annual burning on central Missouri prairie. *Ecol.*43(2):334-336.

Launchbaugh, J.L. 1957. The effects of stocking rate on cattle gains and on native short grass vegetation in west central Kansas. *Kansas Agric. Expt. Sta. Bull.* 394.

Launchbaugh, J.L. 1964. Effects of early spring burning on yields of native vegetation. *J. Range Manage.* 17:5-6.

Launchbaugh, J.L. 1972. Effects of fire on short grass and mixed prairie species. *Proc. Tall Timbers Fire Ecol. Conf.* 12:129-151.

Lavado, R.S., J.O. Sierra, and P.N Hashimoto. 1996. Impact of grazing on soil nutrients in a Pampean grassland. *J. Range Manage.* 49:452-457.

Leetham, J.W., and D.G. Milchunas. 1985. The composition and distribution of soil microarthropods in the short grass steppe in relation to soil water, root biomass, and grazing by cattle. *Pedobiologia* 28:311-325.

Lewis, J.K. 1969. Range management viewed in the ecosystem framework. *In* G.W. Van Dyne (ed), *The Ecosystem Concept in Natural Resource Management*, Academic Press, NY, pp. 97-187.

Manley, J.T., and G.E. Schuman, J.D. Reeder, and R.H. Hart. 1995. Rangeland soil carbon and nitrogen responses to grazing. *J. Soil Water Conserv.* 50:294-298.

Mathews, B.W., L.E. Sollenberger, V.D. Nair, and C.R. Staples. 1994. Impact of grazing management on soil nitrogen, phosphorus, potassium, and sulfur distribution. *J. Environ. Qual.* 23:1006-1013.

Merrill, L.B., and V.A. Young. 1959. Response of curly mesquite to height and frequency of clipping. *Texas Agr. Expt. Sta.,* MP 331.

Milchunas, D.G., and W.K. Lauenroth. 1989. Three-dimensional distribution of plant biomass in relation to grazing and topography and in the short grass steppe. *Okios* 55:82-86.

Milchunas, D.G., and W.K. Lauenroth. 1993. Quantitative effects of grazing on vegetation and soils over a global range of environments. *Ecol. Monogr.* 63:327-366.

Mutch, R.W. 1970. Wildland fires and ecosystems — a hypothesis. *Ecol.*51:1046-1051.

Newberry, J.S. 1873. *Origin of the Prairies.* Geol. Survey of Ohio. Vol. 1, Part 1:26-31.

Ojima, D.S. 1987. *The short-term and long-term effects of burning on tall grass prairie ecosystem properties and dynamics.* Ph.D. dissertation, Colorado State University. Fort Collins, CO.

Ojima, D.S., W.J. Parton, D.S. Schimel, and C.E. Owensby. 1990. Simulated impacts of annual burning on prairie ecosystems. *In* S.L. Collins and L.L. Wallace (eds), *Fire in North American Tall Grass Prairies*, University of Oklahoma Press, Norman, OK, pp. 118-132.

Ojima, D.S., D.S. Schimel, W.J. Parton, and C.E. Owensby. 1994. Long- and short-term effects of fire on nitrogen cycling in tall grass prairie. *Biogeochemistry* 24:67-84.

Owensby, C.E., and K.L. Anderson. 1967. Yield responses to time of burning in the Kansas Flint Hills. *J. Range Manage.* 20:12-16.

Owensby, C.E., and J.B. Wyrill, III. 1973. Effects of range burning on Kansas Flint Hills soil. *J. Range Manage.* 26(3):185-188.

Owensby, C.E., J.N. Ham, A.K. Knapp, D. Bremer, and L.M. Auen. 1996. Water vapor fluxes and their impact under elevated CO_2 in a C_4 tall grass prairie. *Global Change Biol.* 3:189-195.

Owensby, C.E., J.M. Ham, A.K. Knapp, and L.M. Auen. 1999. Biomass production and species composition change in a tall grass prairie ecosystem after long-term exposure to elevated atmospheric CO_2. *Global Change Biol.* 5:497-506.

Redman, R.E. 1978. Plant and soil water potentials following fire in a northern mixed grassland. *J. Range Manage.* 31:443-445.

Reynolds, H.G., and J.W. Bohning. 1956. Effects of burning on a desert grass-shrub range in southern Arizona. *Ecol.*37:769-777.

Rice, C.W., and F.O. Garcia. 1994. Biologically active pools of carbon and nitrogen in tall grass prairie soil. *In* J.W. Doran et al. (eds), *Defining soil quality for a sustainable environment, SSSA Spec. Publ. 35*, SSSA and ASA, Madison, WI, pp. 201-208.

Rice, C.W., F.O. Garcia, C.O. Hampton, and C.E. Owensby. 1994. Soil microbial response in tall grass prairie to elevated CO_2. *Plant and Soil* 165:67-74.

Rice, C.W., T.C. Todd, J.M. Blair, T.R. Seastedt, R.A. Ramundo, and G.W.T. Wilson. 1998. Belowground biology and processes. *In* A.K. Knapp et al. (eds), *Grassland Dynamics*, Oxford University Press, New York, pp. 244-264.

Risser, P., and W.J. Parton. 1982. Ecological analysis of a tall grass prairie: nitrogen cycle. *Ecol.*63:1342-1351.

Sampson, A.W., and E.C. McCarty. 1930. The carbohydrate metabolism of *Stipa pulchra. Hilgardia* 5:60-100.

Sauer, C.O. 1950. Grassland climax, fire, and man. *J. Range Manage.* 3(1):16-21.

Schuster, J.L. 1964. Root development of native plants under three grazing intensities. *Ecol.*45:63-70.

Seastedt, T.R., and R.A. Ramundo. 1990. The influence of fire on belowground processes of tall grass prairie. *In* S.L. Collins and L.L. Wallace (eds), *Fire in North American Tall Grass Prairies*, University of Oklahoma Press, Norman, OK, pp. 99-117.

Sharrow, S.H., and H. Wright. 1977a. Effects of fire, ash, and litter on soil nitrate, temperature, moisture, and tobosagrass production in the Rolling Hills. *J. Range Manage.* 30(4):266-270.

Sharrow, S.H., and H. Wright. 1977b. Proper burning intervals for tobosagrass in west Texas based on nitrogen dynamics. *J. Range Manage.* 30(5):343-346.

Shimek, F. 1911. The Prairies. *Bull. Nat. Hist. of Iowa* 6(1):169-240.

Stewart, O.C. 1951. Burning and natural vegetation in the United States. *Geog. Rev.* 41:317-320.

Towne, G., and C. Owensby. 1984. Long-term effects of annual burning at different dates in ungrazed Kansas tall grass prairie. *J. Range Manage.* 37:392-397.

Trlica, M.J., Jr., and J.L. Schuster. 1969. Effects of fire on grasses of the Texas High Plains. *J. Range Manage.* 22(5):329-333.

Turner, C.L., T.R. Seastedt, and M.I. Dyer. 1993. Maximization of aboveground grassland production: The role of defoliation frequency, intensity, and history. *Ecol. Appl.* 3:175-186.

USDA. 1972. The nation's range resources — a forest-range environmental study. *Forest Resources Rpt.* No. 19.

Ueckert, D.N., T.L. Whigham, and B.M. Spears. 1978. Effect of burning on infiltration, sediment, and other properties in a mesquite-tobosagrass community. *J. Range Manage.* 31:420-425.

White, E.M., W.W. Thompson, and F.R. Gartner. 1973. Heat effects on nutrient release from soils under ponderosa pine. *J. Range Manage.* 26(1):22-24.

Williams, M.A. 1998. *Soil and microbial response in tall grass prairie to elevated CO_2.* M.S. thesis, Kansas State University. Manhattan, KS.

Williams, M.A., C.W. Rice, and C.E. Owensby. 2000. Carbon and nitrogen dynamics and microbial activity in tall grass prairie exposed to elevated CO_2 for 8 years. *Plant and Soil* (submitted).

Williams, R.E., B.W. Allred, R.M. DeNio, and H.E. Paulsen, Jr. 1968. Conservation, development, and use of the world's rangelands. *J. Range Manage.* 21:355-360.

Wolfe, C.W. 1972. Effects of fire on a sandhills grassland environment. *Proc. Tall Timbers Fire Ecol. Conf.* 12:241-256.

SECTION **4**

Using Computer Simulation Modeling to Predict Carbon Sequestration in Grazing Land

CHAPTER **14**

Simulating Rangeland Production and Carbon Sequestration

J.D. Hanson,[1] M.J. Shaffer,[2] and L.R. Ahuja[3]

Introduction

Early range research used soil survey information to classify rangeland sites and to plan their management (Shantz, 1911; Clements, 1934). Interactions of abiotic factors on parent material (Branson et al., 1964) and on slope (Klemmendson, 1964) thus delineated basic ecological and physiological processes occurring in plant communities.

More recent research approaches have used traditional soil physics, plant physiology (Sosebee, 1977), and simulation models to describe the fundamental processes controlling rangeland production (Charley, 1977; Hunt et al., 1983; Innis, 1978; Parton et al., 1987; Hanson et al., 1992). Most of the soils within the Great Plains are relatively shallow and formed in calcareous loamy alluvium or in loamy residuum derived from shale and sandstone. These soils are generally deficient in N, and occasionally other nutrients, for optimum forage production.

Effective management of the Great Plains is based primarily on the ecological relationships between water, soil, and vegetation (Sprock, 1982). Production of forage and dryland crops in the Great Plains is unpredictable because temperature and precipitation exhibit extreme seasonal and long-term variability. Water

[1] Laboratory Director/Research Leader/Rangeland Scientist, USDA, Agricultural Research Service, Northern Great Plains Research Laboratory, Mandan, ND.

[2] Soil Scientist, USDA, Agricultural Research Service, Great Plains Systems Research Unit, Fort Collins, CO. Send correspondence to: Jon D. Hanson, P.O. Box 459, Highway 6 South, Mandan, ND 58554-0459, phone (701) 667-3010, fax (701) 667-3054.

[3] Research Leader/Soil Scientist, USDA, Agricultural Research Service, Great Plains Systems Research Unit, Fort Collins, CO. Send correspondence to: Jon D. Hanson, P.O. Box 459, Highway 6 South, Mandan, ND 58554-0459, phone (701) 667-3010, fax (701) 667-3054.

supplies affect crop and forage production in the Great Plains more than any other single factor (Great Plains Agricultural Council, 1985).

Because of the limited water supply of the short grass steppe, the predominant vegetation type is the blue grama-buffalo grass association (Kuchler, 1964). Lauenroth and Milchunas (1988), Parton et al. (1987), and Sims et al. (1978) compiled detailed descriptions of the vegetation of the Central Plains. Forage production may often be less than 700 kg/ha, and so the carrying capacity for cattle is relatively low. The host of plant species populating native range provides food and habitat for domestic livestock and many species of wildlife.

Complex interactions between system components hinder our understanding of the structure and function of rangelands. The productivity of rangeland and dryland pasture (especially depleted rangelands and abandoned cropland) of the Great Plains must be maximized if future demands for red meat are to be met. This chapter discusses how computer simulation modeling has been used to evaluate the effects of environmental change and of management strategies on improving rangeland production. It begins by describing components of rangeland systems which are important in such modeling.

Rangeland Soil and Hydrology

Effect of rangeland management

Rangeland management practices, such as grazing intensity, affect soil properties and infiltration in a number of ways. The most important impacts are the change in surface cover (vegetated vs. bare areas), compaction, organic matter (OM) and litter content of surface soil, macropores (root channels), plant species composition, surface configuration, and microtopography (surface roughness). In most cases, a combination of all these is involved.

Ample evidence shows that continuous heavy grazing or a high stocking rate reduces infiltration of rainfall into the soil (Gifford and Hawkins, 1978; Weltz and Wood, 1986; Warren et al., 1986; Thurow, 1985). However, the effect varies from soil to soil and for vegetation types. The effect also may vary temporally, with infiltration rates higher in the growing season than in the dormant season (Warren et al., 1986).

Gifford (1977) showed clear evidence of the negative effect of percent bare soil and the positive effect of percent canopy cover on infiltration. The bare soil is generally subject to crusting-sealing by the impact of raindrops and to compaction by trampling (Lull, 1959; Dadkhah and Gifford, 1980). The effect of percent cover on infiltration is almost linear (Gifford, 1977). Thurow (1985) showed that the relationship between the infiltration rate and the total organic cover (including litter and canopy) is also linear. Other than providing surface cover, the vegetation increases infiltration by creating macropores or decayed root channels (Beven and Germann, 1982).

This effect of vegetation, as well as related effects, varies with the type of vegetation. Thurow et al. (1986) found infiltration under oak motte and bunchgrass vegetation to be higher than under sod. Gifford (1977) reported a decrease in the infiltration rate with successional stages of rangeland vegetation, from blue grama to annual weeds and buffalograss. Blackburn (1973) found higher infiltration in coppice dunes under rangeland shrubs than in the interspace between shrubs. Freebairn (1989) showed that surface roughness and clods increased the infiltration rate.

Role of surface cover

Rainfall on a bare soil forms a dense surface seal by breaking down aggregates, washing fine particles into the soil matrix, and causing compaction (Duley, 1939; McIntyre, 1958a, b; Tackett and Pearson, 1965). On drying, this surface seal becomes a crust.

McIntyre (1958a) measured the compacted skin to be only 0.1 mm thick and the wash-in layer to be 1.5 to 2.5 mm thick. However, the hydraulic conductivity of these layers was as much as 2000 times lower than that of the original soil. Other investigators have reported a five-fold or greater reduction in hydraulic conductivity (Hillel, 1960; Tackett and Pearson, 1965; Falayi and Bouma, 1975; Sharma et al., 1981). Whatever the magnitude, the infiltration of water in a sealed-crusted soil can be reduced markedly (Duley, 1939; Mannering and Wiersma, 1970; Morin and Benyamini, 1977).

No comprehensive research has been done on the effects of different levels of cover and different plant architectures on reducing crusting and increasing infiltration. Lange and Mollett (1985) did report the effect of six levels of corn stubble on infiltration. On a clay loam soil, about 60% residue cover was required to avoid a significant reduction in infiltration.

A thin, dense crust will saturate quickly at the start of rainfall or irrigation. Therefore, one can neglect its unsaturated hydraulic conductivity and water storage and characterize only its saturated hydraulic conductivity or hydraulic resistance. Edwards and Larson (1969) measured the change in saturated conductivity of a seal-crust formed under simulated rainfall in bare Ida silt loam (fine silty, mixed [calcareous], mesic Typic Udorthents) soil. The conductivity decreased approximately as an exponential function of rainfall time. The conductivity value was not quite steady after 90 minutes of rainfall, with an intensity of 6 cm/hr.

The work of Amemiya (1968) and Burwell et al. (1968) indicated that soil surface roughness deters sealing-crusting. Following infiltration, the surface cover also directly affects the loss of water by evaporation from the soil surface vs. the amount used by plants for transpiration. The management practices that result in a greater percentage of bare soil cause a greater percentage of loss by evaporation.

Role of root channels or macropores

The term *macropore* here refers to relatively large noncapillary pores or channels, such as interaggregate pores, interpedal voids, drying cracks in clay soils, wormholes, animal burrows, and channels created by plant roots. Most natural soils contain some macropores. Recent reviews of experimental work indicate that water flow and transport in soils containing a network of macropores that are open at the soil surface can deviate substantially from predictions of the current theories for these processes, which are adequate for soils without a large number of macropores (Thomas and Phillips, 1979; Germann and Beven, 1981; White, 1985).

The macropores allow rapid gravitational flow of the free water available at the soil surface during prolonged high-intensity rainfalls or above an impeding soil horizon. Obviously, a less permeable soil matrix will direct more water to macropore flow. Several reports indicate that the macropores formed by earthworms and decayed root channels are preserved in a no-till soil and play a dominant role in the infiltration of water and chemicals from the soil surface (Edwards et al., 1979; Scotter and Kanchanasut, 1981; Dick et al., 1989).

Some recent reports on cropping's effect on water infiltration are interesting. Root growth initially may decrease infiltration rates, but decomposition of roots leaves channels or macropores that result in increased infiltration (Barley, 1954). Disparte (1987) measured higher infiltration rates in plots planted to alfalfa than in unplanted controls. Meek et al. (1989) found 140% to 240% increases in the infiltration rate with time for alfalfa, related to decreases in stand density and number of macropore channels in the profile. The number of channels decreased with depth between 20 and 50 cm. Shirmohammadi and Skaggs (1984) reported a 70% increase in hydraulic conductivity of soil in the top 20 cm by growing fescue and a 40% increase by growing soybeans.

Present models of flow in soils with macropores generally are based on two domains, the soil matrix and macropores. The two domains interact through a common boundary condition or a simple source/sink term. Hoogmoed and Bouma (1980), Beven and Germann (1981), Addiscott (1977), and Hetrick et al. (1982) presented the unsteady, unsaturated models.

Hoogmoed and Bouma (1980) simulated infiltration of water into a cracked dry clay soil by combining existing physical models for vertical and horizontal infiltration. The flow of water into the cracks was assumed to be very rapid. Horizontal infiltration was assumed to occur from cracks into the soil matrix at different depths under a boundary condition of constantly saturated water content at the interface.

Beven and Germann (1981) assumed a laminar flow of water in macropores and developed a kinematic wave model for flow. The model requires knowledge of at least four unknown parameters, besides the soil matrix properties. The model has not yet been validated. Hetrick et al. (1982) developed a bi-continuum layer model that is somewhat similar in principle to Hoogmoed and Bouma's (1980).

Effects of microtopography and surface shaping

Several researchers have pointed out a relationship between the roughness of the soil surface and an increased infiltration rate (Allmaras et al., 1966; Amemiya, 1968; Burwell et al., 1968; Falayi and Bouma, 1975). Increased surface detention enhances infiltration, and rough surfaces deter or delay crusting-sealing of the soil surface (Amemiya, 1968; Burwell et al., 1968). Similar effects should result from shaping the soil surface into ridges and furrows, but this has not been investigated.

Soil Nutrients

Effect on plant production

Forage production in semiarid rangelands is limited most by water and N available on the site (Woodmansee et al., 1978). P may also be a limitation in some areas of the northern Great Plains (Black and Wight, 1979). Although these factors are involved directly in plant growth, predicting production or growth potential from site attributes is untenable because they are not uniform across sites or they are not readily measurable in the field (USDA, 1963). Therefore, indirect approaches have been adopted to predict potential plant production.

Measurements of soil texture and horizon depth are extensively used (Lentz and Simonson, 1987a; Passey et al., 1982). Soil subgroups provide the most meaningful level of soil classification for correlation with broad plant associations (Passey et al., 1982). Additional information is necessary before correlation of vegetative production with soils grouped at the family, series, or phase level is feasible.

After studying the correspondence of soil properties and classification units, Lentz and Simonson (1987b) concluded that *Soil Taxonomy* could provide a sensitive means for classifying and distinguishing soils having even minor differences in growth potential. They suggested that additional detailed studies of soil-plant community relations were needed to quantitatively describe how soil morphological properties interact to determine levels of temperature, soil aeration, and nutrient status.

Effect of grazing on soil nitrogen

Grazing by herbivores affects the soil N cycle directly by removing plant N and depositing manure and urine. Harvesting N from plants delays N input into soil organic matter (SOM). Reduced inputs, in turn, limit the total available soil N for plant uptake. Alternatively, light to moderate grazing may stimulate plant growth; thus, the amount of soil inorganic N that plants take up may increase. Overall, grazing probably accelerates the soil N cycle. A faster turnover time for soil organic N could lead to soil inorganic N's leaching if the rate of soil inorganic N's production exceeds the rate of plant uptake. On the short grass prairie, however, there is no evidence for nitrate leaching without fertilization.

In a grassland ecosystem without herbivores, aboveground residues return to the soil surface or are blown away by wind. SOM builds up to some steady-state level typical for the soil texture, annual precipitation, annual temperature, and level of annual inputs. Soil inorganic N reaches plants through microbially mediated processes, and leaching is probably at a minimum.

Under these conditions, soil organic N should remain at some steady-state level for the given grassland ecosystem (Fig. 14.1). Dead root N, litter N, and green shoot N also should remain at some steady-state level for the given system. Plants should take up soil inorganic N as fast as possible, and the N should stay at some steady-state level within the soil system, with very little leaching. As a result of these cumulative conditions, total N mineralized should be at a minimum, while total soil N should be maximized for the given steady-state level of the ecosystem.

Under light or moderate grazing, and depending on the grassland ecosystem in question, plant production should increase. Increasing plant production also increases inputs from litter and dead roots. Trampling facilitates contact of aboveground residue with the soil surface and stimulates microbially mediated decomposition. Manure returns high quality substrate to the soil surface. Soil inorganic N is still being taken up as quickly as it becomes available and, with increased decomposition rates, more soil inorganic N is available within the soil system. Under light or moderate grazing, therefore, total N mineralized and soil organic N should increase, while total soil N decreases. At the same time, dead root N, litter N, and green shoot N should increase.

In grassland ecosystems, heavy grazing results ultimately in decreased plant productivity, increased runoff and soil erosion, and structural changes to the plant community. Loss of inputs implies loss of SOM, and the loss of community structure implies that plant uptake of N changes. Decreased plant uptake of N would be followed by an increased potential for soil inorganic N to leach from the soil profile, although little evidence indicates leaching without fertilization.

The severe changes in the grassland community resulting from heavy grazing should result in maximum losses of soil organic N and soil inorganic N. Plant N

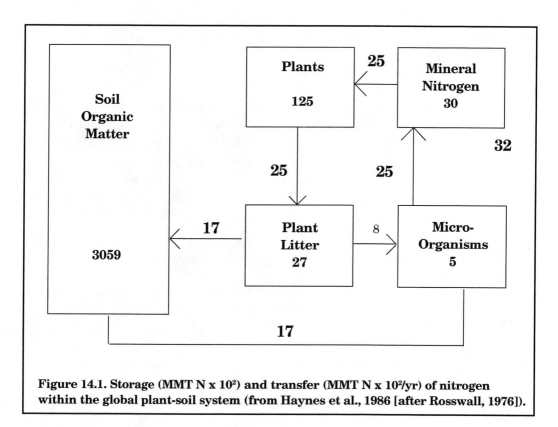

Figure 14.1. Storage (MMT N x 10²) and transfer (MMT N x 10²/yr) of nitrogen within the global plant-soil system (from Haynes et al., 1986 [after Rosswall, 1976]).

probably would decrease under heavy grazing over time; this would reduce the quality of inputs and returned manure, thereby perpetuating the cycle of degradation. Total N mineralized should reach a maximum, while total soil N should fall to its minimum level for the system.

Importance of nitrogen to soil organic matter

Forage quality in a grazed ecosystem links animal production to SOM. The more N-rich the forage, the more pounds of animal product will be realized. In a rangeland situation, where all the N comes from the soil, from plant residues, and from manure and urine, and where there are no additional inputs of N to the system, SOM-associated N is the major contributor of plant available N. N-rich SOM implies N-rich forage.

Organic N may occur in live roots, live soil fauna, soil microorganisms, organic or inorganic end products of N metabolism from the entire soil population, and accumulated SOM. Amino- and amide-N groups are important in the bonding between clay particles and SOM.

Ammonia is produced by enzymatic digestion and becomes readily available to microorganisms. It subsequently becomes available to higher plants, following nitrification. Of primary importance to the plant is the release of N (as well as P and S) from SOM through the action of soil microorganisms and the binding of micronutrients by polysaccharides (keeping them readily available for plant uptake).

Release and immobilization of N, P, and S occur during decomposition of SOM. Polysaccharides (long gummy chains of sugars) and fulvic acid (the acid-soluble portion of the soil humus that is high in N, P, and S) are the two main chemical components of the SOM active fraction (Weil, 1992). High levels of polysaccharides may enhance either denitrification or N mineralization, depending upon the soil's condition, and may increase microbial immobilization of N by serving as C substrates for microbial biomass.

Models of the effects of grazing on the nitrogen cycle

Simulating the soil N cycle in a grazed ecosystem demonstrates the feedback effects between forage quality and SOM-associated N. The tall grass prairie version of ELM (Innis, 1978) was used to simulate the impact of grazing on N cycling for a site in northeast Oklahoma. Data collected from the Osage site indicated that total SOM and total N were similar on grazed and ungrazed treatments. ELM modeling indicated that moderate seasonal steer grazing increased N mineralization from SOM and increased the formation of soil organic N (Risser and Parton, 1982).

The CENTURY model (Parton et al., 1987, 1988) was used to examine how plant physiological responses to herbivory on and off prairie dog colonies (Wind Cave National Park, South Dakota) provide feedbacks to nutrient availability by controlling the C supply to soil heterotrophs. The model showed that, when grazing limited aboveground production, soil inorganic N dramatically increased, because the mineralized vegetation did not use N; and the decreased inputs of plant C (dead roots), in turn, decreased the microbial use of N. With increasing grazing intensity, losses of surplus N, via leaching or denitrification of soil inorganic N, caused the decline of total SOM and its C-to-N ratio.

The GRASS model (Coughenour et al., 1984) was used for the short grass ecosystem in the Serengeti National Park, Tanzania, to simulate effects of grazing on soil N and N mineralization. The model showed that mean soil mineral N increased with grazing intensity to a maximum and then declined with the heaviest grazing intensities. Net mineralized microbial N declined with light grazing, increased to a maximum at moderate intensities, and declined again for the heaviest intensity. Plant N uptake followed mineralization and peaked at a moderate grazing intensity (Seagle et al., 1992).

Field experiments on the effects of grazing on the nitrogen cycle

Simulating soil dynamics is always easier and quicker than taking actual soil samples and completing the lab analyses. However, without field verification, models become meaningless. Few studies have measured the N cycle under grazing.

Soils from the uppermost horizon of Canadian rough-fescue grassland for three treatments (no grazing, light grazing, and very heavy grazing) were analyzed for their N fractions. Total N remained the same between exclosures and grazing treatments but, under grazing, ammonia-N and soil inorganic N increased and mineralizable N decreased. This combination of losses and gains in N fractions resulted in a change in the quality of SOM and the amount of N available to the plant for uptake (Dormaar et al., 1990; Dormaar and Willms, 1990).

Plant Growth Models

Models of varying resolution describing plant growth and development have been produced in the past decade. These models range in complexity from regression models (West and Lauenroth, 1982) to extremely complex process models (Innis, 1978; Jones et al., 1980; Reynolds et al., 1980; White, 1984; van Keulen and Wolf, 1986; Hanson et al., 1992).

Some models are based on a metaphor that considers the plant to be some physical or biochemical system, such as an enzymatic system (Thornley, 1972; Thoughton, 1977). Other models were developed using a systems' modeling approach which makes them particularly useful in describing and synthesizing the conceptual structure of the system (Ares and Singh, 1974). These models are often nonlinear and usually incorporate maximum reduction techniques (van Bavel et al., 1973; Connor et al., 1974; Reed et al., 1976; Gilmanov, 1977; Cunningham and Reynolds, 1978; Detling, 1979, Skiles et al., 1982; Hanson et al., 1987, 1988, 1992; Coughenour et al., 1984).

Different approaches used to structure plant-growth models within any particular system include linear vs. nonlinear (Patten, 1972), empirical vs. mechanistic, and deterministic vs. stochastic (Hanson et al., 1985). Using these different approaches, models of a selected ecosystem emphasize those factors limiting plant growth (Singh et al., 1980).

Plant growth models generally assume that the processes which determine plant production (photosynthesis, respiration, C allocation, etc.) do not differ appreciably between species of various ecosystems (Hanson et al., 1988). Available water and N usually limit plant growth on arid and semiarid grasslands. Date (1973) reported N availability was most responsible for determining the structure and productivity of grassland communities. Reuss and Innis (1977) stressed that plant available N almost universally limits plant growth and production on grasslands.

Available water also dynamically influences the structure and production of rangelands (Lauenroth, 1979). Soil moisture depends not only on the amount of precipitation, but also on soil's water-holding characteristics and on rainfall's distribution (Duncan and Woodmansee, 1975). Water stress reduces plant growth, but little mechanistic information is available which describes the detailed physiological recovery of post-drought plants (Chu et al., 1979).

Models of Grazing Behavior

In addition to the obvious effects of N and water availability, rangeland production models must consider the effect of large domestic herbivores. These grazers alter the plant community's structure by consuming plants and plant parts, trampling vegetation, and recycling nutrients.

Arid and semiarid grassland models must also deal with the landscape's spatial heterogeneity. Unless stocking rates are high or various management techniques are employed, cattle do not graze pastures uniformly. Rather, they tend to graze selectively and leave large areas of the pasture effectively ungrazed (Herbel and Anderson, 1959, Klipple and Costello, 1960; Launchbaugh and Owensby, 1978; Hart et al., 1993). Models developed to simulate grasslands under grazing must consider this patchy grazing behavior (Hanson et al., 1993). Parton and Risser (1980) found that the ELM model (Innis, 1978) would not work correctly without considering spatial distribution of the vegetation.

The irregular grazing patterns in pastures may be induced by the variable distribution of nutrients throughout the grassland. Schimel et al. (1985) found that C, N, P, and clay increased from the top to bottom of a slope; the lower slope position seemed to have a larger soil N pool, which appeared to have a slower turnover rate than those of corresponding upslope soils. Other obvious factors that need to be considered in modeling grazing behavior on grasslands include fences, salt blocks, water tanks or holes, plant species' composition, and type of grazer.

Importance of soil organic matter

SOM can be argued to be the primary indicator of rangeland health (Cole et al., 1987). Information has been collected on loss of OM after cultivation (Bauer and Black, 1981; Campbell and Souster, 1982; Hunt, 1977; Tiessen et al., 1984), yet little is known about the effects of resting cultivated lands or converting them to grasslands (Berg, 1988). Nutrients trapped in SOM for thousands of years account for much of the cropping potential of the U.S. Great Plains region. These nutrients have been released and transformed into corn, wheat, hay, beef, and numerous other agricultural commodities over previous decades. Haas et al. (1957) reported average losses of 42% and 36% of organic C (OC) and N, respectively, following 30 to 43 years of various cropping systems at 11 locations in the Great Plains. Ongoing current research is quantifying the losses and gains of C caused by changing land use in the Great Plains (Burke et al., 1991; Cole et al., 1992a,b; Parton et al., 1988, 1989).

When abandoned farmlands and marginal croplands are transformed to grasslands, many soil properties known to enhance plant growth increase (Dormaar and Smoliak, 1985). However, even 49 years after establishing crested wheatgrass, some croplands in Canada have not regained the nutrient levels found in adjacent native sites (Dormaar et al., 1978). In pastures restored to grass-clover vegetation following topsoil removal, microbial biomass and enzyme concentrations increase more rapidly than OC or total N (Ross et al., 1982). These data appear to indicate that some of the mechanisms for grassland renovation are in place, but actual recovery time will vary (Sandor et al., 1986).

Grasslands managed under grazing may or may not have similar patterns of nutrient losses or gains as those that are plowed and subsequently cropped. The fertility of rangeland soils probably varies more than the fertility of agricultural soils when soil physical characteristics, moisture content, OM, and soil fauna are taken into account (Heady, 1975). However, unless soil nutrients are added through fertilization, the range plant being grazed and the agronomic crop being raised for harvest similarly depend on available SOM as a nutrient source.

Importance of Human Activities

People are a major component of current ecosystems of the Great Plains. Agriculture, in the form of ranching and farming, has been present for over a hundred years. In building ecosystem theory, ecologists tend to relegate human effects to the category of *outside disturbance*. Human influences and ecological systems usually are considered separate entities. However, human interactions with our surroundings are pervasive. Humans interact with natural systems in many ways and at all levels of organization. Interactions can be subtle or profound, harmful or benign. We therefore are an integral component of present ecological systems (Forman and Godron, 1986). Ecosystem studies of the human component have not been sufficient (Naveh, 1982).

Purposes of Simulation Models

Computer simulation models are written for a variety of reasons. Usually, the author of a model declares its purpose. The user of the model is responsible for applying it only for its intended purposes, which include, among others, manipulating answers to "what if" questions, training, investigating, testing, innovating, predicting, and making decisions.

Answering "what if" questions

Many times the real system which is being modeled is too time-consuming or expensive to manipulate to answer "what if" questions. Indeed, the real system may not even exist.

Training

Models can be used in "gaming" situations to impart knowledge to those who do not fully understand the system being modeled. By perturbing the driving variables or parameters of the model, the user can see the effects on certain indicator variables within the model.

Investigating the unknown

Models can be used to formulate a guess at how a poorly understood real system is functioning. By applying models in this way, the user is forced to gain an improved understanding of what is really happening in the system being modeled.

Testing hypotheses

Models have been used to test hypotheses of how a system works. For example, as early as 1978, ELM was used to test Tex's Hypothesis which stated that *prairie ecosystems cannot be seriously damaged by heavy domestic cattle grazing during the second year of two successive good years* (Innis, 1978).

Innovating

Models can and most often do stimulate new ideas and experimental approaches, especially when the model is being verified and tested against actual data.

Predicting

Models are developed for interpolation and extrapolation of data and thereby can be used to extend our current knowledge and understanding of how systems function. Models are especially useful for investigating second- and third-order interactions among system components.

Supporting decisions

Models can suggest priorities for applied research and development and, if used cautiously, can help managers in decision making.

Rangeland Production Modeling (SPUR2)

SPUR2 (Simulation of Production and Utilization of Rangelands, Version 2), a newer and enhanced version of SPUR (Wight and Skiles, 1987), is a general grassland ecosystem model that simulates the cycling of C and N through several compartments, including standing green, standing dead, live roots, dead roots, seeds, litter, and SOM. It also simulates competition between plant species, the impact of grazing on vegetation, rangeland hydrology, plant production, beef production, and forage removal by wildlife and grasshoppers (Hanson et al., 1992).

SPUR2 has been modified to simulate the direct effects of CO_2 on plant production (Hanson et al., 1993). Required initial conditions include the initial biomass content for each compartment and parameters that characterize the species to be simulated (Hanson et al., 1988). SPUR2 allows the user to simulate cow-calf operations, stocker steers, up to ten wildlife species, and two species of grasshoppers.

Daily inputs of precipitation, maximum and minimum temperatures, solar radiation, and daily wind drive the model. These variables come either from extant weather records or from a stochastic weather generator. The soils/hydrology component calculates upland surface runoff volumes, peak flow, snowmelt, upland sediment yield, and channel stream flow and sediment yield.

Soil-water tensions, used to control various aspects of plant growth, are generated from a soil-water balance equation. Surface runoff is estimated by the Soil Conservation Service's curve number procedure, and soil loss is computed with the modified universal soil-loss equation. The snowmelt routine employs an empirical relationship between air temperature and energy flux of the snowpack.

SPUR2 has been used to simulate the effects of climate change on grassland ecosystem processes and cattle production on U.S. rangelands (Baker et al., 1993; Hanson et al., 1993). Because of its mechanistic, process-oriented structure, SPUR2 is well suited for examining the interactions between management decisions and climatic influences on short-term ecological processes and for evaluating possible adaptive management strategies.

Cow-calf model

The Colorado Beef Cattle Production Model (CBCPM) has been incorporated into SPUR2. CBCPM is a second-generation beef-cattle production model that was a modification of the Texas A&M Beef Model (Sanders and Cartwright, 1979). CBCPM is a herd-wide, life cycle simulation model and operates at the level of the individual animal.

The biological routines of CBCPM simulate animal growth, fertility, pregnancy, calving, death, and demand for nutrients. Currently, 14 genetic traits related to growth, milk, fertility, body composition, and survival can be studied. The user can define the size and age distribution of the cow herd, the breeding season, the calving season, and the weaning date and can specify a particular culling strategy.

Intake of grazed forage is calculated by FORAGE, a deterministic model that interfaces CBCPM and SPUR2 (Baker et al., 1992). The model is driven by weight from the animal growth curve, the animal's demand for grazed forage, and the quantity and quality of forage available for each time step of the simulation. FORAGE determines the intake of grazed forage by simulating the rate of intake and grazing time of each animal in the time step.

Steer model

The basic structure of the steer model has been adapted from the Texas A&M University (TAMU) Beef Simulation Model (Sanders and Cartwright, 1979). Notter (1977) incorporated further modifications into the TAMU Beef Simulation Model. The grazing season is defined by Julian dates of turnout to pasture and removal from the pasture. The initial physical and physiological status of the steers is inferred from their age and weight at turnout. Supplemental feed can be offered between input Julian dates. Steers consume all supplemental feed before eating any of the available herbage. Forage intake and diet selection in SPUR2 are calculated by the FORAGE interface.

Wildlife model and insect components

SPUR2 allows the user to specify up to 10 different wildlife species. Dry matter intake (DMI) is calculated as

$$DMI = WT^{0.9} * k \, 10^3 \quad (27)$$

where WT is the mature body weight of the simulated wildlife species and k is the daily intake of dry matter per unit of metabolic weight (mg/kg). The model assumes the animal eats 25 mg/kg body weight (Minson, 1990). The user defines site and forage preferences. Wildlife is not limited physically from grazing specific sites. Data for mature body weights in the user interface were obtained from Chapman and Feldhamer (1982).

The grasshopper population-dynamics submodel is called on a daily basis from SPUR2. The model simulates daily activities of two types of spring emerging grasshoppers, "grass feeders" that feed exclusively on grass species present and "mixed feeders" that feed on both grasses and forbs in proportion to their relative frequency. Grasshoppers are more competitive grazers than cattle and can reduce carrying capacity drastically. Their demand for forage is likely to be highest when forage production is lowest.

SPUR2 simulates a dynamic grasshopper population. The user must specify the emergent date and the density for grasshopper larvae. Grasshopper density within the year and subsequent grasshopper damage exponentially decrease. Therefore, if control is necessary, action should taken early in the season. SPUR2 allows control using either carbaryl or malithion. The HOPPER model has been used successfully to simulate grasshopper populations on rangeland when coupled with a simple plant production model (Berry and Hanson, 1991).

Carbon Sequestration Modeling

Simulation of C sequestration must include consideration of the important processes, sources, and sinks involved in the C cycle on rangelands.

C cycling and sequestration in rangeland soils depend highly on the rooting environment associated with rangeland grasses and other plants. C recycled from surface litter, animal manure, and wastes such as sewage sludge can play important roles in some cases.

Figure 14.2 shows the dominant processes associated with C cycling under rangeland conditions. Death and sloughing of plant root materials, especially grasses, form the primary C input into the soil. C losses also can occur from soil erosion and from net mineralization of fresh root and other materials, added from plant production and other sources.

OMNI

The OMNI computer program for OM/N cycling was developed as a major component submodel of the Root Zone Water Quality Model (RZWQM) (RZWQM Development Team, 1998). OMNI is a state-of-the-art model for C and N cycling in soil systems. An attempt was made to unify the form of the process-rate equations governing C and N cycling in soils and to provide a firm theoretical basis for environmental interactions and future extensions. Significant use was made of concepts and principles found in existing models such as NTRM (Shaffer and Lar-

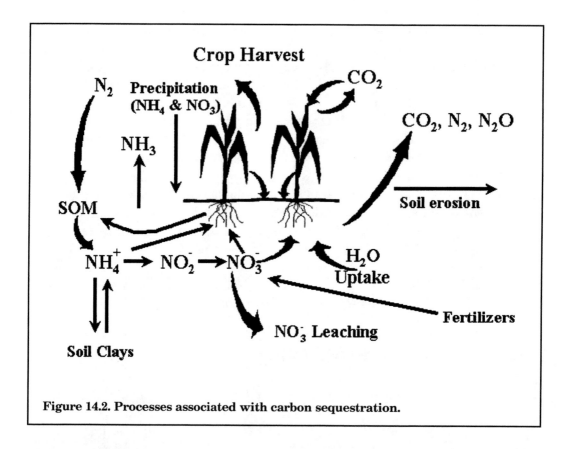

Figure 14.2. Processes associated with carbon sequestration.

son, 1987), Phoenix (Juma and McGill, 1986), CENTURY (Schimel et al., 1990), and Frissel's N model (Frissel and van Veen, 1981).

General flow diagrams of the soil C-N cycle appear in Figures 14.3 and 14.4. OMNI simulates all the major pathways illustrated, including mineralization-immobilization of plant litter, manure, and other organic wastes; mineralization of the soil humus fractions; inter-pool transfers of C and N; denitrification (production of N_2 and N_2O); gaseous loss of ammonia (NH_3); nitrificiation of ammonium to produce nitrate-N; production and consumption of methane gas (CH_4) and CO_2; and microbial biomass (MBM) growth and death. In OMNI, growth and subsequent death of microorganisms drive most of the processes and are a function of environmental variables such as soil temperature and water content, soil pH, soil oxygen levels, and solution concentrations (or activities) of nutrients.

Carbon and nitrogen pools

In the model, OM is distributed over five computational pools and is decomposed by three microbial biomass populations (Fig. 14.4). The OM pools consist of slow and fast pools for crop residues and other organic amendments and fast, medium, and slowly decaying SOM, respectively. The fast and medium SOM pools

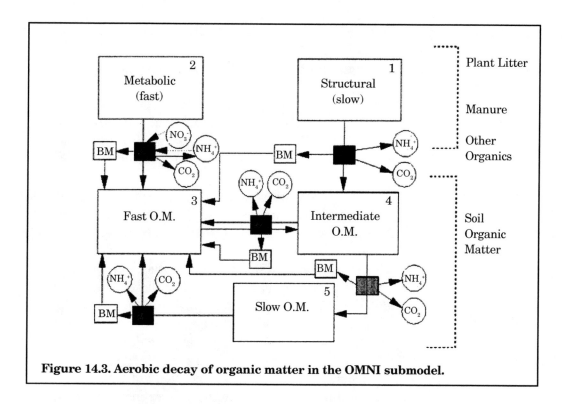

Figure 14.3. Aerobic decay of organic matter in the OMNI submodel.

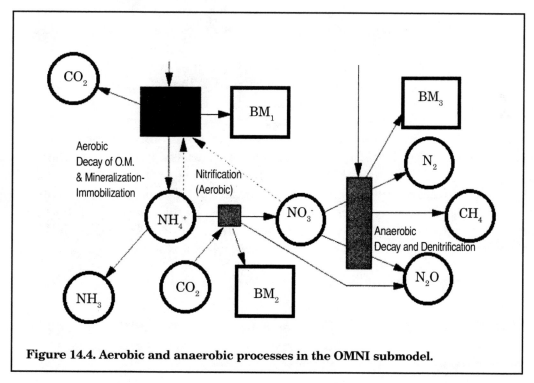

Figure 14.4. Aerobic and anaerobic processes in the OMNI submodel.

approximately correspond to the potentially mineralizable N pool (NO) the literature frequently mentions. For use in OMNI, the data input about NO should be determined on soils that do not contain fresh additions of crop litter, manure, or other forms of OM.

This partitioning of OM is based, in part, on work of Parton et al. (1983), Juma and McGill (1986), and others. The MBM populations include two heterotrophic groups (soil fungi and facultative bacteria, populations 1 and 3) and one autotrophic group (nitrifiers, population 2). Populations 1 and 2 are strict aerobes, while population 3 is primarily anaerobic.

For computational (mass balance) and conceptual purposes, C sink/source compartments for CO_2 and CH_4 gases also are included. These gases behave as by-products as well as sources during various parts of the C and N cycle. N concentrations are simulated in organic form as organic residues, SOM, and microbial biomass according to specified OM pool and microbial C:N ratios; in mineral form as NO_3-N and NH_4-N; as urea fertilizer; and as general N sinks (N_2 and N_2O).

Soil humus is depleted by microbial decay and built up by the addition of dead biomass and inter-pool transfers. Biomass populations are depleted through death and built up by biomass assimilation/growth during the decay of OM and nitrification.

Growth of heterotrophs occurs primarily by aerobic decay of OM. However, some facultative bacteria grow during denitrifying activity or methane production under anaerobic conditions. The autotrophs grow exclusively because of nitrifying activity under aerobic conditions. Urea is converted to ammonium (NH_4^+) by the enzymatic process of hydrolysis. The C source-sink storage slots increase as a result of MBM respiration, receiving CO_2 from aerobic respiration and CH_4 from anaerobic respiration. CO_2 behaves as a C source during nitrification, and CH_4 is a C source during aerobic OM decay.

Mineral N in solution exists as NO_3^- and NH_4^+. NH_4^+ is the primary form in the sense that NO_3^- is formed only by microbial oxidation of NH_4^+ through the process of nitrification. Solution NH_4^+ may be adsorbed onto the surfaces of soil clays by the process of cation exchange, making it temporarily unavailable for leaching. Conversion of organic N to mineral N first produces NH_4^+, then NO_3^-. The process of denitrification and immobilization removes NO_3^-. NH_4^+ is formed by the hydrolysis of urea and by transformations of organic N along mineralizing pathways. Conversely, NH_4^+ is removed via NH_3 volatilization, nitrification, and immobilization.

Miscellaneous processes

Many processes operate simultaneously on the various N and C species in OMNI. However, rate equations independently model only a defining subset of those processes. For mass balance consistency, the remaining processes are modeled as functions of the specified independent rates.

For example, OM decay is calculated by specified first-order rate equations. Aerobic microbial heterotrophic growth, on the other hand, is calculated as a function of the decay based on the OM and MBM C:N ratios and specified assimilation efficiency factors. Similarly, the augmentation of the fast OM pool is calculated as a function of MBM death, which an independent set of rate equations models.

This conceptual model of OM and N cycling in OMNI is relatively straightforward. However, mass balance must be maintained in the system when sources become limited. For example, when the demand for mineral N (NH_4^+ and NO_3^-) during immobilization exceeds the supply (storage plus mineralization), various rates must be adjusted to bring demand into balance with available supply. Not to do so would result in negative mineral N quantities (physically meaningless) and produce numerical problems.

Similar adjustments are required when total OM decay (via aerobic decay and denitrification-related anaerobic decay) exceeds available OM supply and when calculated death rates exceed available biomass population. The algorithms for handling these situations are somewhat complex.

CENTURY

The CENTURY model was developed to simulate monthly biogeochemical cycling of C, N, P, and S in both natural grassland and managed agro-ecosystems (Parton et al, 1987, 1988). Driving variables and major parameters for the model include surface soil physical properties; monthly precipitation and temperature; plant N and lignin contents; and land use.

The model's simplistic design allows it to be used to predict regional trends in key ecosystem processes, such as C and N fluxes, including primary production and the fate of soil C and N across a wide range of climates and soil types. However, the model is mechanistic enough to explore regional responses to climate change (Schimel et al., 1990, 1991; Burke et al., 1991).

GEM

The objectives of the GEM model are to predict seasonal and year-to-year biomass dynamics of primary producers, microbes, and soil fauna and to predict N availability in grasslands and the effects of the level of CO_2 and climate change on these dynamics (Hunt et al., 1991). Information needed to run this model includes daily precipitation, weekly mean, maximum, and minimum air temperature, wind speed, relative humidity, and monthly soil temperature.

The model's structure allows for investigation of the effects of climate change and elevated CO_2 on feedbacks through the trophic structure of the grassland ecosystem. GEM has been used to predict the ecosystem-level effects of climate change in a short grass steppe (Hunt et al., 1991).

References

Addiscott, T.M. 1977. A simple computer model for leaching in structured soils. *J. of Soil Sci.* 28:554-563.

Allmaras, R.R., R.E. Burwell, W.E. Larson, and R.F. Holt. 1966. Total porosity and random roughness of the interrow zone as influenced by tillage. *Conserv. Res. Rep.* 7 USDA-ARS, Washington, DC.

Amemiya, M. 1968. Tillage-soil water relations of corn as influenced by weather. *Agron. J.* 60:534-537.

Ares, J., and J.S. Singh. 1974. A model of the root biomass dynamics of a short grass prairie dominated by blue grama (*Bouteloua gracilis*). *J. Appl. Ecol.* 11:727-744.

Baker, B.B., J.D. Hanson, R.M. Bourdon, and J.B. Eckert. 1993. The potential effects of climate change on ecosystem processes and cattle production on U.S. rangelands. *Climate Change* 25:97-117.

Barley, K.P. 1954. Effects of root growth and decay on the permeability of a synthetic sandy soil. *Soil Sci.* 78:205-210.

Bauer, A., and A.L. Black. 1981. Soil carbon, nitrogen, and bulk density comparisons in two cropland tillage systems after 25 years and in virgin grassland. *Soil Sci. Soc. Am. J.* 45: 1166-1170.

Berg, W.A. 1988. Soil nitrogen accumulation in fertilized pastures of the Southern Plains. *J. Range Manage.* 41:22-25.

Berry, J.S. and J.D. Hanson. 1991. A simple, microcomputer model of rangeland forage growth for management decision support. *J. of Prod. Agric.* 4:491-500.

Beven, K.J., and P. Germann. 1981. Water flow in soil macropores. II. A combined flow model. *J. Soil Sci.* 32:15-30.

Beven, K.J. and P. Germann. 1982. Macropores and water flow in soils. *Water Resour. Res.* 18:1311-1325.

Black, A.L., and J.R. Wight. 1979. Range fertilization: Nitrogen and phosphorus uptake and recovery over time. *J. Range Manage.* 32:349-353.

Blackburn, W.H. 1973. *Infiltration rate and sediments production of selected plant communities and soils in five rangelands in Nevada.* Agric. Exp. Sta., College of Agric., Univ. of Nevada, Reno, NV.

Branson, F.A., R.F. Miller, and I.S. McQueen. 1964. Effects of two kinds of geologic materials on plant communities and soil moisture. *In* W. Keller and T.S. Ronningen (eds), *Forage Plant Physiology and Soil—Range Relationships*, ASA Special Publ. No. 5, ASA, Madison, WI, pp. 165-175.

Burwell, R.E., L.L. Sloneker, and W.W. Nelson. 1968. Tillage influences water intake. *J. Soil Water Conserv.* 23:185-187.

Cambell, C.A., and W. Sousster. 1982. Loss of organic matter and potentially mineral-izable nitrogen from Saskatchewan soils due to cropping. *Can. J. Soil Sci.* 62:651-656.

Charley, J.L. 1977. Mineral cycling in rangeland ecosystems. *In* R.E. Sosebee (ed), *Rangeland Plant Physiology*, Forage Sci. Service No. 4. Soc. Forage Manage, Denver, CO, pp. 215-256.

Chu, A.C.P., H.G. McPerson, and G. Halligan. 1979. Recovery growth following water deficits of different duration in prairie grass. *Aust. J. Plant Physiol.* 6:255-263.

Clements, F. E. 1934. The relic method in dynamic ecology. *J. Ecol.* 22:39-68.

Cole, C.V., J. Williams, M. Shaffer, and J. Hanson. 1987. Nutrient and organic matter dynamics as components of agricultural production systems models. *Soil Fertility and Organic Matter as Critical Components of Production Systems.* SSSA Spec. Pub. No. 19, pp. 147-166.

Connor, D.J., J.F. Brown, and M.J. Trlica. 1974. Plant cover, light interception, and photosynthesis of short grass prairie: A functional model. *Photosynthetica* 8:18-27.

Coughenour, M.B., S.J. McNaughton, and L.L. Wallace. 1984. Modeling primary production of perennial graminoids—Uniting physiological processes and morphometric traits. *Ecol. Modelling* 23:101-134.

Cunningham, G.L., and J.F. Reynolds. 1978. A simulation model of primary production and carbon allocation in the creosotebush (*Larrea tridentata* [DC] Cov.). *Ecology* 59:37-52.

Dadkhah, M., and G.F. Gifford. 1980. Influence of vegetation, rock cover and trampling on infiltration rate and sediment production. *Water Resour. Bull.* 16:979-987.

Date, R.A. 1973. Nitrogen: A major limitation in the productivity of natural communities, crops and pastures in the Pacific area. *Soil Biol. Biochem.* 5:5-18.

Detling, J.K. 1979. Processes controlling blue grama production on *Bouteloua gracilis* biomass dynamics on the North American short grass prairie. *Oecologia* 38:157-191.

Dick, W.A., R.J. Roseberg, E.L. McCoy, W.M. Edwards, and F. Haghiri. 1989. Surface hydrologic response of soils to no-tillage. *Soil Sci. Soc. Am. J.* 53:1520-1526.

Disparte, A.A. 1987. *Effect of root mass density on infiltration among four Mediterranean dryland forages and two irrigated forage legumes.* M.S. thesis, University of California, Riverside, CA.

Dormaar, J. F., and S. Smoliak. 1985. Recovery of vegetative cover and soil organic matter during revegetation of abandoned farmland in a semiarid climate. *J. Range Manage.* 38:487-491.

Dormaar, J.F., A. Johnson, and S. Smoliak. 1978. Long-term soil damages associated with seeded strands of crested wheat grain in southeastern Alberta, Canada. *Proc. First International Rangeland Congress.* Denver, CO. pp. 623-625.

Duley, F.L. 1939. Surface factors affecting the rate of intake of water by soils. *Soil Sci. Soc. Am. Proc.* 9:60-64.

Duncan, D.A., and R.G. Woodmansee. 1975. Forecasting from precipitation in California's annual rangeland. *J. Range Manage.* 28:327-329.

Edwards, W.R., and W.E. Larson. 1969. Infiltration of water into soils as influenced by surface seal development. *Trans. ASAE* 12:463-465, 470.

Edwards, W.M., R.R. van der Ploeg, and W. Ehlers. 1979. A numerical study of the effects of noncapillary-size pores upon infiltration. *Soil Sci. Soc. Am. J.* 45:851-856.

Falayi, O., and J. Bouma. 1975. Relationships between the hydraulic conductance of surface crusts and soil management in a typic Hapludalf. *Soil Sci. Soc. Am. Proc.* 39:957-963.

Forman, R.T.T., and M. Godron. 1986. *Landscape Ecology.* John Wiley, New York.

Freebairn, D.M. 1989. *Rainfall and tillage effects on infiltration of water into soil.* Ph.D. thesis, University of Minnesota, Minneapolis, MN.

Frissel, M.J., and J.A. van Veen. 1981. Chapter 13, Simulation model for nitrogen immobilization and mineralization. *In* Iskandar (ed), *Modeling Wastewater Renovation. Land Treatment,* John Wiley & Sons, New York, pp. 359-381.

Germann, P., and K.J. Beven. 1981. Water flow in soil macropores. 1. An experimental approach. *J. Soil Sci.* 32:1-13.

Gifford, G.F. 1977. Vegetation allocation for meeting site requirements. *Development of Strategies for Rangeland Management.* Westview Press.

Gifford, G.F., and R.H. Hawkins. 1978. Hydrologic impact of grazing on infiltration: A critical review. *Water Resour. Res.* 14:305-313.

Gilmanov, T.G. 1977. A comparison of potential photosynthesis, productivity and yield of plant species with differing photosynthetic metabolism. *Aust. J. Plant Phys.* 1:107-117.

Great Plains Agricultural Council. 1985. *Research needs for efficient use of water by plants in the Great Plains.* GPAC Publication No. 114. Denver, CO.

Hanson, J.D. 1991. Integration of the rectangular hyperbola for estimates of daily net photosynthesis. *Ecol. Modelling* 58:209-216.

Hanson, J.D., G.L. Anderson, and R.H. Haas. 1992. Combining remote sensing techniques with simulation modeling for assessing rangeland resources. *Geocarto Internat.* 7:99-104.

Hanson, J.D., B.B. Baker, and R.M. Bourdon. 1993. Comparison of the effects of different climate change scenarios on rangeland livestock production. *Agric. Sys.* 41:487-502.

Hanson, J.D., W.J. Parton, and G.S. Innis. 1985. Plant growth and production of grassland ecosystems: A comparison of modeling approaches. *Ecol. Modelling* 29:131-144.

Hanson, J.D., W.J. Parton, and J.W. Skiles. 1983. SPUR plant growth component. *In* J.R. Wight (ed), *SPUR—Simulation of Production and Utilization on Rangelands: A Rangeland Model for Management and Research,* USDA Misc. Pub. 1431., USDA, Washington, DC, pp. 68-73.

Hanson, J.D., J.W. Skiles, and W.J. Parton. 1987. SPUR: Simulating plant growth on rangeland. *In* J.R. Wight and J.W. Skiles (eds), *SPUR: Simulation of Production and Utilization of Rangelands, Documentation and User Guide,* ARS-63, USDA, Washington, DC, pp. 57-74.

Hanson, J.D., J.W. Skiles, and W.J. Parton. 1988. A multispecies model for rangeland plant communities. *Ecol. Modelling* 44:89-123.

Hart, R.H., J. Bissio, M.J. Samuel, and J.W. Waggoner, Jr. 1993. Grazing systems, pasture size, and cattle grazing behavior, distribution and gains. *J. Range Manage.* 46:81-87.

Herbel, M.C.H., and K.L. Anderson. 1959. Response of true prairie vegetation on major Flint Hills range sites to grazing treatments. *Ecol. Monogr.* 29:171-186.

Hetrick, D.M., J.T. Holdeman, and R.J. Luxmoore. 1982. *AGTEHM: Documentation of modifications to terrestrial ecosystem hydrology model (TEHM) for agricultural applications.* ORNL/TM-7856, Oak Ridge Nat. Lab. Oak Ridge, TN.

Hillel, D. 1960. Crust formation in loessial soils. *Trans. Int. Congr. Soil Sci. 7th* 1:330-339.

Hoogmoed, W.B., and J. Bouma. 1980. A simulation model for predicting infiltration into cracked clay soil. *Soil Sci. Soc. Am. J.* 44:458-461.

Hunt, H.W. 1977. A simulation model for decomposition in grasslands. *Ecol.* 58:469-483.

Hunt, H.W., J.W.B. Stewart, and C.V. Cole. 1983. A conceptual model for interaction among carbon, nitrogen, sulfur, and phosphorus in grasslands. *In* C.B. Bolin and R.B. Cook (eds), *The Major Bio-geochemical Cycles and Their Interaction*, Scope Vol. 21, John Wiley & Sons, New York, pp. 303-325.

Hunt, H.W, M.J. Trlica, E.J. Redente, J.C. Moore, J.K. Detling, T.G. Kittel, D.E. Walter, M.C. Fowler, D.A. Klein, and E.T. Elliot. 1991. Simulation model for the effects of climate change on temperate grassland ecosystems. *Can. J. Plant Sci.* 71:609-617.

Jones, J.W., L.G. Brown, and J.D. Hesketh. 1980. COTCROP: A computer model for cotton growth and yield. *In* J.D. Hesketh and J.W. Jones (eds), *Predicting Photosynthesis for Ecosystem Models,* Vol. II, CRC Press, Boca Raton, FL, pp. 209-242.

Klipple, G.E., and D.F. Costello. 1960. *Vegetation and cattle response to different intensities of grazing on short grass ranges on the Central Great Plains.* Tech. Bull. 1216. USDA. Washington, D.C.

Klemmedson, J.O. 1964. Topo-function of soils and vegetation in range landscape. *In* W. Keller and T.S. Ronningen (eds), *Forage Plant Physiology and Soil-Range Relationships*, ASA Special Publ. No. 5, ASA, Madison, WI.

Kuchler, A.W. 1964. *Potential natural vegetation of the coterminous United States.* Am. Geographical Soc. Special Publ. 36. New York.

Lange, P.M., and J.B. Mollet. 1985. Effect of the amount of surface maize residue on infiltration and soil loss from a clay loam soil. *South Afr. J. Plant and Soil.* 1:(3)97-98.

Lauenroth, W.K. 1979. Grassland primary production: North American grasslands. *In* N. French (ed), *Perspectives in Grassland Ecology*, Springer, New York, pp. 3-21.

Lauenroth, W.K., and D.G. Milchunas. 1988. The short grass steppe. *In* R.T Coupland (ed), *Ecosystems of the World, Vol. 8, Natural Croplands*, Elsevier Scientific Pub. Co., New York.

Launchbaugh, J.L., and C.E. Owensby. 1978. *Kansas rangelands: their management based on a half century of research.* Kans. Agric. Exp. Stn. Bull. 622.

Lentz, R.D., and G.H. Simonson. 1987a. Correspondence of soil properties and classification units with sagebrush communities in southeastern Oregon: I. comparisons between mono-taxa soil-vegetation units. *Soil Sci. Soc. Am. J.* 51:1263-1271.

Lentz, R.D. and G.H. Simonson. 1987b. Correspondence of soil properties and classification units with sagebrush communities in southeastern Oregon: II. Comparisons within a multi-taxa soil-landscape unit. *Soil Soc. Am. J.* 51:1271-1276.

Lockeretz, W. 1978. The lessons of the Dust Bowl. *Am. Sci.* 66:560-569.

Lull, H.W. 1959. Soil compaction on forest and range lands. U.S.F.S., Washington, D.C.

Mannering, J.V., and D. Wiersma. 1970. The effect of rainfall energy on water infiltration into soils. *Proc. Indiana Acad. Sci.* 79:407-412.

McIntyre, D.S. 1958a. Permeability measurements of soil crusts formed by raindrop impact. *Soil Sci.* 85:185-198.

McIntyre, D.S. 1958b. Soil splash and the formation of surface crusts by raindrop impact. *Soil Sci.* 85:261-266.

Meek, B.D., E.A. Rechel, L.M. Carter, and W.R. DeTar. 1989. Changes in infiltration under alfalfa as influenced by time and wheel traffic. *Soil Sci. Soc. Am. J.* 53:238-241.

Morin, J., and Y. Benyamini. 1977. Rainfall infiltration into bare soils. *Water Resour. Res.* 13:813-817.

Naveh, Z. 1982. Landscape ecology as an emerging branch of human ecosystem science. *Adv. Ecol. Res.* 12:189-237.

Parton, W.J., D.W. Anderson, C.V. Cole, and J.W.B. Stewart. 1983. Simulation of soil organic matter formations and mineralization in semiarid ecosystems. *In* Lowrance et al. (eds), *Nutrient Cycling in Agricultural Ecosystems*, *Special Publication 23*, University of Georgia, College of Agriculture Experiment Station, Athens, GA.

Parton, W.J., and P.G. Risser. 1980. Impact of management practices on the tallgrass prairie. *Oecologia (Berlin)* 46:223-234.

Parton, W.J., D.S. Schimel, C.V. Cole, and P.S. Ojima. 1987. Analysis of factors controlling soil organic matter levels in Great Plains grasslands. *Soil Sci. Soc. Am. J.* 51:1173-1179.

Parton, W.J., J.S. Singh, and D.C. Coleman. 1978. A model of production and turnover of roots in short grass prairie. *J. Appl. Ecol.* 47: 515-542.

Passey, H.B., V.K. Hugie, E.W. Williams, and D.E. Ball. 1982. *Relationships between soil, plant community, and climate on rangelands of the Intermountain West.* USDA-SCS Tech. Bull. No. 1669. U.S. Government Printing Office. Washington D.C.

Patten, B.C. 1972. A simulation of the short grass prairie ecosystem. *Simulation* 19:177-186.

Reed, K.L., E.R. Hammerly, B.E. Dinger, and P.G. Jarvis. 1976. An analytical model for field measurement of photosynthesis. *J. Appl. Ecol.* 13:925-942.

Reuss, J.O., and G.S. Innis. 1977. A grassland nitrogen flow simulation model. *Ecol.* 58:379-388.

Reynolds, J.F., B.R. Strain, G.L. Cunningham, and K.R. Knoerr. 1980. Predicting primary productivity for forest and desert ecosystem models. *In* J.D. Hesketh and J.W.

Jones (eds), *Predicting Photosynthesis for Ecosystem Models,* Vol. I, CRC Press, Boca Raton, FL, pp. 169-208.

RZWQM Development Team. 1998. RZWQM: Simulating the effects of management on water quality and crop production. *Agric. Sys.* 57:161-195.

Ross, D.J., T.W. Spier, K.R. Tate, A. Cairns, D.S. Meyrick, and E.A. Pansier. 1982. Restoration of pasture after top-soil removal: Effect on soil carbon and nitrogen mineralization, microbial turnover, and enzyme activities. *Soil Biol. Biochem.* 14:575-581.

Sandor, J.A., P.L. Gersper, and W.J. Hawley. 1986. Soils in prehistoric agricultural terracing sites in New Mexico: Organic matter and bulk density changes. *Soil Sci. Soc. Am. J.* 50:173-177.

Schimel, D.S., T.G. Kittel, A.K. Knapp, T.R. Seastedt, W.J. Parton, and V.B. Brown. 1991. Physiological interactions along resource gradients in a tallgrass prairie. *Ecology* 72:672-684.

Schimel, D.S., W.J. Parton, T.G. Kittel, D.S. Ojima, and C.V. Cole. 1990. Grassland biogeochemistry: Links to atmospheric processes. *Climatic Change* 17:13-25.

Schimel, D., M.A. Stillwell, and R.G. Woodmansee. 1985. Biogeochemistry of C, N and P in a soil catena of the short grass steppe. *Ecology* 66:276-282.

Scotter, D.R., and P. Kanchanasut. 1981. Anion movement in a soil under pasture. *Aust. J. Soil Res.* 19:299-307.

Shaffer, M.J., and W.E. Larson, 1987. *NTRM, a Soil-Crop Simulation Model for Nitrogen, Tillage, and Crop-Residue Management, Conservation Research Report 34-1.* USDA-ARS. USDA Technical Information Service, Washington, D.C.

Shaffer, M.J., K. Rojas, D.G. DeCoursey, and C.S. Hebson. 1999. Chapter 5, Nutrient Chemistry Processes (OMNI). *Root Zone Water Quality Model.* Water Resources Pub., Inc. Englewood, CO. (In press.)

Shantz, H.L. 1911. *Natural vegetation as an indicator of the capabilities of land for crop production in the Great Plains area.* USDA Government Printing Office. Washington D.C.

Sharma, P.P., C.J. Gantzer, and G.R. Blake. 1981. Hydraulic gradients across simulated rainformed soil surface seals. *Soil Sci. Soc. Am. J.* 45:1031-1034.

Shirmohammadi, A., and R.W. Skaggs. 1984. Effect of soil surface conditions on infiltration for shallow water table soils. *Trans. Am. Soc. Agric. Eng.* 27:1780-1787.

Sims, P.L., J.S. Singh, and W.K. Lauenroth. 1978. The structure and function of ten Western North American grasslands. I. Abiotic and vegetational characteristics. *J. of Ecol.* 66:251-285.

Singh, J.S., M.J. Trlica, P.G. Risser, R.E. Redmann, and J.K. Marshall. 1980. Autotrophic subsystem. *In* A.I. Breymeyer and G.M. Van Dyne (eds), *Grassland, Systems Anaylsis and Man, International Biological Programme 19,* Cambridge University Press, Cambridge, pp. 59-200.

Skiles, J.W., J.D. Hanson, and W.J. Parton. 1982. Simulation of above- and belowground nitrogen and carbon dynamics of Bouteloua gracilis and Agropyron smithii. *In* W. K. Lauenroth et al. (eds), *Analysis of Ecological Systems: State-of-the-Art in Ecological Modelling,* Elsevier Scientific Pub. Co., New York, pp. 467-473.

Sosebee, R.E. 1977. *Rangeland Plant Physiology. Range Sci. Series No. 4.* Soc. Range Manage. Denver, CO 80204.

Sprock, H. 1982. *Rangeland Soil Survey of Weld County, Colorado, Northern part.* USDA-SCS. U. S. Government Printing Office, Washington, DC.

Tackett, J.L., and R.W. Pearson. 1965. Some characteristics of soil crusts formed by simulated rainfall. *Soil Sci.* 99:407-413.

Thomas, G.W., and R.E. Phillips. 1979. Consequences of water movement in macropores. *J. Environ. Qual.* 8:149-152.

Thornley, J.H.M. 1972. A balanced quantitative model for root:shoot ratios in vegetative plants. *Ann. Bot.* 36: 431-441.

Thoughton, A. 1977. The rate of growth and partitioning of assimilates in young grass plants: A mathematical model. *Ann. Bot.* 41:553-565.

Thurow, T.E. 1985. *Hydrologic relationship with vegetation and soil as affected by selected livestock grazing systems and climate on the Edwards Plateau.* Ph.D. Thesis, Range Science Dept., Texas A & M. College Station, TX.

Thurow, T.E., W.H. Blackburn, and C.A. Taylor. 1986. Hydrologic characteristics of vegetation types as affected by livestock grazing systems, Edwards Plateau, Texas. *J. Range Manage.* 39:506-510.

Tiessen, H., J.W.B. Stewart, and C.V. Cole. 1984. Pathways of phosphorus transformations in soils of differing pedogenesis. *Soil Sci. Soc. Am. J.* 48:853-858.

USDA. 1963. *Range Research Methods. USDA Miscellaneous Publ. 940.* USDA-Forest Service. Washington, D.C.

van Bavel, C.H.M., DeMichele, D.W. and Ahmed, J. 1973. *A model of gas and energy exchange regulation by stomatal action in plant leaves.* Texas Agric. Exp. Stn. MP-1078.

van Keulen, H., and J. Wolf (eds). 1986. *Modeling of Agricultural Production: Weather, Soils and Crops.* Pudoc Wageningen. Centre for Agricultural Publishing and Documentation. The Netherlands.

Warren, S.D., W.H. Blackburn, and C.A. Taylor, Jr. 1986. Soil hydrologic response to number of pastures and stocking density under intensive rotation grazing. *J. Range Manage.* 39:501.

Weltz, M., and M.K. Wood. 1986. Estimating macroporosity in a forest watershed by use of a tension infiltrometer. *Soil Sci. Soc. Am. J.* 50:578-587.

West, W.H., and W.K. Lauenroth. 1982. Estimating long-term forage production on rangelands. *In* W.K. Lauenroth et al. (eds), *Analysis of Ecological Systems: State-of-the-Art in Ecological Modelling,* Elsevier Scientific Pub. Co., New York, pp. 443-450.

White, E.G. 1984. A multispecies simulation model of grassland producers and consumers. II. Producers. *Ecol. Modelling* 24:241-262.

White, R.E. 1985. The influence of macropores on the transport of dissolved and suspended matter through soil. *Adv. in Soil Sci.* 3:95-120.

Woodmansee, R.G., J.L. Dodd, R.A. Bowman, F.E. Clark, and C.E. Dickman. 1978. Nitrogen budget for a short grass prairie ecosystem. *Oecologia* 34:363-376.

CHAPTER **15**

Modeling Soil C Responses to Environmental Change in Grassland Systems

W.J. Parton,[1] J.A. Morgan,[2] R.H. Kelly,[3] and D. Ojima[4]

Introduction

This chapter focuses on the dynamics of soil organic C (SOC) in the carbon (C) cycle in grasslands. Recent years have seen much discussion of the potential impact of altered environmental conditions on storage of C in grasslands and other natural ecosystems (Parton et al., 1995; Ojima 1993 a, b; Hunt et al., 1991; VEMAP, 1995). This chapter discusses how potential changes in climatic factors like precipitation and air temperature, induced by greenhouse gases, may alter C dynamics in grasslands, and it considers the effect of increased atmospheric CO_2 and the combined effect of increased CO_2 and altered climatic conditions.

This chapter also includes a review of the literature about the direct effect of atmospheric CO_2 on grasslands; a general description of the factors that control soil C dynamics; and an analysis of the results of modeling the effects of environmental change on grassland dynamics. The modeling section reviews other grassland models but provides detailed results from the CENTURY ecosystem model (Parton et al., 1987) and primarily highlights results for the U.S. Great Plains region. The CENTURY model has been tested extensively, using observed soil C and plant production data from grasslands around the world (Burke et al., 1991; Parton et al., 1993; Gilmanov et al., 1997), with the most extensive testing completed for the U.S. Great Plains region.

[1] Sr. Rsch. Sci., Natural Resource Ecology Lab., CO State Univ., Fort Collins, CO 80523-1499, phone (970) 491-1987, fax (970) 491-1965, e-mail billp@nrel.colostate.edu.

[2] Rsch. Ldr., High Plains Grasslands Research Station, 8408 Hildreth, Cheyenne, WY 82009, phone (307) 772-2433; and Supervisory Plant Physiologist, USDA-ARS Rangeland Resources, 16 Crops Research Lab, 1701 Centre Avenue, Fort Collins, CO 80526; phone (970) 498-4216; e-mail morgan@lamar.colostate.edu.

[3] Rsch. Assoc., Natural Resource Ecology Lab., and Project Mgr., Shortgrass Steppe LTER, Dept. of Forest Sciences, CO State Univ., Fort Collins, CO 80523-1499; e-mail robink@cnr.colostate.edu.

[4] Natural Resource Ecology Lab., CO State Univ., Fort Collins, CO 80523-1499; e-mail dennis@nrel.colostate.edu.

Factors that Control Soil Carbon Dynamics

One of the unique features of grasslands is in the distribution of C, with over 95% of it in soil organic matter (SOM). Aboveground plant organs typically account for 40 to 200 g C/m^2. Total root C (0- to 30-cm depth, 80% to 90% of the total roots) ranges from 200 to 400 g C/m^2, while soil C levels (0 to 100 cm) range from 3000 to 20,000 g C/m^2.

The typical depth distribution of C in plant root material and SOM for a short-grass steppe, and grasslands in general, is striking in its rapid decrease with depth (Fig. 15.1) (Mosier, unpublished data). Most SOM forms as a result of decomposition of dead grass roots and resultant stabilization through the microbial biomass into organic matter (OM). This stabilized OM, combined with resistant OM from plant material with a high lignin content and material protected within soil aggregates, makes up a relatively large pool of OM with a slow turnover time. The decrease in soil organic content (SOC) with increasing soil depth results from an exponential decline in root biomass and associated litter and exudates moving down through the profile.

For most grasslands, over 90% of the live roots are in the top 30 cm of the soil (Yonker at al., 1991, Gill et al., 1999), and 92% of the variability in surface (5 cm) total soil C can be explained by a total root biomass covariant (Kelly et al., 1996). The relationship between live plant and SOM C is similar for most grasslands, although the trend is toward a lower root:shoot ratio in more productive grasslands (20% increase from a short grass steppe to a tall grass prairie) and a higher root biomass in cold grassland systems (Kelly et al., submitted).

An important structural feature of grasslands is a rapid increase in the age of SOM with increasing soil depth (see Fig. 15.1c). Radiocarbon testing estimated that SOM's age (conventional ^{14}C age) for a native site in Iowa increases from less than 100 years in the 0- to 10-cm layer to over 1500 years at 50 cm (Harden, unpublished data). The details of how soil C's age increases with depth are understood poorly. Possible factors include decreasing soil temperature with depth, increasing soil anaerobic conditions with depth, and slow movement of resistant OM down the soil profile. Since the older soil C at depth highly resists change, we focus here on potential changes in SOM in the top 20 to 30 cm of the soil, resulting from changes in atmospheric CO_2 and climatic factors.

Figure 15.2's simplified flow diagram of soil C dynamics in grasslands, based on the CENTURY ecosystem model, shows controls on SOM. The major factors that control soil C dynamics include C inputs to the soil, the abiotic decomposition factor (DEFAC), and soil texture. Plant production, which is a function of climate and soil nutrient availability, controls C inputs to the soil. Grazing and fire also may alter inputs. Soil temperature and soil water status control the abiotic decomposition index (DEFAC) (Fig. 15.3), with decomposition increasing with increasing soil temperature and more favorable soil water status.

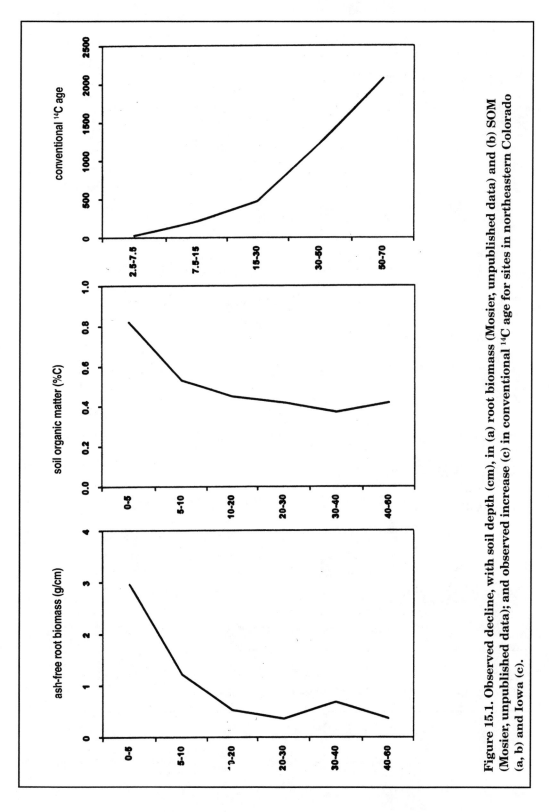

Figure 15.1. Observed decline, with soil depth (cm), in (a) root biomass (Mosier, unpublished data) and (b) SOM (Mosier, unpublished data); and observed increase (c) in conventional ^{14}C age for sites in northeastern Colorado (a, b) and Iowa (c).

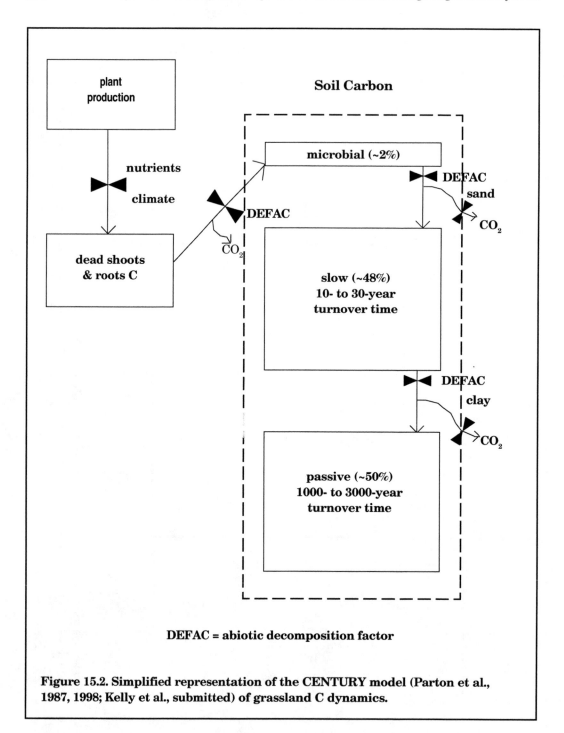

DEFAC = abiotic decomposition factor

Figure 15.2. Simplified representation of the CENTURY model (Parton et al., 1987, 1998; Kelly et al., submitted) of grassland C dynamics.

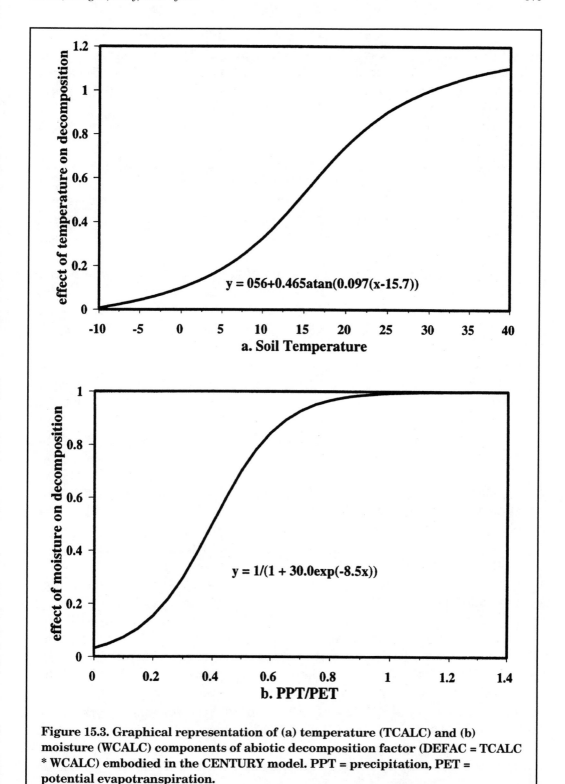

Figure 15.3. Graphical representation of (a) temperature (TCALC) and (b) moisture (WCALC) components of abiotic decomposition factor (DEFAC = TCALC * WCALC) embodied in the CENTURY model. PPT = precipitation, PET = potential evapotranspiration.

The soil temperature curve CENTURY uses (Fig. 15.3a) shows that decomposition is most rapid at relatively low soil temperatures (Q_{10} = 4 at 5°C vs. Q_{10} = 1.6 at 25°C). The greatest increase in decomposition occurs as the ratio of available water to potential evapotranspiration rate increases from 0.2 to 0.6, with little impact for ratios above 0.6.

Like other widely used models of C dynamics (van Veen and Paul 1985; Jenkinson 1990), CENTURY divides the total SOM into multiple pools to account for different C to N ratios and rates of turnover. The SOM in CENTURY is divided into three different pools, with approximately 2%, 48%, and 50% of the soil C in the microbial, slow, and passive pools, respectively. In a 50- to 100-year time period, the most important changes in soil C result from changes in the microbial and slow pools, because they are coupled more tightly to the ambient environment than are chemically stabilized passive materials.

Soil texture has an important impact on physical and chemical protection of OM, with higher levels of OM and physical-chemical protection occurring in soils with high clay and silt content. Holding other factors constant, higher respiration losses for sandy soils result in lower soil C levels, while clay soils have low respiration loss and high soil C levels.

We used CENTURY's simulated patterns of soil C, plant production, and abiotic decomposition index (Fig. 15.4a,b,c,d) to demonstrate the controls on soil C levels for the U.S. Great Plains region. These model experiments were conducted using the Vegetation/Ecosystem Modeling and Analysis Project (VEMAP) 0.5 x 0.5 degree-scale database of climate and soils (Kittel et al., 1995) as input to a version of CENTURY designed to run across a grid (Schimel et al., 1996, 1997).

The CENTURY model has been tested extensively using observed soil C and plant production data for the Great Plains region (Burke et al., 1991; Sala et al., 1988). Schimel et al. (1998 VEMAP validation paper) demonstrated that the model accurately represents regional patterns of soil C in the U.S.

Comparison of the simulated patterns of NPP and DEFAC show that each has its highest values in the Southeastern U.S., and values decrease as you move northwest across the Great Plains. For CENTURY VEMAP simulations, a high correlation was observed between net primary productivity (NPP) and DEFAC for the U.S. (unpublished data). Comparison of VEMAP sand content data with simulated soil C numbers shows that soil C patterns in the Great Plains correlate with sand content (higher soil C with low sand content [unpublished data]).

A detailed analysis of these simulated patterns shows that soil C tends to increase with increasing plant production, but the increase is not direct, since decomposition rates are also greater in the high productivity zones. Burke et al. (1989) showed that, while mean annual temperature and annual precipitation have a substantial impact on observed soil C (more soil C with higher precipitation and lower soil temperatures), soil texture exerts dominate control on regional patterns in soil C content.

Follett, Kimble, and Lal, editors

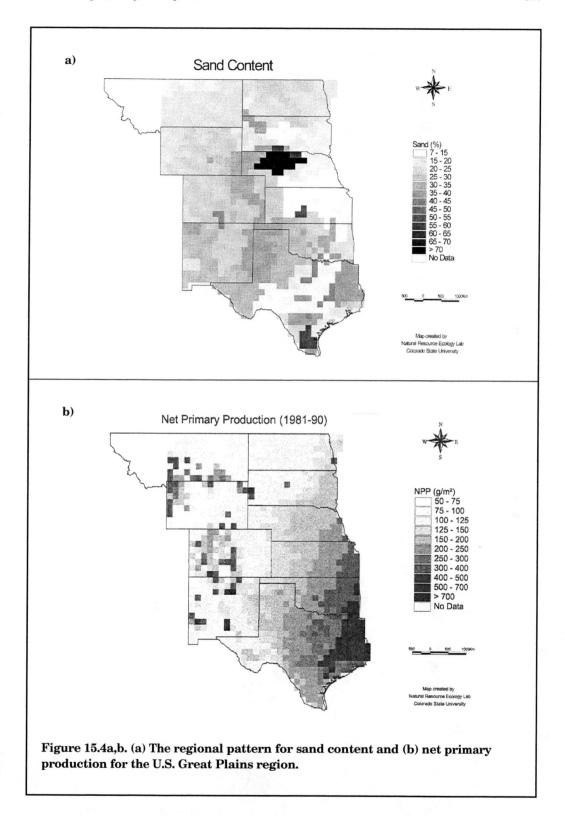

Figure 15.4a,b. (a) The regional pattern for sand content and (b) net primary production for the U.S. Great Plains region.

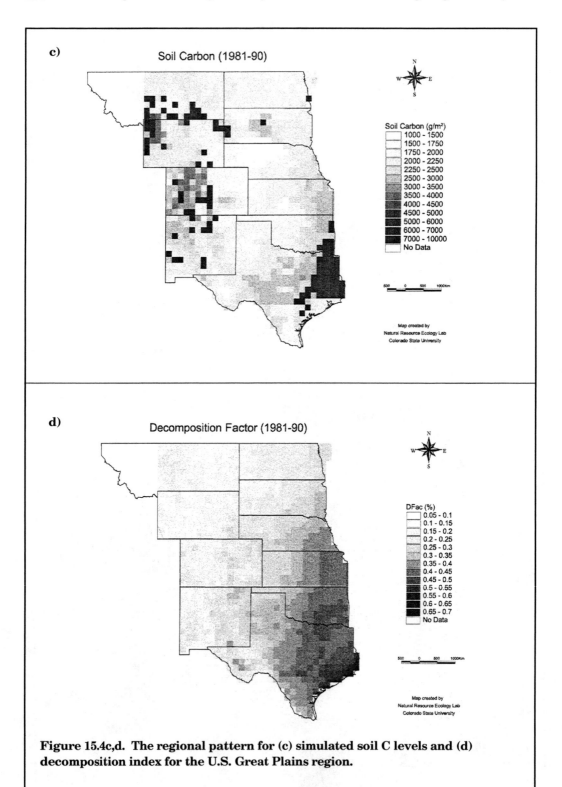

Figure 15.4c,d. The regional pattern for (c) simulated soil C levels and (d) decomposition index for the U.S. Great Plains region.

Impacts of Atmospheric Carbon Dioxide on Grasslands

An important consideration in evaluating grasslands' ability to sequester C must be how increased atmospheric CO_2 affects the C cycle, both through its influence on plant photosynthesis and associated plant processes and through the subsequent cycling of that C through the entire plant-soil system. This section reviews some of the latest findings on these effects of CO_2, primarily in grassland systems.

C_3 plants

CO_2 is the C substrate for photosynthesis in plants. For plants using the C_3 pathway, a group comprising approximately 95% of the world's plant species, atmospheric concentrations of CO_2 are, at present, well below levels required for photosynthetic saturation (Drake et al., 1997).

Increasing atmospheric CO_2 concentrations, which have risen from approximately 270 ppm in pre-industrial times to over 360 ppm today, are expected to increase both photosynthesis and productivity in plants. However, the degree of stimulation of photosynthesis due to increased atmospheric CO_2 depends on a complicated interaction of CO_2 with nutrients, water, and climate, particularly temperature and light (Poorter, 1993). Large differences exist among species in their capacity to respond to elevated CO_2, but an average productivity increase is 37% in plants subjected to growth at CO_2 concentrations approximately twice the present atmospheric levels of 360 ppm (Poorter, 1993).

C_4 plants

Unlike that of C_3 plants, photosynthesis in C_4 plants is nearly saturated at present atmospheric CO_2 concentration, suggesting little potential growth responses to rising CO_2 levels. Plants with the C_4 photosynthetic metabolism represent a much smaller group than C_3 plants, less than 5% of the world's species.

Grasses of this group, however, are particularly important in the world's tropical and subtropical grasslands (Ehleringer et al., 1997). Despite a photosynthetic apparatus that is nearly saturated at present ambient CO_2 concentrations, Wand and Midgley (1999) reported that production increased an average 15% in C_4 grasses when grown at doubled CO_2 concentrations, a response strikingly similar to the 23% average enhancement noted for C_3 grasses.

Some grassland studies indicate that CO_2 enrichment may result in comparable growth enhancements between C_3 and C_4 plants (Hunt et al., 1996; Morgan et al., 1998) or, on occasion, even greater growth responses in C_4 than C_3 grasses (Owensby et al., 1993). The significant growth enhancement in C_4 plants under elevated CO_2 is due primarily to indirect effects of CO_2 on plant-water relations, although some evidence shows a direct photosynthetic enhancement in some C_4 grasses (Ghannoum et al., 1997; LeCain and Morgan, 1998).

Stomata are sensitive to CO_2 and tend to close as CO_2 concentrations rise (Drake et al., 1997), irrespective of photosynthetic class. In grasslands, this change in stomatal conductance can reduce canopy level evapotranspiration (Ham et al., 1995), retard the rate of soil water depletion (Kirkham et al., 1991; Owensby et al., 1993; Jackson et al., 1994; Morgan et al., 1998), enhance plant and soil water relations (Knapp et al., 1994; Morgan et al., 1994, 1998; Wilsey et al., 1997), and lengthen the growing season (Chiarielle and Field, 1996).

Grasslands almost always are characterized by periods of water shortage (Campbell et al., 1997), and these periods may be especially important in the large responses of C_4 grasses to CO_2. Therefore, CO_2 enrichment should increase water use efficiencies and NPP of most grasslands, and these responses will likely be expressed more readily in arid and semi-arid regions.

Acclimation

Many studies indicate that, over time, plants' physiological traits, like photosynthesis, tend to acclimate to CO_2 enrichment, leading to a lessened effect over time (Drake et al., 1997; Sage, 1994). This downward regulation is particularly common in environments in which essential nutrients for plant growth, like N, are limiting (Morgan et al., 1994; Sage. 1994). Nevertheless, even when acclimation is apparent, growth still generally is enhanced above the rate of present ambient CO_2 concentrations (Drake, 1996).

Acclimation which occurs in suboptimal soil environments, investigated mostly in regard to low available N (Canadell et al., 1996), suggests that reduced sensitivity of plants to CO_2 over time may be due to the inability of soil nutrients to keep pace with enhanced growth under elevated CO_2. However, good evidence shows that plants grown at elevated CO_2 become more efficient in use of N through changes in their metabolism (Berntson and Bazzaz, 1996; Conroy, 1992; Drake et al., 1997). Both soil and plant metabolic responses are likely to figure into plants' acclimation to CO_2 enrichment.

Above- and belowground growth

In most grasslands, the proportion of plant biomass in perennial belowground organs dominates the vegetative compartment (Tate and Ross, 1997). While CO_2 may vary considerably in how it affects partitioning of growth between above- and belowground plant organs, in most natural ecosystems, particularly when nutrients and/or water are limiting, CO_2 enrichment appears to cause increased partitioning of biomass to belowground organs (Canadell et al., 1996; Morgan et al., 1994; Owensby et al., 1993; Rogers et al., 1994, 1996).

Soil water

The above discussion indicates a strong probability that the flow of NPP and C into soils will be enhanced in future CO_2-enriched grasslands. However, the storage and cycling of that C is less certain. A myriad of complex direct and indirect responses of plants and soils to elevated CO_2, and their interactions, over time, ultimately will determine the effects on grassland soil C.

In general, elevated CO_2 will affect the dynamics of C cycling through its impact on soil water and through increased rhizodeposition (Cardon, 1996). CO_2-induced enhanced efficiency of water use, due to stomatal closure, will be especially important in characteristically water-limited grasslands. Altered stomatal conductance may affect the dynamics of wetting and drying of grassland soils, resulting in periods of enhanced soil water content. Soil water content is an important driver of soil microbial activities, so these responses to elevated CO_2 will be important for soil C and N cycling (Hungate et al., 1996; Hunt et al., 1991).

Soil microbes

Root symbionts like mycorrhizae and N-fixing organisms represent an additional potential C sink to consider in CO_2 enriched environments. One might expect that these sinks would be important in helping plants adapt in future CO_2-enriched atmospheres, and evidence from several studies supports this (Canadell et al., 1996; Diaz, 1996).

Changes in mycorrhizae or N-fixation capacity may be due either to alterations in the specific infection rate or fixation activity, or simply to larger root systems (Canadell et al., 1996). Increases in NPP tend to be greater in nodulated and mycorrhizae-infected species (Diaz, 1996), suggesting a potentially greater response to elevated CO_2 than that of species with no associated symbionts. This greater response is expected, given the importance of mycorrhizae in nutrient acquisition and plant water relations (Allen, 1991) and the generation of available N through fixation.

The ability of grassland soil resources to sustain from CO_2 enrichment, in the long term, and the enhanced productivity observed in short-term studies, will depend on the complex interactions among soil biology, hydrology, and plants' responses. While researchers generally agree that elevated CO_2 will increase inputs of labile C compounds into SOM and that soil microbial biomass also will be enhanced (Berntson and Bazzaz, 1996), they also hypothesize that soil microbes may enhance (Zak et al., 1993) or constrain (Díaz et al., 1993) NPP responses to elevated CO_2.

Changes in grassland plant communities from CO_2 enrichment are likely to affect important ecosystem driving factors like soil water availability, available N, or C inputs and to interact in ways that are difficult to predict from our present knowledge, which has come mostly through studies of single plant species over short time periods. Modeling can provide a useful tool for extrapolating beyond our experimental boundaries and for exploring the likely long-term consequences of CO_2 enrichment on grassland C cycling.

Modeling the Impacts of Environmental Change on Grassland Dynamics

This section examines modeling studies that simulate the potential impacts of environmental changes on grassland dynamics. A recent review by Parton et al. (1996) shows how grassland models have been used extensively during the last 25 years to evaluate changes in grassland dynamics resulting from altered climatic conditions, elevated atmospheric CO_2 levels, and different grassland management practices (e.g., grazing intensity, seasonality of grazing, fertilization, changes in grass species, etc). Here we focus on more recent (post-1990) studies that evaluate the effects of climatic change, increased atmospheric CO_2, and the combined effect of CO_2 and climatic change on grassland dynamics.

The ecosystem variables we consider include plant production, soil C storage, soil water status, and abiotic factors affecting decomposition (DEFAC). We primarily present results from the CENTURY model (Parton et al., 1993) but also include a review of other modeling studies. Most studies before 1995 evaluated the effect of altered climatic conditions, while studies after 1995 include the combined effects of increased atmospheric CO_2 and altered climatic conditions. We present new results from the daily time-step version of the CENTURY model (DAYCENT) (Parton et al., 1998) for sites in Colorado and Kansas, and results from a recent CENTURY transient climate change run (1990 to 2100) using the Hadley Centre for Climate Prediction and Research General Circulation Model (GCM [*http://www.meto.govt.uk/sec5/CR_div/index_climate.html*]).

CENTURY 1.0

Schimel et al. (1990) used an earlier version of the CENTURY model (Version 1.0; Parton et al., 1987) to simulate the ecosystem-level impacts of different GCM-predicted climatic changes for selected sites in the U.S. Great Plains. Simulation studies indicate that climate change scenarios result in increased plant production and N cycling, while SOM storage decreases. Predicted increases in air temperature (2 to 4°C) have the largest impact on system response. The increase in N availability causes plant production to increase, but this increase in C inputs to the soil is insufficient to compensate for increased decomposition of SOM.

These results were improved upon in a later paper (Schimel et al., 1991) which evaluated climate change impacts across the Great Plains using the Goddard Institute for Space Studies' (GISS [Hansen et al., 1984]) GCM double CO_2 scenarios. The model's results showed that plant production increased and soil C decreased for most of the region, with the exception of Texas and Oklahoma, where plant production decreased by >10% as a result of increased drought stress.

Burke et al. (1991) did a more detailed analysis of GISS climatic change impacts for the southern Great Plains and found results similar to Schimel et al.'s (1991). They showed that the 50-year simulated decreases in soil C (100 to 300 g C/m) were small compared to the observed decreases in soil C resulting from agricultural land use changes (cultivation-induced decreases of 1000 to 2000 g C/m^2).

Jenkinson

Jenkinson's (1990) SOM model was used to evaluate the impact of increasing air temperature (3°C) on soil C levels for the natural ecosystems of the world. His simulated C losses were much greater than the values CENTURY simulated. These larger losses probably occurred because he did not include the temperature-induced increase in N availability on plant production and the resultant increase of C inputs into the soil.

CENTURY 4.0

Version 4 of the CENTURY model (Parton et al., 1993) has been tested using observed plant production and soil C and N data from tropical and temperate grasslands. It was used to assess the potential impact of different GCM projections on plant production, soil C levels, and DEFAC for all of the major grasslands in the world (Ojima et al., 1993a; Parton et al., 1995).

Projected changes in climate from the 2XCO$_2$ GCM scenarios occurred as a 50-year linear increase (no change after 50 years), and the direct impacts also were considered. The major direct effects of increasing atmospheric CO_2 are decreasing water loss due to transpiration (stomatal conductance decreasing with increasing atmospheric CO_2) and increase in the C:N ratio of live plant shoots (Owensby et al., 1998; Ojima et al., 1993a). Increasing atmospheric CO_2 generally resulted in an increase in soil C storage (Table 15.1) for global grassland systems except the dry savannas. This primarily results from the CO_2-induced increase in plant pro-

Table 15.1. Simulated change in aboveground plant production, represented as an average change during a 50- to 75-year period following initiation of altered climate conditions and CO_2 levels (doubling in 50 years).

Sites	Land Cover Area (Mha)	Current Plant Production (gC/m²/yr)	% Change in Plant Production				
			$2xCO_2$	CCC CC	CCC $2xCO_2$ and CC	GFDL CC	GFDL $2xCO_2$ and CC
Extreme continental steppe	209.5	59.7	29.4	−24.8	−2.3	−27.7	−3.3
Dry continental steppe	294.3	36.4	11.0	−17.3	4.0	7.9	34.6
Humid temperate	395.8	133.8	16.9	13.4	34.1	11.6	31.5
Mediterranean	16.1	78.8	15.3	25.1	25.1	7.0	27.0
Dry savanna	510.9	54.6	15.0	22.1	22.1	−1.0	16.2
Savanna	799.0	190.9	13.0	14.1	14.1	6.7	21.4
Humid savanna	170.8	339.6	9.9	8.1	8.1	2.4	14.6

Climate change scenarios were derived from the Canadian Climate Center (CCC) and the Geophysical Fluid Dynamics Laboratory (GFDL) GCM 2 x CO_2 simulations (Ojima et al., 1993). CC = climate change. Following Table 2, Environmental Change in Grasslands (Parton et al., 1994).

duction (Table 15.2). Increased CO_2 also increases DEFAC (causes wetter soil conditions), but the increase in production is greater than the increase in DEFAC.

The Canadian Climate Centre (CCC) (Boer, McFarlane, and Lazare, 1992) and Geophysical Fluid Dynamics Laboratory (GFDL) (Manabe and Wetherald, 1990; Wetherald and Manabe, 1990) climate change scenarios result in a decrease in soil C for the majority of the grassland biomes, with large decreases (>10%) for the dry and extreme continental steppe systems. This decrease for most grasslands results in more rapid increase in DEFAC than in plant production with the climate change scenarios.

The larger decrease for the continental steppe systems results from large increases in DEFAC and substantial decreases in plant production (as great as 25%) resulting from increased drought stress (higher temperatures and reduced precipitation). Adding the CO_2 effect to the climate change runs results in decreased soil C losses and increased plant production. The combined effect of increased atmospheric CO_2 and climate change is a net loss of 0.2 Pg soil C, compared to 0.4 Pg for climate change alone.

Table 15.2. Simulated change in soil C, represented as an average change during a 50- to 75-year period following initiation of altered climate conditions and CO$_2$ levels (doubling in 50 years).

Sites	Land Cover Area (Mha)	Current Soil C (Pg to 20 cm)	2xCO$_2$	CCC CC	CCC 2xCO$_2$ and CC	GFDL CC	GFDL 2xCO$_2$ and CC
					% Change in SOM		
Extreme continental steppe	209.5	12.65	1.42	−11.64	−12.21	−9.08	−9.05
Dry continental steppe	294.3	7.72	1.03	−13.16	−12.92	−14.11	−13.10
Humid temperate	395.8	28.89	2.45	−4.39	−1.95	−4.19	−2.12
Mediterranean	16.1	0.25	2.94	−9.23	−6.44	−9.14	−6.23
Dry savanna	510.9	12.79	−0.09	−0.37	0.04	0.49	1.03
Savanna	799.0	39.46	3.44	−1.61	1.62	−1.13	2.31
Humid savanna	170.8	3.76	8.55	−4.36	3.56	−3.44	5.59

Climate change scenarios were derived from the Canadian Climate Center (CCC) and the Geophysical Fluid Dynamics Laboratory (GFDL) GCM 2 x CO$_2$ simulations (Ojima et al., 1993). CC = climate change, Pg = 10^6 g. Following Table 3, Environmental Change in Grasslands (Parton et al., 1994).

TEM 4.0

The Terrestrial Ecosystem Model (TEM version 4), a process-based ecosystem model (Raich et al., 1989; McGuire et al., 1992, 1993, 1996a,b; Melillo et al., 1993, 1995), is similar to the CENTURY model in many ways and includes the interactive feedback of nutrient availability and climate on plant production and the direct effects of atmospheric CO$_2$ on plant production. The global model results (Melillo et al., 1993) show that climate change alone causes decreased soil C levels and increased plant production. Adding the direct effect of increasing atmospheric CO$_2$ substantially reduces the loss of soil C and increases plant production.

VEMAP

The VEMAP model comparison (VEMAP 1995) compared biogeochemistry models to evaluate the impact of CO$_2$-induced GCM climate change on ecosystems' dynamics in the continental U.S., using a .5x.5 degree grid. The models were set up to use the same land use, climatic change scenarios, and current climatic conditions. The biogeochemistry model's results for the grassland regions in the Great Plains showed that soil C and plant production would increase for the central and northern Great Plains, but plant production and soil C storage tended to decrease for the southern Great Plains. This general pattern was most pronounced for the TEM and CENTURY models' results using the GFDL and the United Kingdom Meteorological Office [UKMO (Wilson and Mitchell 1987)] GCM scenarios.

DAYCENT Model Carbon Dioxide Results

The major factors that have been included in models that consider direct CO_2 effects are:

1. stomatal conductance response
2. alteration of photosynthetic rate
3. changing allocation to roots vs. shoots, and
4. modified nutrient content of live leaves.

DAYCENT (Parton et al., 1996; Kelly et al., submitted) has been used extensively to simulate the effects of increased atmospheric CO_2 concentration on system C dynamics, and it allows the user to prescribe the magnitude of each of these effects. We conducted an illustrative set of simulations for two sites representative of two climatic extremes of the U.S. Great Plains.

Konza Prairie Research Natural Area, a tall grass prairie Nature Conservancy site in central Kansas, is characteristic of subhumid grasslands lying at the eastern edge of the Great Plains grasslands. The Central Plains Experimental Range located in northeast Colorado is a short grass steppe lying at the northern boundary of Great Plains grasslands. Identical effects of CO_2 doubling were applied for each site.

Both grasslands are dominated by C_4 grasses which can respond to CO_2 (Hunt et al., 1996; Morgan et al., 1994, 1998a,b; Owensby et al., 1993b), including with limited direct photosynthetic enhancement at elevated CO_2 (Morgan et al., 1994; LeCain and Morgan, 1998). However, since growth enhancements in these grasslands under elevated CO_2 appear driven primarily by improved water relations via stomatal closure (Kirkham et al., 1991; Morgan et al., 1994, 1998a,b; Owensby et al., 1993), and also because photosynthetic acclimation of C_3 prairie grasses severely limits their photosynthetic response to elevated CO_2 (Morgan et al., 1994, and unpublished data), we assumed for modeling purposes that CO_2 enrichment reduced stomatal conductance by 30% but had no effect on photosynthesis.

We also assumed that root:shoot ratios increased 50% under elevated CO_2, given the hypothesis that CO_2 enrichment leads to increased root:shoot ratios in N-limited systems like the short grass steppe, as a plant mechanism for achieving balanced growth (Hunt et al., 1998; Morgan et al., 1994; Rogers et al., 1994, 1996). The live ratio of C:N was increased by 30% to reflect the common observation of reduced tissue N concentration of plants grown in CO_2-enriched atmospheres (Hunt et al., 1996; Morgan et al., 1994; Owensby et al., 1993a; Read et al., 1997).

For each site, we created a long-term equilibrium simulation at ambient CO_2 levels before the 50-year post-doubling test period. Models for both sites showed a slight increase in total production over the experimental 50-year period. The increase in production was almost entirely belowground, annually averaging 41% and 26% greater than an ambient CO_2 control for Konza and CPER, respectively, over the 50-year post-doubling period (Fig. 15.5a). Concomitant with the increase in production was an increase in total SOM, averaging 6% and 2% higher than ambient CO_2 control for Konza and CPER, respectively (Fig. 15.5b).

As total SOM increased, the modeled chemical composition of the SOM also changed over time, toward a higher ratio of C:N. This change in OM quality may have long-term implications for ecosystem productivity. The lower quality implies a lower overall capacity to supply mineral N to the system.

The abiotic decomposition factor (DEFAC) was increased for both CPER (12%) and Konza (11%). The dynamics of the change in DEFAC are particularly interesting (Fig. 15.6). While the difference between ambient and doubled CO_2 simulation results for Konza remained relatively constant over time (CV = 21%), results for CPER were far more variable (CV = 71%).

The greater overall variability for CPER is due to both greater interannual variation and to a change in modeled response at 28 years after CO_2 doubling. Overall, the enhancement of DEFAC by high CO_2 decreases, and the amplitude of the interannual variation increases (including single-year decreases in DEFAC to levels lower than simulated under ambient CO_2). Thus it appears that, in years with relatively low precipitation, DEFAC may be lower under high CO_2 conditions than at ambient levels.

New VEMAP Results

The most recent VEMAP model comparisons use the transient GCM results from the HEDLEY CENTRE and CCC models. CENTURY's simulations of NPP, soil C, and DEFAC after 50 years of climatic change (2050 values minus 1980 levels) using the HEDLEY CENTRE GCM show soil C storage tending to increase in the southern and central Great Plains and decrease in the northern Great Plains (Fig. 15.7a,b,c).

The decrease in soil C in the northern Great Plains primarily results from the increase in DEFAC resulting from increased soil temperatures. In the central Great Plains, soil C content increased because of increased plant production combined with a small increase in DEFAC. In the southern Great Plains, soil C increased slightly as a result of a big decrease in DEFAC (drier conditions reduced DEFAC) and a decrease in plant production. The HEDLEY CENTRE GCM results indicate that precipitation was increasing by 25% for most of the Great Plains and this increase resulted in higher NPP for most of the region. These results show the differential impact of climatic changes on plant production and DEFAC and the combined effect of both factors on soil C levels.

A summary of the observed data and model results shows that the major factors which affect grassland soil C levels are soil texture, decomposition rates, and C input to the soil. Some of the initial modeling work assessing climate change

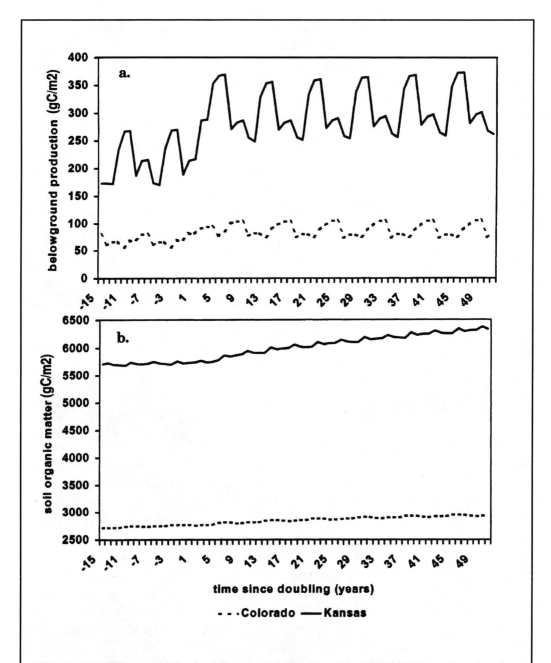

Figure 15.5. Effects of CO_2-doubling on production (a) and SOM (b) as simulated by the CENTURY ecosystem model (Parton et al., 1987, 1998; Kelly et al., submitted) for an arid Colorado site and a subhumid Kansas site.

suggested that grassland soil C levels could decrease as much as 5 to 10% with 2X CO_2 climate change scenarios, with most of the decrease due to increases (2 to 4°C) in the soil temperatures. The more recent model's results, which include the direct impact of increased atmospheric CO_2 on plant growth, suggest the earlier estimates of C loss were exaggerated and that, for the U.S. Great Plains, soil C levels may increase slightly, although soil C may decrease in the North.

Increasing atmospheric CO_2 concentration enhances grassland plant production and allocation of C to the root system. These factors are primarily responsible for the predicted increases in soil C. Another possible factor in new GCM scenarios is the predicted increase in precipitation associated with the increase in air temperature, which may result in increased plant production and inputs of C to the soil.

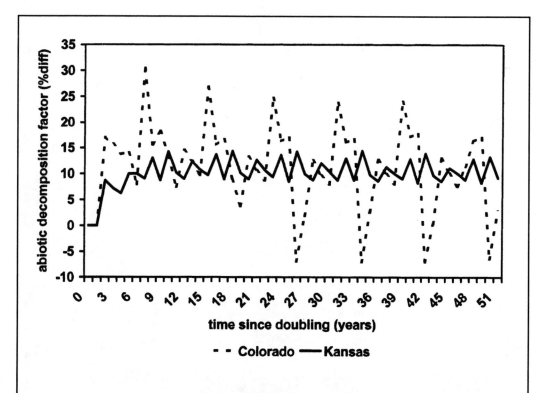

Figure 15.6. Effects of CO_2-doubling on the abiotic decomposition factor (DEFAC) as simulated by the CENTURY ecosystem model (Parton et al., 1987, 1998; Kelly et al., submitted) for an arid Colorado site and a subhumid Kansas site.

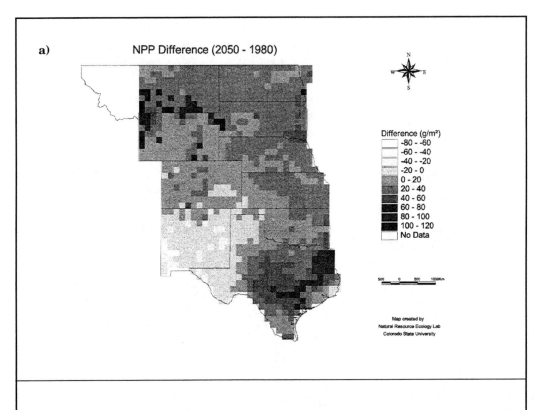

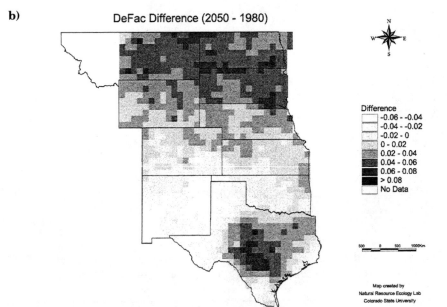

Figure 15.7a,b. Simulated changes in (a) net primary production and (b) decomposition index (DEFAC) under the Hedley Centre transient GCM scenario for 50 years of climate change.

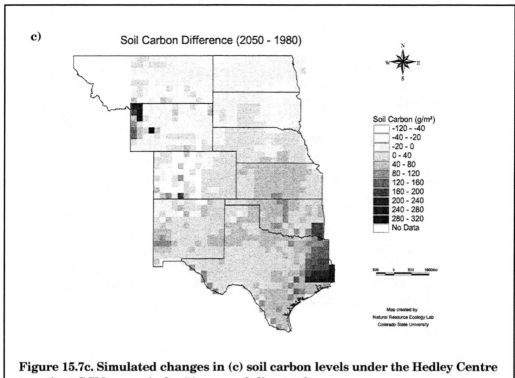

Figure 15.7c. Simulated changes in (c) soil carbon levels under the Hedley Centre transient GCM scenario for 50 years of climate change.

Conclusion

Current field research and modeling studies suggest that key research efforts should evaluate the effects of the interactions between potential climatic changes and increasing atmospheric CO_2 concentration. Increasing atmospheric CO_2 for grasslands generally results in increased total plant production, higher root:shoot ratios, reduction of transpiration water loss and N content of the live shoots, and increased soil water contents during the growing season. Available soil N generally reduces the response of plants to increased atmospheric CO_2 levels, but whether long-term exposure to higher CO_2 levels will increase N inputs to soil through N fixation remains unclear.

We are just becoming able to evaluate how increased atmospheric CO_2 levels affect plant growth. Current modeling studies have shown both grasslands' sensitivity to the combined impacts of climatic change and increased atmospheric CO_2 concentration, and our lack of knowledge about that sensitivity.

We also know too little about plant communities' potentially complex responses to global change. These responses may include competitive interactions regarding water, nutrients, and light, as well as functional differences in re-

sponses to CO_2, temperature, and water. Functional differences are likely to dominate how these plant communities respond.

Various modeling systems describe what happens to plants' resposes which
plant production drives. This simplistic treatment, although vital in the development of modeling studies, ignores changes in species' composition. Also emerging
is evidence of a potentially significant response of C_4 grasses to CO_2, including
their photosynthetic acclimation (Morgan et al., in review). Tissue quality may
decline, resulting in a declining N supply in the system. This implies a potentially
important role of legumes in future CO_2-enriched environments. It also underscores our need to understand these processes formally, so that we can describe
how plant systems will respond to CO_2 enrichment.

Acknowledgments

NASA and USFS provided funds for VEMAP simulations. DOE funded the
Great Plains Assessment for transient model simulations. The USDA-ARS funded the NSF SGS LTER programs and collaborative research.

References

Alcamo, J., G.J.J. Kreileman, J.C. Bollen, G.J. van den Born, R. Gerlagh, M.S. Krol,
A.M.C. Toet, and H.J.M. de Vries. 1996. Baseline scenarios of global environmental
change. *Global Environ. Change* 6:261-303.

Allen, M.F. 1991. *The Ecology of Mycorrhizae.* Cambridge University Press, Cambridge, UK.

Bazzaz, F.A. 1990. The response of natural ecosystems to the rising global CO_2 levels.
Annu. Rev. of Ecol. Syst. 21:167-196.

Berntson, G.M., and F.A. Bazzaz. 1996. Belowground positive and negative feedbacks
on CO_2 growth enhancement. *Plant and Soil* 187:119-131.

Boer, G.J., N.A. McFarlane, and M. Lazare. 1992. Greenhouse gas-induced climate
change simulated with the CCC second-generation general circulation model. *J.
Climate* 5:1045-1077.

Burke, I.C., T.G.F. Kittel, W.K. Lauenroth, P. Snook, C.M. Yonker, and W.J. Parton.
1991. Regional analysis of the central Great Plains. *Bioscience* 41:685-692.

Campbell, B.D., D.M. Stafford-Smith, and G.M. McKeon. 1997. Elevated CO_2 and water supply interactions in grasslands: a pastures and rangelands management perspective. *Global Change Biol.* 3:177-187.

Canadell, J.G., L.F. Pitelka, and J.S.I. Ingram. 1996. The effects of elevated [CO_2] on
plant-soil carbon belowground: A summary and synthesis. *Plant and Soil* 187:391-
400.

Cardon, Z.G. 1996. Influence of rhizodeposition under elevated CO_2 on plant nutrition and soil organic matter. *Plant and Soil* 187:277-288.

Chiariello, N.R., and C.B. Field. 1996. Annual grassland responses to elevated CO_2 in multiyear community microcosms. *In* C.H. Körner and F.A. Bazzaz (eds). *Carbon Dioxide, Populations, and Communities*, Academic Press, San Diego, pp. 139-157.

Conroy, J.P. 1992. Influence of elevated atmospheric CO_2 concentrations on plant nutrition. *Aust. J. of Bot.* 40:445-456.

Diaz, S. 1996. Effects of elevated CO_2 at the community level mediated by root symbionts. *Plant and Soil* 187:309-320.

Díaz, S., J.P Grime, J. Harris, and E. McPherson. 1993. Evidence of a feedback mechanism limiting plant response to elevated carbon dioxide. *Nature* 364:616-617.

Drake, B.G., M.A. Gonzàlez-Meler, and S.P. Long. 1997. More efficient plants: A consequence of rising atmospheric CO_2? *Annu. Rev. of Plant Physiol. and Plant Mol. Biol.* 48:607-637.

Ehleringer, J.R., T.E. Cerling, and B.R. Helliker. 1997. C_4 photosynthesis, atmospheric CO_2, and climate. *Oecologia* 112:285-299.

Farrar, J.F., and S. Gunn. 1996. Effects of temperature and atmospheric carbon dioxide on source-sink relations in the context of climate change. *In* E. Zamski and A.A. Schaffer (eds), *Photoassimilate Distribution in Plants and Crops*, Marcel Dekker, New York.

Fredeen, A.L., G.W. Koch, and C.B. Field. 1995. Effects of atmospheric CO_2 enrichment on ecosystem CO_2 exchange in a nutrient and water limited grassland. *J. of Biogeogr.* 22:215-219.

Ghannour, O., S. von Caemmerer, E.W.R. Barlow, and J.P. Conroy. 1997. The effect of CO_2 enrichment and irradiance on the growth, morphology and gas exchange of a C_3 (*Panicum laxum*) and a C_4 (*Panicum antidotale*). *Austr. J. of Plant Phys.* 24:227-237.

Gill, R.A., I.C. Burke, D.G. Milchunas, and W.K. Lauenroth. 1999. Relationship between root biomass and soil organic matter pools in the shortgrass steppe of eastern Colorado. *Ecosystems* 2:226-236.

Gilmanov, T.G., W.J. Parton, and D.S. Ojima. 1997. Testing the CENTURY ecosystem level model on data sets from eight grassland sites in the former USSR representing wide climatic/soil gradient. *Ecol. Modelling* 96:191-210.

Hungate, B.A., F.S. Chapin III, H. Zhong, E.A. Holland, and C.B. Field. 1996. Stimulation of grassland nitrogen cycling under carbon dioxide enrichment. *Oecologia* 109:149-153.

Hunt, H.W., E.T. Elliott, J.K. Detling, J.A. Morgan, and D.-X. Chen. 1996. Responses of a C_3 and a C_4 perennial grass to elevated CO_2 and temperature under different water regimes. *Global Change Biol.* 2:35-47.

Hunt, H.W., J.A. Morgan, and J.J. Read. 1998. Simulating growth and root-shoot partitioning in prairie grasses under elevated atmospheric CO_2 and water stress. *Ann. of Bot.* 81:489-501.

Hunt, H.W., M.J. Trlica, E.F. Redente, J.C. Moore, J.K. Detling, T.G.F. Kittel, D.W. Walter, M.C. Fowler, D.A. Klein, and E.T. Elliott. 1991. Simulation model for the effects of climate change on temperate grassland ecosystems. *Ecol. Modelling* 53:205-246.

Jackson, R.B., O.E. Sala, C.B. Field, and H.A. Mooney. 1994. CO_2 alters water use, carbon gain, and yield for the dominant species in a natural grassland. *Oecologia* 98:257-262.

Jenkinson, D.S. 1990. The turnover of carbon and nitrogen in soil. *Philoso. Trans. of the R. Soc. of London B* 329:361-368.

Kelly, R.H., I.C. Burke, and W.K. Lauenroth. 1996. Soil organic matter and nutrient availability responses to reduced plant inputs in shortgrass steppe. *Ecology* 77:2516-2527.

Kelly, R.H., W.J. Parton, M.D. Hartman, L.K. Stretch, D.S. Schimel, and D.S. Ojima. 2000. Intra and interannuual variability of ecosystem processes in shortgrass steppe. *J. of Geophys. Res.* (in revision).

Kirkham, M.B., H. He, T.P. Bolger, D.J. Lawlor, and E.T. Kanemasu. 1991. Leaf photosynthesis and water use of big bluestem under elevated carbon dioxide. *Crop Sci.* 31:1589-1594.

Kittel, T.G.F, N.A. Rosenbloom, T.H. Painter, D.S. Schimel, and VEMAP Modeling Participants (1995) The VEMAP integrated database for modeling United States ecosystem/vegetation sensitivity to climate change. *J. of Biogeogr.* 22(4-5):857-862.

Knapp, A.K., J.T. Fahnestock, and C.E. Owensby. 1994. Elevated atmospheric CO_2 alters stomatal responses to variable sunlight in a C_4 grass. *Plant, Cell and Environ.* 17:189-195.

Koch, G.W., and H.A. Mooney. 1996. Response of terrestrial ecosystems to elevated CO_2: A synthesis and summary. *In* G.W. Koch and H.A. Mooney (eds), *Carbon Dioxide and Terrestrial Ecosystems*, Academic Press, Inc., San Diego, CA, pp. 415-429.

LeCain, D.R., and J.A. Morgan. 1998. Growth, gas exchange, leaf nitrogen and carbohydrate concentrations in NAD-ME and NADP-ME C_4 grasses grown in elevated CO_2. *Physio. Plant.* 102:297.

Manabe, S., and R.T. Wetherald. 1990. *In* J.F.B. Mitchell, S. Manabe, V. Meleshko, T. Tokioka, Equilibrium climate change and its implications for the future, *in* J.T. Houghton, G.J. Jenkins, and J.J. Ephraums (eds), *Climate Change: The IPCC Scientific Assessment*, Cambridge University Press, Cambridge, UK, pp. 131-172.

McGuire, A.D., L.A. Joyce, D.W. Kicklighter, J.M. Melillo, G. Esser, and C.J. Vorosmarty. 1993. Productivity response of climax temperate forests to elevated temperature and carbon dioxide: a North American comparison between two global models. *Climate Change* 24:287-310.

McGuire, A.D., D.W. Kicklighter, and J.M. Melillo. 1996b. Global climate change and carbon cycling in grasslands and conifer forests. *In* J.M. Melillo and A.I. Breymeyer (eds), *Global Change: Effect on Coniferous Forests and Grasslands,* SCOPE volume 56, John Wiley and Sons, Chichester, West Sussex, England.

McGuire, A.D., J.M. Melillo, D.W. Kicklighter, and L.A. Joyce. 1996a. Equilibrium responses of soil carbon to climate change: Empirical and process-based estimates. *J. Biogeogr.* 22:785-796.

McGuire, A.D., J.M. Melillo, D.W. Kicklighter, A.L. Grace, B. Moore III, and C.J. Vorosmarty. 1992. Interactions between carbon and nitrogen dynamics in estimating net primary productivity for potential vegetation in North America. *Global Biogeochemical Cycles* 6:101-124.

Melillo, J.M., D.W. Kicklighter, A.D. McGuire, W.T. Peterjohn, and K.M. Newkirk. 1995. Global change and its effects on soil organic carbon stocks. *In* R.G. Zepp and C.H. Sonntag (eds), *Role of Nonliving Organic Matter in the Earth's Carbon Cycle,* John Wiley & Sons, New York, pp.175-189.

Melillo, J.M., A.D. McGuire, D.W. Kicklighter, B. Moore III, C.J. Vorosmarty, and A.L. Schloss. 1993. Global climate change and terrestrial net primary production. *Nature* 363:234-240.

Milchunas, D.G., and W.K. Lauenroth. 1992. Carbon dynamics and estimates of primary productivity by harvest, ^{14}C dilution and ^{14}C turnover. *Ecology* 73:593-607.

Mooney, H.A., B.G. Drake, R.J. Luxmoore, W.C. Oechel, and L.F. Pitelka. 1991. Predicting ecosystem responses to elevated CO_2 concentrations. *BioScience* 41:96-104

Morgan, J.A., D.R. LeCain, A.R. Mosier, D.G. Milchunas, W.J. Parton, and D. Ojima. 1998. Carbon dioxide enrichment enhances photosynthesis, water relations and growth in C_3 and C_4 shortgrass steppe grasses. *ESA Abstracts*, ESA, Baltimore, MD, p. 196.

Morgan, J.A., D.R. LeCain, J.J. Read, H.W. Hunt, and W.G. Knight. 1998. Photosynthetic pathway and ontogeny affect water relations and the impacts of CO_2 on *Bouteloua gracilis* (C_4) and *Pascopyrum smithii* (C_3). *Oecologia* 114:483-493.

Morgan, J.A., H.W. Hunt, C.A. Monz, and D.R. LeCain. 1994. Consequences of growth at two carbon dioxide concentrations and temperatures for leaf gas exchange of *Pascopyrum smithii* (C_3) and *Bouteloua gracilis* (C_4). *Plant, Cell and Environ.* 17:1023-1033.

Morgan, J.A., R.H. Skinner, and J.D. Hanson. 2000. Re-growth and biomass partitioning in forages under nitrogen and CO_2 regimes: Differences among species of three functional groups. *Crop Science* (in review)

Morgan, J.A., W.G. Knight, L.M. Dudley, and H.W. Hunt. 1994. Enhanced root system C-sink activity, water relations and aspects of nutrient acquisition in mycotrophic *Bouteloua gracilis* subjected to CO_2 enrichment. *Plant and Soil* 165:139-146.

Norby, R.J. 1994. Issues and perspectives for investigating root responses to elevated atmospheric carbon dioxide. *Plant and Soil* 165:9-20.

Norby, R.J., E.G. O'Neill, W.G. Hood, and R.J. Luxmoore. 1987. Carbon allocation, root exudation and mycorrhizal colonization of *Pinus exhinata* seedlings grown under CO_2 enrichment. *Tree Physiol.* 3:203-210.

Ojima, D.S., W.J. Parton, D.S. Schimel, T.G.F. Kittel, and J.M.O. Scurlock. 1993a. Modeling the effects of climatic and CO_2 changes on grassland storage of soil C. *Water, Air, and Soil Pollu.* 70:643-657.

Ojima, D.S., B.O.M. Dirks, E.P. Glenn, C.E. Owensby, and J.M.O. Scurlock. 1993b. Assessment of C budget for grasslands and drylands of the world. *Water, Air, and Soil Pollu.* 70:95-109.

Overdiek, D. 1993. Elevated CO_2 and the mineral content of herbaceous and woody plants. *Vegetatio* 104/105:403-411.

Owensby, C.E., P.I. Coyne and L.M. Auen. 1993. Nitrogen and phosphorus dynamics of a tall grass prairie ecosystem exposed to elevated carbon dioxide. *Plant, Cell and Environ.* 16:8434-850.

Owensby, C.E., P.I. Coyne, J.M. Ham, L.M. Auen, and A.K. Knapp. 1993. Biomass production in a tall grass prairie ecosystem exposed to ambient and elevated CO_2. *Ecol. Apps.* 3:644-653.

Pan, Y., J.M. Melillo, A.D. McGuire, D.W. Kicklighter, L.F. Pitelka, K.Hibbard, L.L. Pierce, S.W. Running, D.S. Ojima, W.J. Parton, D.S. Schimel, and other VEMAP members including H. Fisher, T. Kittel, R. McKeown, and N. Rosenbloom. 1998. Modeled responses of terrestrial ecosystems to elevated atmospheric CO_2: A comparison of simulations by the biogeochemistry models of the Vegetation/Ecosystem Modeling and Analysis Project (VEMAP). *Oecologia* 114:389-404.

Parton, W.J., D.S. Schimel, C.V. Cole, and D.S. Ojima. 1987. Analysis of factors controlling soil organic matter levels in Great Plains grasslands. *Soil Sci. Soc. of Am. J.* 51:1173-1179.

Parton, W.J., D.S. Schimel, and D.S. Ojima. 1994. Environmental change in grasslands: assessment using models. *Climatic Change* 28:111-141.

Parton, W.J., J.M.O. Scurlock, D.S. Ojima, T.G. Gilmanov, R.J. Scholes, D.S. Schimel, T. Kirchner, J.C. Menaut, T. Seastedt, E.G. Moya, A. Kamnalrut, and J.I. Kinyamario. 1993. Observations and modeling of biomass and soil organic matter dynamics for the grassland biome worldwide. *Global Biogeochem. Cycles* 7:785-809.

Parton, W.J., J.M.O. Scurlock, D.S. Ojima, D.S. Schimel, D.O. Hall, and SCOPEGRAM Group Members. 1995. Impact of climate change on grassland production and soil carbon worldwide. *Global Change Biol.* 1:13-22.

Parton, W.J., D.S. Ojima, and D.S. Schimel. 1996. Models to evaluate soil organic matter storage and dynamics. *In* M.R. Carter and B.A. Stewart (eds), *Structure and Organic Matter Storage in Agricultural Soils*, CRC Press, Inc., Boca Raton, FL, pp. 421-448.

Parton, W.J., M.D. Hartman, D.S. Ojima, and D.S. Schimel. 1998. DAYCENT and its land surface submodel: description and testing. *Global and Planetary Change* 19:35-48.

Polley, H.W., H.S. Mayeux, H.B. Johnson, and C.R. Tischler. 1997. Viewpoint: Atmospheric CO_2, soil water, and shrub/grass ratios on rangelands. *J. of Range Manage.* 50:278-284.

Polley, H.W., J.A. Morgan, B.D. Campbell, and M. Stafford-Smith. 2000. Crop ecosystem responses to climatic change: rangelands. *In* K.R. Reddy and H.F. Hodges (eds), *Climate Change and Global Crop Productivity*, CAB International, New York, pp. 293-314.

Poorter, H. 1993. Interspecific variation in the growth response of plants to an elevated ambient CO_2 concentration. *Vegetatio* 104/105:77-97.

Raich, J.W., E.B.Rastetter, J.M. Melillo, et al. (1991) Potential net primary productivity in south America: Application of a global model. *Ecol. Apps.* 4:399-429.

Read, J.J., J.A. Morgan, N.J. Chatterton, and P.A. Harrison. 1997. Gas exchange and carbohydrate and nitrogen concentrations in leaves of *Pascopyrum smithii* (C_3) and *Bouteloua gracilis* (C_4) at different carbon dioxide concentrations and temperatures. *Ann. of Bot.* 79:197-206.

Rogers, H.H., G.B. Runion, and S.V. Krupa. 1994. Plant responses to atmospheric CO_2 enrichment with emphasis on roots and the rhizosphere. *Environ. Pollution* 83:155-189.

Rogers, H.H., S.A. Prior, G.B. Runion, and R.J. Mitchell. 1996. Root to shoot ratio of crops as influenced by CO_2. *Plant and Soil* 187:229-248.

Sage, R.F. 1994. Acclimation of photosynthesis to increasing atmospheric CO_2: The gas exchange perspective. *Photosyn. Res.* 39:351-368.

Sala, O. E., W. J. Parton, L. A. Joyce, and W.K. Lauenroth. 1988. Primary production of the central grassland region of the United States. *Ecol.* 69:40-45.

Schimel, D.S., T.G.F. Kittel, and W.J. Parton. 1991. Terrestrial biogeochemical cycles: global interactions with the atmosphere and hydrology. *Tellus* 43AB:188-203.

Schimel, D.S., W.J. Parton, T.G.F. Kittel, D.S. Ojima, and C.V. Cole. 1990. Grassland biogeochemistry: links to atmospheric processes. *Climatic Change* 17:13-25.

Schimel, D.S., B.H. Braswell, and W. Pulliam. 1996. Climate and nitrogen controls on the geography and timescales of terrestrial biogeochemical cycling. *Global Biogeochem. Cycles* 10:677.

Schimel, D.S., VEMAP Participants, and B.H. Braswell. 1997. Continental scale variability in ecosystem processes: Models, data, and the role of disturbance. *Ecol. Monogr.* 67:251-271.

Soussana, J.F., E. Casell, and P. Loiseau. 1996. Long-term effects of CO_2 enrichment and temperature increase on a temperate grass sward. *Plant and Soil* 182:101-114.

Stitt, M. 1991. Rising CO_2 levels and their potential significance for carbon flow in photosynthetic cells. *Plant, Cell and Environ.* 14:741-762.

Stulen, I., and J. den Hertog. 1993. Root growth and functioning under atmospheric CO_2 enrichment. *Vegetatio* 104/105:99-115.

Tate, K.R., and D.J. Ross. 1997. Elevated CO_2 and moisture effects on soil carbon storage and cycling in temperate grasslands. *Global Change Biol.* 3:225-235.

van Ginkel, J.H., A. Gorissen, and J.A. van Veen. 1997. Carbon and nitrogen allocation in Lolium perenne in response to elevated atmospheric CO_2 with emphasis on soil carbon dynamics. *Plant and Soil* 188:299-308.

van Veen, J.A., R. Merckx, and S.C. van der Geijn. 1989. Plant and soil related controls of the flow of carbon from root through the soil microbial biomass. *Plant and Soil* 115:179-188.

VEMAP members. 1995. Vegetation/ecosystem modeling and analysis project: Comparing biogeography and biogeochemistry models in a continental-scale study of terrestrial ecosystem responses to climate change and CO_2 doubling. *Global Biogeochem. Cycles* 9:407-437.

Vorosmarty, C.J., B. Moore III, A.L. Grace, et al. 1989. Continental scale model of water balance and fluvial transport: an application to south America. *Global Biogeochem. Cycles* 3:241-265.

Wand, S.J.E., and G.F. Midgley. 1999. Responses of wild C_4 and C_3 grass (*Poaceae*) species to elevated atmospheric CO_2 concentrations: a test of current theories and perceptions. *Global Change Biol.* 5:723-741.

Wetherald, R.T., and S. Manabe. 1990. *In* U. Cubasch and R.D. Cess, Processes and Modeling, *in* J.T. Houghton, G.J. Jenkins, and J.J. Ephraums (eds), *Climate Change: The IPCC Scientific Assessment*, Cambridge University Press, Cambridge, UK., pp. 69-91.

Wilsey, B.J., J.S. Coleman, and S.J. McNaughton. 1997. Effects of elevated CO_2 and defoliation on grasses: A comparative eocsystem approach. *Ecol. Apps.* 7:844-853.

Wilson, C.A., and J.F.B. Mitchell. 1987. A doubled CO_2 climate sensitivity experiment with a global climate model including a simple ocean. *J. Geophys. Res.* 92 (D11):13,315-13,343.

van Veen, J.A., and E.A. Paul. 1981. Organic carbon dynamics in grassland soils. I.Background information and computer simulation. *Can. J. of Soil Sci.* 61:185-201.

Zak, D.R., K.S. Pregitzer, P.S. Curtis, J.A. Teeri, R. Fogel, and D.L. Randlett. 1993. Elevated atmospheric CO_2 and feedback between carbon and nitrogen cycles. *Plant and Soil* 51:105-117.

SECTION 5

SECTION 5

Summary and Overview and Research and Development Priorities

CHAPTER **16**

The Potential of U.S. Grazing Lands to Sequester Soil Carbon

R.F. Follett,[1] J.M. Kimble,[2] and R. Lal[3]

Introduction

Previous work (Lal et al.,1998) described greenhouse gas processes, global warming concerns, and the potential of U.S. cropland soils to sequester C and mitigate the greenhouse effect. That work was based on the potential C sequestration that results from land conversion, land restoration, and intensification through the use of conservation tillage, improved water and fertility management, and improved cropping systems. The estimates were that 75 to 208 (mean = 142) MMTC/yr could be sequestered in cropland soils.

On the other side of the ledger, annual CO_2-C emissions resulting from U.S. agriculture — from production inputs (fuel and fertilizer), soil erosion, and normal soil oxidative processes associated with cropland agriculture — were estimated at about 43 MMTC/yr. Thus, soil C sequestration in cropland can offset emissions from U.S. agriculture and in addition provide a significant contribution to offset CO_2-C emission from other sectors of the U.S. economy.

Because of the extensive area that grazing lands encompass in the U.S., the question now requiring an answer is the potential of U.S. soils under grazing lands to sequester C and help mitigate the greenhouse effect. Sequestering soil C in grazing lands is important for enhancing soil and water quality and reducing the rate of emissions of radiatively active gases (greenhouse gases) to the atmosphere. In contrast to most cropland, grazing lands can sequester soil C both as soil organic C (SOC) and soil inorganic C (SIC). U.S. grazing lands therefore can

[1] Supervisory Soil Scientist, USDA-ARS, P.O. Box E, Fort Collins, CO 80522.

[2] Research Soil Scientist, USDA-NRCS-NSSC, Fed. Bldg. Rm 152 MS 34, 100 Centennial Mall North, Lincoln, NE 68506-3866, phone (402) 437-5376, fax (402) 437-5336, e-mail john.kimble@nssc.nrcs.usda.gov.

[3] Prof. of Soil Science, School of Natural Resources, The Ohio State University, Columbus, OH, e-mail lal.1@osu.edu.

act as a full counterpart to U.S. croplands (Lal et al., 1998) in helping to mitigate the greenhouse effect.

The potential for grazing lands soils to store significant amounts of C is high because:

1. Grazing lands have comparatively low current rates of management inputs, but high potential rates of SOC sequestration where such management inputs as fertilizer, pesticides, improved species, etc., can be justified economically (especially for pasture lands).
2. Arid and semiarid grazing lands have positive potential to sequester SIC.
3. Grazing lands involve an extremely large area.

This chapter quantifies the relative losses and gains of soil C from the various processes at work in U.S. grazing lands, and it provides an overall estimate of their potential to sequester C.

Chapter 2 (Sobecki et al.) describes grazing lands as we refer to them here. In comparison to the area of about 155 Mha that cropland soils occupy (Lal et al., 1998), privately owned grazing lands occupy about 212 Mha — about 161 Mha rangeland and 51 Mha pasture. Also, Hart et al. (1980) report 1.6 Mha of high-altitude wet mountain meadows (elevation $\geq$ 1830 m) in the western U.S. mountain meadow sites vary from narrow strips along small streams to large areas of seasonally flooded plains along major rivers.

In addition to privately owned grazing land are large areas of grazed federal and state (publicly owned) land. The areas of these additional grazing lands are uncertain and may total from less than half to more than two-thirds more additional land used for grazing (Sobecki et al., Ch. 2; Lal et al., 1998). The area of publicly owned grazing land thus is estimated at between 106 and 141 Mha (average = 124 Mha).

Because of the uncertainty, we will consider that there are 124 Mha of publicly owned and 212 Mha of privately owned grazing lands in the U.S., or a total of 336 Mha. Our focus will be on the 212 Mha privately owned lands because they generally have a higher C sequestration potential, due to their soil and climate characteristics and to the fact that economic incentives are easier to apply to them.

Processes Affecting Carbon Sequestration or Loss in Grazing Lands

In terrestrial ecosystems, SOC is the largest pool and globally contains over 1550 Pg C, followed by the SIC pool that contains 750 to 950 Pg C (Batjes, 1996; Eswaran et al., 1993; Schlesinger, 1995). Terrestrial vegetation is reported to contain an additional 600 Pg C (Houghton, 1995; Schimel, 1995). The atmosphere is reported to contain 750 Pg C (Sarmiento and Wofsy, 1999). Thus, the soil C pool is about four times larger than the terrestrial vegetation pool and three times larger than the atmospheric C pool.

The net annual increase in atmospheric CO_2-C is estimated to be about 3.3 $\pm$ 0.2 Pg C/yr (Sarmiento and Wofsy, 1999; Houghton, 1999; Houghton et al., 1998). Consequently, even a small percentage change in the amount of C stored or released from these large terrestrial C stocks easily affects the annual net exchange with atmospheric CO_2. Computer simulation models are becoming more common to help describe C dynamics and evaluate estimates of the amounts of SOC that potentially can be sequestered. This book describes these models, the CENTURY (Ch. 15, Parton et al.) and SPUR2 (Ch. 14, Hanson et al.) and other rangeland modeling schemes (Ch. 14, Hanson et al.).

C can be sequestered in organic (SOC) and inorganic (SIC) forms. When sequestered as SOC, plants capture the CO_2-C through photosynthesis and then physical, biological, and/or chemical processes stabilize various by-products during decomposition. If the CO_2-C is sequestered as SIC, then primarily chemical processes are involved, although there is evidence that biogenic processes also occur (Monger et al., 1991). The greatest potential for SIC sequestration in U.S. soils is in the arid and semiarid western U.S. because of the dryer climate, large land areas, and availability of noncarbonate sources of calcium.

Soil organic carbon

The net magnitude of the processes and amounts of SOC sequestered depends on ecosystem properties such as climate, plant species, plant structure and function, partitioning of C between tops and roots within the plant, and total amount of plant material produced. The consumption of plant material and cycling of C and nutrients by herbivores, and the trampling, camping, and other activities of grazing animals also affect the amounts and types of residual plant material (litter, roots, and other plant derived material) available to enter the soil C cycle.

The soil organic carbon pool cycle

The conceptual diagram in Figure 16.1 summarizes the SOC addition and loss processes discussed here. Roman numerals (i through xvii) in Figure 16.1 appear in the narrative to help clarify the discussion.

Conceptually, the change in the SOC pool size is the net result of additions minus losses. The arrows with solid lines in Figure 16.1 are part of the "addition/input" processes of C into the SOC pool. Arrows with dashed lines are part of the "losses/output" process, whereby C is removed from the SOC pool.

Atmospheric CO_2-C is captured through photosynthesis where it (i) becomes the C that is in green plants. The C in green plants begins its entry into the SOC

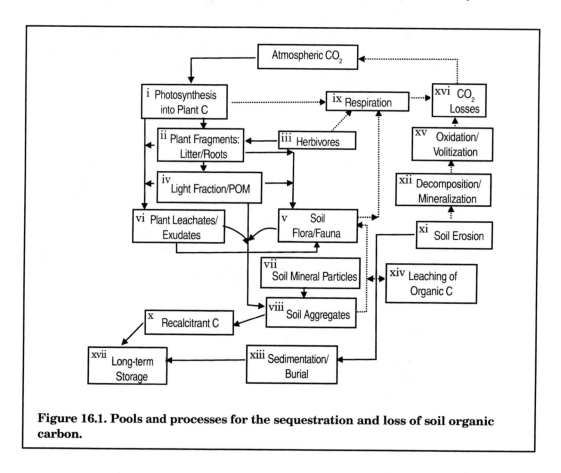

Figure 16.1. Pools and processes for the sequestration and loss of soil organic carbon.

pool in the form of litter, plant root material, and root exudates (ii). Plant material also is consumed by herbivores (iii) and then enters into the SOC pool from animal excreta. Litter sized plant material becomes mechanically (mostly) reduced to smaller fractions, with the fractions often referred to as *light fraction* and/or *particulate organic matter* (POM) (iv). Some of the C and essential elements in these fractions become a source of nutrition for soil flora and fauna (v), including bacteria, fungi, and micro- and macrofauna.

Root biomass and microbial processes are central to the sequestration of SOC (Ch. 6, Reeder et al.). Buried litter and roots feed into an active C pool, comprised of microbial biomass and microbial metabolites. Turnover times are roughly 1.5 years for the active fraction, 25 years for the slow fraction, and over 1000 years for the passive (recalcitrant) pools (Paul and Clark 1996). The turnover time of 25 years is the time for achieving a practical upper limit for C sequestration in soil.

Lal et al. (1998) estimated that achieving this limit could require 50 years. We estimate 25 to 50 years as the time limit on sinks reported in Table 16.1.

Plant residue leachates and microbial and plant root exudates (vi) are an additional source of energy for microbes. These exudates, along with fungal hyphae, earthworm casts, and other biologically binding mechanisms, contribute to the encasing of the light fraction and POM sized plant material with clay and other soil mineral particles (vii). In addition, chemical binding by amorphous iron and aluminum compounds, and mineral cations and anions, results in complex organo-mineral compounds that resist microbial attack. Micro- and macroaggregates (viii) are formed around POM as the microbial and chemical binding of the C to soil mineral particles continues.

Because living plants, roots, animals, and soil organisms respire (ix), much of the C that initially was photosynthesized and entered the soil returns to the atmosphere as CO_2. If the net amount of C entering the SOC pool exceeds that lost back to the atmosphere, the SOC pool increases. Some of the C entering the SOC pool will be physically and/or chemically protected in recalcitrant C forms (x) that have mean resident times (MRTs) in the soil of hundreds to thousands of years (Follett et al., 1997b; Paul et al., 1997). This recalcitrant C is part of a long-term C storage pool (xvii).

SOC fluxes

Understanding net SOC fluxes requires considering both gains and losses of SOC (Ch. 3, Follett). Losses of SOC as CO_2 (xvi) to the atmosphere occur through respiration (ix) and directly by nonbiological oxidation or volatilization (xv). Soil erosion (xi) generally has been considered to decrease SOC, at least at the point on the landscape where erosion occurs. Lal et al. (1998) estimate that perhaps 20% of the SOC that is transported across the landscape by soil erosion may be lost to the atmosphere through microbial decomposition/mineralization (v and xii) of the C exposed by the breakdown of soil aggregates, etc.

Another 10% eventually is transported to the ocean and other water bodies and is removed for a long period of time from the active oxidative processes, and the remaining 70% is redistributed across and within the landscape. Thus, it also is important to recognize that terrestrial sedimentation and deposition (xiii) results in the (often deep) burial of vast quantities of OC, as Stallard (1998) described. Such C deposits, especially if they are deep, should be expected to have very long MRTs and enter the long-term C storage pool (xvii). Combined colluvial, alluvial, and aeolian sedimentation and the eventual remobilization of these deposits is extremely complex to quantify because the remobilization rates and the processes are either poorly understood or difficult to measure (Stallard, 1998).

Leaching (xiv) of dissolved OC (DOC) from various soil C pools can move C deeper below the soil profile and eventually into groundwater. Such leaching generally is considered a loss mechanism, although it can result in the C not being exposed to oxidative processes for long periods of time and can occur concurrently with soil erosion, sedimentation, and/or deposition. However, DOC also may be transported and/or remain within the soil profile or be transported by interflow back to the soil surface.

Because DOC is likely to contain soluble compounds such as sugars, other soluble carbohydrates, amino acids, and perhaps proteins that are readily available sources of C to microbial populations (v), DOC is susceptible to consumption and respiration (ix) that would return it to the atmosphere as CO_2. Oxidation and volatilization (xv), such as from the effect of fire on SOC, are even more difficult to evaluate and, in addition, may result in the deposition of charcoal from the burned plant material. Currently the importance of charcoal for C sequestration and its susceptibility to microbial decomposition is understood poorly.

Fire is thought by some to destroy SOC, with a subsequent nutrient loss and exposure of the soil surface. Long-term burning studies in the Kansas Flint Hills failed to substantiate that conclusion (Ch. 13, Rice and Owensby). Grasslands generally have been subjected to repeated fires during their evolution, and compensating mechanisms probably prevent catastrophic changes in soil physical and chemical properties. Most fires do not affect the belowground C and, as Rice and Owensby (Ch. 13) describe, may help stimulate belowground C sequestration.

Root biomass turnover in grassland systems insures a continual, renewed supply of SOC. OM is essential to soil aggregation, water infiltration capacity, and decreased soil erosion. Only C and N are appreciably lost due to volatilization (Fig. 16.1) as a result of fire, and the amounts lost largely depend on fire intensity (Ch. 13, Rice and Owensby). The amounts and availability of other nutrients that remain can increase following burning, and competition from woody species killed by fire is decreased. Increased plant growth often occurs on burned areas, and the net effect on SOC content of mineral soils in more humid tall grass prairies is negligible.

The effects of fire vary greatly with type of vegetation and climatic zone. Negligible and/or positive response to burning is not universal. Mixed and short grass prairie fires can result in lower plant productivity, apparently because of loss of plant residues, decreased soil water availability, and increased evaporation because of the loss of surface residue. Forage production may be decreased in burned areas only the first season, but not by the end of the second season. Lal (Ch. 10) reports literature showing low water infiltration rates and hydrophobic soil conditions following burning of oak and juniper vegetation types. On the other hand, certain areas, particularly in humid regions, may require fire to maintain the integrity of their ecosystems.

Energy fluxes

Energy flux through soil microbial biomass (SMB) determines if the system is building or depleting the SOC pool and if the rate of substrate depletion is equal to the rate of substrate addition (input of litter, roots, and other plant materials) annually. The energy flux through the SMB drives the decomposition of organic residues (Smith and Paul, 1990) and soil organic matter (SOM). If decomposition exceeds C inputs, SOM will decline.

SMB is generally in a resting state, with periodic flushes of activity and growth. Much of the yearly throughput of energy (plant material) is used for population maintenance. However, with no or limited inputs of C, the SMB uses available supplies of OC in the soil, and the size of the SMB pool itself will decrease. Inputs of C into the soil system must be maintained or increased for SOC levels to remain stable or increase.

A study Follett reported (1997) illustrates this relationship. Soil was collected (0- to 10-cm depth) from long-term experimental plots near Akron, Colorado, and Sidney, Nebraska, and incubated in a constant temperature room with no additional C substrate added. By day 842 of the long-term incubation, the SMB remaining in the soil (averaged across all treatments and sites) decreased to only 17% of that which was present at the beginning of the study. Similarly, the SOC remaining in the soil at day 842 decreased to an average of 67% of that which was present at the beginning of the study.

Soil inorganic carbon

Chapter 4 (Monger and Martinez) discusses the processes and potential to sequester SIC in grazing land soils. The conceptual diagram in Figure 16.2 summarizes some of that information. As before, roman numerals (i through ix) appear in Figure 16.2 and in the narrative to clarify the discussion.

The accumulation of carbonate (primarily $CaCO_3$) occurs in soils with an alkaline to neutral pH. The atmosphere and also root and microbial respiration are the source of the CO_2 for SIC sequestration (i). Respiration and buildup of high concentrations of gaseous (CO_2) in the soil atmosphere, from plant root and soil microbial respiration in the presence of water, contributes to the formation of carbonic acid and bicarbonate (HCO_3^-) in the soil solution (ii). Increased CO_2 in the presence of water drives the reaction (ii) to the right, while the addition of H^+ ions drives the reaction to the left.

The source of the Ca^{+2} ions that react with the carbonic acid system is important, because the source largely determines whether SIC sequestration as $CaCO_3$ occurs. If the Ca^{+2} ions are derived from noncarbonate minerals, such as from the weathering of calcium silicates (iii) or from atmospheric additions of noncarbonate Ca^{+2}, then reaction of the Ca^{+2} ions with the carbonic acid system in alkaline

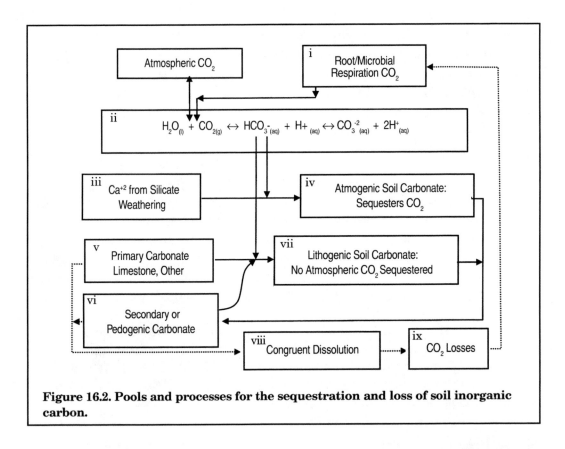

Figure 16.2. Pools and processes for the sequestration and loss of soil inorganic carbon.

soils will result in SIC sequestration as "atmogenic" carbonate (iv). However, if the Ca^{+2} ions are derived from pre-existing primary (v), pre-existing secondary or pedogenic (vi), or even pre-existing atmogenic (iv) carbonates that are dissolved and then re-precipitated as "lithogenic" carbonate (vii), then there is no net sequestration of SIC.

Calcium carbonate ($CaCO_3$) minerals can continue to react with the carbonic acid system, but there is no net increase in the sequestration of SIC. Under certain conditions, congruent dissolution (viii) of carbonates out of the soil profile can occur, but this process will not be discussed here (see Ch. 4). However, congruent dissolution plays an important role because it results in the loss of CO_2 to the atmosphere (ix).

Potential for Carbon Emissions from and Sequestration in U.S. Grazing Land

The amount of SOC or SIC sequestered in soils is the net difference between the amount of C added to the soil system and the amount of C lost from the soil.

Thus, with time, the amount of C present in the soil will either decrease or increase, as a result of the processes described in this book. In addition, emission losses of C can occur that are associated with soil losses, land management practices, and the manufacture and use of production inputs (i.e., fertilizer and fuel).

Estimates of these losses which appear here complement the previous estimates Lal et al. (1998) made for cropland. Data in this chapter about mined land and part of the C emissions resulting from production inputs and farm energy uses are from Table 2 in Lal et al. (1998). The Conservation Reserve Program (CRP) data in this chapter are from more recent information than the data in that 1998 report. These land uses involve grasses and grazed areas (even if grazed only by wildlife), and improved estimates of C sequestered under CRP are now available (Follett et al., 2000).

Emission Losses from Grazing Land

Emission losses of CO_2 from grazing land that occur nationally in the U.S. include C emissions resulting from soil erosion (wind and water), acid precipitation on soil layers containing carbonates, reaction of acid fertilizer with carbonates in the soil that result from application of N fertilizer to pastures, and production inputs for operating machinery used directly for grazing land management and indirectly for fertilizer production and application to improved pastures. Cumulatively, we estimate these emissions are 12.0 to 19.5 MMTC/yr, with a mean of 15.8 MMTC/yr, as Table 16.1, Section A, and Figure 16.3 show.

Soil erosion

Soil erosion by wind and water are major sources of soil C loss from grazing lands (Table 16.1). Lal (Ch. 9) estimates that annual sediment losses on U.S. grazing lands are 1760 and 610 MMT soil for wind and water erosion, respectively. The mean C content (enrichment ratio) in suspended sediments is perhaps two to three times higher than in the topsoil itself; thus large amounts of SOC and SIC are exposed to oxidation and gaseous loss mechanisms. Moreover, soil erosion promotes the loss of SIC as well as SOC, because erosion exposes $CaCO_3$ to wind and water transport and dissolution, and the microbiotic crust and root-microbial infrastructure to oxidative and other associated processes. As Monger and Martinez (Ch. 4) discuss, carbonate, in part, is precipitated by roots and microbes, and the loss of this biopedologic system may curtail $CaCO_3$ precipitation.

Estimates by Lal (Ch. 9) are that the losses resulting from wind and water erosion from U.S. grazing land total between 8.7 and 14.6 MMT of CO_2-C/yr (Table 16.1, Sections A.1a to A.1b). For additional discussion of the processes and factors causing soil erosion and the role of soil erosion management to decrease its impacts, see Lal et al. (1998).

Table 16.1. Estimated potential losses and benefits to the mitigation of atmospheric CO_2 resulting from conversion, restoration, and intensified management of U.S. grazing lands.

	Area Mha	Rate of C Sequestration kg C/ha/yr	Quantity Sequestered MMTC/yr	Cumulative C Sequestered MMTC/yr
A. Emissions Losses from Grazing Lands				−12.0 to −19.5
1. Soil Erosion				
a. Wind Erosion of soils and calciferous layers			−5.7 to −9.5	
b. Water Erosion of soils and calciferous layers			−3.0 to −5.1	
2. Acid Precipitation on Calcic Soils/Layers	228.8	(0.003 to 0.03)	−0.0007 to −0.007	
3. CO_2 emissions from lime for acidity from fertilizer N application (0.43 kg C/kg N) pastures; N rate = 200 -300 kg/ha	10.2	(86 to 129)	−0.9 to −1.3	
4. Production/Energy Inputs (Fert Pastures)			−2.4 to −3.6	
B. Land Conversion and Restoration				17.6 to 45.7
1. Land Use				
a. Conservation Reserve Program				
1. Estimated current area	12.75	600 to 900	7.6 to 11.5	
2. Potential additional area	1.98	600 to 900	1.2 to 1.8	
b. Land/Soil Restoration				
1. Eroded Soils	123	50 to 200	6.2 to 24.6	
2. Mined Lands	0.63	1000 to 3000	0.6 to 1.9	
2. Land Conversion from Forest to Pasture (assume area = 1% of current pasture)	0.51	100 to 200	0.05 to 0.1	
3. Land Conversion from Cropland to Pasture				
a. Estimated current area (4%)	2.4	400 to 1200	1.0 to 2.9	
b. Potential additional area (4%)	2.4	400 to 1200	1.0 to 2.9	
C. Nonintensively Managed Grazing Lands				−4.1 to 13.9
1. Soil Inorganic Carbon	262.4	0.12 to 13.0	0.032 to 3.4	
2. Soil Organic C (Rangelands)	54	(100) to 100	−5.4 to 5.4	
3. Pastureland (25% currently unmanaged)				
a. SOC (12.5% remains unmanaged)	6.38	50 to 200	0.32 to 1.27	
b. SOC (12.5% improved management)	6.38	150 to 600	0.96 to 3.82	
4. Cold Region Systems	—	(200) to 200	—	
D. Improve/Intensify Management				16.0 to 50.4
1. Improved Rangeland Management	107	50 to 150	5.4 to 16.0	
2. Improved Pastureland Management				
a. Fertility management = liming, and fertilizer-N; rate = 200 to 300 kg/ha				
1. Estimated current area (20%)	10.2	100 to 200	1.0 to 2.1	
2. Potential additional area (10%)	5.1	100 to 200	0.5 to 1.0	
b. Application of manure from confined livestock with a N-rate = 250 kg/N ha				
1. Estimated current area (25%)	12.75	200 to 500	2.6 to 6.4	
2. Potential additional area (10%)	5.1	200 to 500	1.0 to 2.6	
c. Planting improved plant species				
1. Estimated current area (5%)	2.6	100 to 300	0.3 to 0.8	
2. Potential additional area (10%)	5.1	100 to 300	0.5 to 1.5	
3. Grazing management on Pasture				
a. Estimated current area (20%)	10.2	300 to 1300	3.1 to 13.3	
b. Potential additional area (10%)	5.1	300 to 1300	1.5 to 6.6	
4. N-fertilization of Mountain Meadows fertilizer-N; rate = 100 to 200 kg/ha				
a. Estimated current area (20%)	0.32	100 to 200	0.03 to 0.07	
b. Potential additional area (10%)	0.16	100 to 200	0.02 to 0.03	
		Total Gain =	29.5 to 110.0	
		Total Mean Gain =	69.8	
		Net Gain =	17.5 to 90.5	
		Net Mean =	54.0	

Follett, Kimble, and Lal, editors

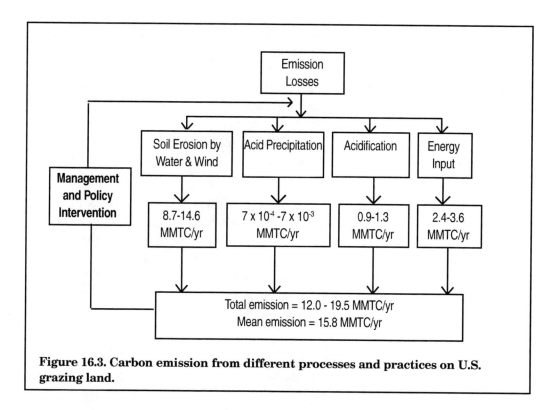

Figure 16.3. Carbon emission from different processes and practices on U.S. grazing land.

Acid precipitation

The potential for the acidity in precipitation to cause CO_2 emissions from the dissolution of $CaCO_3$ is highest in soils with carbonate at or near the land surface. Acidity in precipitation occurs when emissions of sulfate, chloride, nitrate, ammonia, or hydrogen exceed those of basic cations such as calcium or magnesium. Once in the atmosphere, such ions can travel and return to the earth's surface as dry deposition or as wet deposition with precipitation (rain, snow, etc.) and result in excess acidity and available H^+ ions that react with carbonates or bicarbonates (Ch. 4, Monger and Martinez). The presence of H^+ drives the reactions in Equation 1 or 2 to the right.

$$2H^+ + CO_3^{-2}{}_{(aq)} \longleftrightarrow H^+{}_{(aq)} + HCO_3^-{}_{(aq)} \qquad \text{(Equation 1)}$$

$$H^+ + HCO_3^-{}_{(aq)} \longleftrightarrow H_2O_{(l)} + CO_{2(g)} \qquad \text{(Equation 2)}$$

However, in order for these reactions to occur, enough acidity is necessary not only to dissolve $CaCO_3$ and produce HCO_3^- (Eq. 1), but also to cause HCO_3^- to convert to CO_2 and H_2O (Eq. 2). Such dissolution might occur not only where acidic rain contacts $CaCO_3$, but also where microbiotic crusts grow and excrete organic acids on exposed $CaCO_3$ (Ch. 4, Monger and Martinez).

To evaluate the potential of acid precipitation to result in CO_2 emissions, we obtained chemical analyses, data provided by the National Acid Deposition Program (NADP, 1999), and determined the range of excess acidity in the precipitation that fell during 1997 on nine different sites across the area of calcic soils Machette identified (1985). Assuming that one mole of CO_2 was emitted for each mole of excess H^+ in the precipitation (Eq 3), we calculated the range of CO_2-C emissions that Table 16.1 (item A.2.) shows to range from 0.003 to 0.03 kg C/ha/yr.

Moles of excess acidity (H^+) = moles $(NH_4^+ + NO_3^- + SO4^{-2} + Cl^{-1})$
- moles $(Ca^{+2} + Mg^{+2})$ (Equation 3)

Converting the annual rates of C losses or gains to regional values requires multiplying them by the area affected and then converting them to units of MMTC/yr (Eq. 4). In this case, we assumed that three-fourths of the total grazing land ecoregion area Monger and Martinez report (Ch. 4) has carbonates vulnerable to dissolution by acidic rain (i.e., 228.8 Mha).

Soil C Loss/Gain (MMTC/yr) = Area (Mha) *
rate of change (kg C/ha/yr) * 0.001   (Equation 4)

Calculations for the effects of acid precipitation show that annual emissions from calcic soil areas amount to between 0.0007 and 0.007 MMTC/yr (Table 16.1, Section A2), amounts that are largely inconsequential in comparison to the other values. If we calculate CO_2 emissions based on pH, we get similar results. For example, if we assume (1) that rainwater has a pH of 6 and thus will yield 10^{-6} moles of H^+ per liter, (2) that in a particular year carbonate soils receive 20 cm of rainfall, and (3) that one mole of H^+ will produce one mole of CO_2, CO_2 emissions theoretically would total 0.024 kg C ha/yr. For an area of 228.8 Mha, this would total 0.005 MMTC/yr.

In a second case, if we assume that (1) rainwater has a pH of 5 and thus will yield 10^{-5} moles of H^+ per liter, and (2) one mole of H^+ will produce one mole of CO_2, CO_2 emissions theoretically would total 0.37 kg C/ha/yr. For an area of 228.8 Mha, this would total 0.085 MMTC/yr, and again we conclude that CO_2 emission from acid precipitation is inconsequential in comparison to other grazing land sources.

Acid-forming fertilizer

Fertilizers that supply or produce ammonia when added to the soil develop an acid residue. The major effect of ammonium ions occurs when they are nitrified. Upon oxidation of ammonium (NH_4^+) to nitrate (NO_3^-), acidity is increased, as the nitrification reactions in Equations 5 and 6 show.

$2 NH_4^+ + 3O_2 <---> 2 NO_2^- + 2H_2O + 4 H^+ + energy$ (Equation 5)

$2 NO_2^- + O_2 <---> 2 NO_3^- + energy$ (Equation 6)

These nitrification reactions represent enzymatic oxidation mediated by bacterial groups that obtain energy from transforming NH_4^+ ions to NO_3^- ions in eastern U.S. pasture systems and, in the process, release excess H^+ ions, which can be neutralized with liming materials such as $CaCO_3$ or $MgCO_3$. Ammonium nitrate and urea often are used to fertilize pastures. For both of these materials, about 1.8 kg $CaCO_3$ is required to neutralize the acidity from 1 kg of N (Brady, 1974). Schnabel et al. (Ch. 12) indicate that 0.43 kg C are released per kg N applied as ammonium nitrate, and we use this constant in our calculations (Table 16.1, Section A.3).

The exact areas receiving commercial fertilizer annually are not known. In addition, pastures often are fertilized with confined animal manure instead of commercial fertilizer or otherwise are not fertilized at all. Consequently, we assumed that 20% of pastureland would receive annual applications of commercial N fertilizer. (We made these estimates in consultation with one or more authors of Ch. 12.) We estimated average annual rates of N fertilizer application at 200 to 300 kg N/ha, applied in forms that would result in average excess acidity that was the same as that for ammonium nitrate. Estimated C emission from N-fertilizer additions to pastures is 0.9 to 1.3 MMTC/yr (Table 16.1, Section A.3).

Production/energy inputs

We based our estimate of the C emissions that result annually from energy required to manufacture fertilizer and pesticides on an average annual amount of fertilizer N use (Table 16.1, Section A.3) of 1.0 to 1.5 MMT of N. Converting the energy consumption (USDA-ERS, 1994) that results from N fertilizer used (as in Lal et al., 1998) shows 0.82 kg C emission per kg N, or 0.82 to 1.23 MMTC emitted as a result of N fertilizer use on grazing land, compared to 9.4 MMTC for N use for all U.S. agriculture (i.e., 8.7 to 13.1% of the 9.4 MMTC). We multiplied these percentages by the total C emissions (27.9 MMTC) associated with production inputs and agricultural energy uses previously reported (Lal et al.,1998). Thus, we estimate C-emission from production/energy inputs for grazing lands are 2.4 to 3.6 MMTC/yr (Table 16.1, Section A.4.).

Land Conversion and Restoration

Land conversion and restoration in the U.S. include that encouraged by conservation programs enacted by the U.S. Congress and, for the purposes of this chapter, include the effects of the Conservation Reserve Program (CRP) and programs and practices to control soil erosion and reclaim mined lands. In addition, conversion and restoration also can occur because of economic considerations of individual producers and industries. We estimate that land conversion and restoration in the U.S. on soils of grazing lands sequester 18.6 to 46.9 MMTC/yr, with a mean of 32.8 MMTC/yr, as Table 16.1, Section B, and Figure 16.4 show.

Land use

The land use effects that we consider are the CRP and land uses specifically required to restore eroded soils and mined lands. For a more complete discussion of the CRP, its background, and its relationship to previous conservation programs established by the U.S. Congress, see Follett (1997). For eroded lands and mined lands, see Lal et al. (1998).

Conservation Reserve Program (CRP)

CRP enrollment was 12.75 Mha (as of Oct.1, 1999) (Harte, 1999) and at 100% enrollment would include another 1.98 Mha and total 14.73 Mha. Totals change with additional signups and policy decisions about the program. Beneficial effects accrue from placing and/or keeping land under permanent cover, such as with the CRP — decreased soil erosion, increased wildlife habitat, and the sequestration of atmospheric CO_2 as SOC.

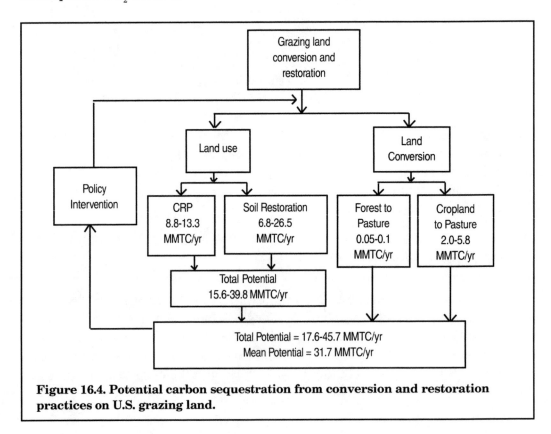

Figure 16.4. Potential carbon sequestration from conversion and restoration practices on U.S. grazing land.

Reeder et al. (1998) in Wyoming emphasized the need to evaluate C and N in the entire soil solum and observed that, 5 years after reestablishing grass on a sandy loam soil, both total and potential net mineralized C and N in the surface soil had increased to levels equal to or greater than those in the A horizon of adjacent native range. On the clay loam soil, however, significant increases in SOC were observed only in the surface 2.5 cm of N-fertilized grass plots.

In a recent study, Follett et al. (2000) sampled 14 sites in 9 states across the historic grasslands region in the central U.S. Carefully paired CRP and cropped fields were sampled and the difference in the amount of SOC was divided by the number of years that the CRP field had been under permanent cover. The base level of SOC was assumed to be the amount of SOC in the cropped fields, so that differences in amount of SOC (i.e., CRP minus cropped) were considered to be the amount of SOC sequestered.

The study showed significant results for the top 3 profile depths across this 5.6 Mha region. Amounts of SOC sequestered at 0 to 5, 0 to 10, and 0 to 20 cm were 560, 740, and 910 kg C/ha/yr, respectively. These values are similar to those Gebhart et al. (1994) obtained previously, measuring SOC sequestration rates of between 800 and 1100 kg C/ha/yr across 5 paired sites within this same region. Other authors (Paustian et al., 1995; Lal et al., 1998) have used estimates for this same region ranging from <100 to 650 kg C/ha/yr.

A steady state, where outputs equals inputs, eventually will be reached (Lal et al., 1998). However, based on values Follett et al. (2000) and Gebhart et al. (1994) experimentally obtained in two separate studies across a total of 19 paired sites, we estimate SOC sequestration rates for Equation 4 at 600 to 900 kg C/ha/yr across a potential enrollment area of 14.73 Mha, with a total potential for annual SOC sequestration of 9.8 to 14.5 MMTC/yr for about 25 years for the CRP land in the U.S. (Lal et al., 1998) (Table 16.1, Section B1a).

Land and soil restoration

The area of privately owned, highly eroded land (HEL) under grazing land management is about 99.8 Mha rangelands and 22.9 Mha pastures (Lal, Ch. 9) and 0.63 Mha mined lands (Lal et al., 1998). Collectively, such lands continue to be important CO_2 emission sources, because of their extent and potential vulnerability to further degradation. The soils on these lands are depleted in SOC and consequently offer considerable opportunity to sequester C when effectively treated and/or managed.

Effective management practices for semiarid or arid rangeland soils are often more difficult to devise because of these soils' heterogeneity. Bird et al. (Ch. 5) recommend that management inputs be targeted to those parts of the landscape with high potential to enhance productivity and C storage due to higher resource availability. An holistic understanding of rangeland systems and relationships between soil structure, SOC, and nutrient and water distribution on a variety of spatial scales can enhance the success of such targeting greatly. Increases in available water (precipitation or higher amounts of run-on) and applied nutrients (manures) and other management techniques can be applied judiciously and economically to establish improved vegetative cover and restore soil quality and higher levels of SOC.

Restored grazing lands soils may sequester C at a rate of 50 to 200 kg C/ha/yr (Table 16.1, Section B1b). Although these rates are modest, the vast land area involved (123 Mha) results in a potential for soil C sequestration (Eq. 4) of 6.2 to 24.6 MMTC/yr. Also, as soil C levels approach steady state, rates of sequestration will decrease.

In the case of mined lands, sufficient economic resources generally have resulted from the extracted minerals (including coal) to fund reclamation. The large potential rates of C sequestration in mined soils (1000 to 3000 kg C/ha/yr), even though for a relatively small land area (Table 16.1, Section B1b2), reflect the potential where substantial resource inputs (including nutrients, plant species mixtures, replacement of topsoil, and land shaping) are available. Reclamation of mined lands is estimated to sequester 0.6 to 1.9 MMTC/yr.

Land conversion from forest to pasture

As Chapter 12 discusses (Schnabel et al.), animal based agriculture in the humid eastern U.S. is possible where forests are cleared and grasses and/or legumes are planted to create highly productive pastureland. For pastures in humid temperate regions, this requires management of grazing frequency, inputs of lime and fertilizer, and occasional replanting to provide a favorable environment.

Much of the pasture land in the humid U.S. previously was forested or could support forest. Following hardwood forest clearing, increases in SOC are reported to occur at the soil surface and also at a depth of 12.5 to 20 cm, perhaps from the incorporation of C-rich forest surface residue at the time of forest conversion (Ch. 12, Schnabel et al.). The current rate of conversion of forest to pasture equals about 1% of current pasture area (we made our estimates in consultation with one or more authors of Ch. 12) or about 0.51 Mha/yr, with estimated rates of SOC sequestration of 100 to 200 kg C/ha/yr. Consequently, the contributions to the total amount of SOC sequestrated by U.S. pastureland is between 0.05 and 0.1 MMTC/yr (Table 16.1, Section B2).

Such a low estimate for the rate of SOC may be questioned. With adequate management inputs, C-sequestration could likely triple. This evaluation does not consider aboveground biomass and the loss of C that results from clearing forest biomass.

We do not advocate the conversion either of forest to pasture or pasture to forest. Because comparative information is minimal and because other national goals must be considered and evaluated carefully, neither system should be considered preferential to the other. Land use considerations include potential rates of C sequestration, whether sequestration is aboveground or belowground, whether C sequestration rates in either system can be improved and enhanced, and whether sequestration is desired for the purpose of producing energy crops, enhancing soil quality, or increasing agronomic and livestock productivity.

Land conversion from cropland to pasture

The CRP program illustrates the potential to sequester C in soil by converting cropland to grass cover (pasture). Gebhart et al. (1994) reported for the Great Plains that 21% of the C lost by decades of intensive tillage was recovered within 5 years under CRP, with C sequestration rates of 800 to 1100 kg C ha/yr. Recently, Follett et al. (2000) observed sequestration rates of 900 kg C/ha/yr across a broad region of the historic grasslands in the central part of the U.S. As Chapter 12 notes (Schnabel et al.), more small and medium sized dairy producers are converting cropland to pasture or alternative grazing systems to reduce production costs by relying more on grazing to feed their livestock. We estimate (in consultation with one or more authors of Ch. 12) that about 2.4 Mha/yr of cropland are converted to pasture and that the C sequestration rates are 400 to 1200 kg C/ha/yr. Thus, such converted lands sequester between 0.96 and 2.9 MMTC/yr. However, policy intervention could make such conversion more economically viable, so at least twice that much land (4.8 Mha/yr) potentially could be converted, sequestering 1.9 to 5.8 MMTC/yr (Table 16.1, Sec B3).

C Sequestration on Nonintensively Managed Grazing Land

Estimating C sequestration in soils of grazing lands with minimal management inputs is difficult. It requires recognition that these large land areas contribute to the overall C fluxes between the atmosphere and the soil, but rates are relatively low. Collectively, nonintensively managed grazing lands have rates of C sequestration estimated at −4.1 to 13.9 MMTC/yr, with an average of 4.8 MMTC/yr (Table 16.1, Section C).

Soil inorganic carbon

Chapter 4 (Monger and Martinez) addresses the issue of annual sequestration rates of carbonate C in U.S. grazing land, and from that we draw our estimates. The three conditions required for SIC sequestration are (1) the land is limited to the Pedocal zone of low rainfall and high pH (i.e., the process does not occur in humid regions), (2) a noncarbonate source of Ca^{+2} ions such as from Ca-silicates or gypsum is available, and (3) the HCO_3^- supply, generated by biotic respiration, is sufficiently large. Based upon rates of SIC sequestration in Chapter 4 (Monger and Martinez), an area of 262.4 Mha (i.e., the total grazing land ecoregion minus limestone terrane), and Equation 4, we estimate rates of C sequestration on unmanaged rangeland soils at from 0.032 to 3.4 MMTC/yr (Table 16.1, Section C1).

Soil organic carbon in rangeland

As Chapter 11 discusses (Schuman et al.), if a consistent management and environment persists, SOC reserves in any ecosystem eventually approach a steady state. A shift in management or environmental conditions would be necessary to change the current condition on unmanaged rangelands.

Chapter 2 (Sobecki et al.) discusses the broad-scale ecological status of U.S. grazing lands and, to the extent the data allow, uses a steady-state model to estimate soil C under given vegetation types. Knowledge is rapidly increasing about the effects of various processes, management, and other factors on soil C sequestration, many of which such models as the SPUR2 (Ch. 14, Hanson, et al.) and the CENTURY (Ch. 15, Parton et al.) describe.

At present, about 54 Mha of U.S. rangelands have no serious ecological or management problems, and Schuman et al. (Ch. 11) assumed no net increases in SOC reserves in these areas. We assume, because of annual variations in climate, grazing requirements, and other conditions, that rates of sequestration would range from –100 to 100 kg C/ha/yr (Table 16.1, Section C2).

Innovative research is needed to determine economical and practical techniques to increase C sequestration on these soils. In addition, tall grass and likely short grass prairie species can respond to elevated atmospheric CO_2 levels by sequestering additional soil C (Ch. 13, Rice and Owensby).

Soil organic carbon in pastures

We estimate that as much as 25% of U.S. pasturelands is nonintensively managed, not in equilibrium with the environment, and sequestering SOC at rates of 50 to 200 kg C/ha/yr. Thus, using Equation 4, these soils currently sequester from 0.64 to 2.55 MMTC/yr. Considering the potential of pastureland to sequester SOC with improved intensive management, it is realistic to estimate that the rates of soil C sequestration in these areas could double or even triple with adoption of one or more improved practices (i.e., fertility, grazing, and/or improved plant species).

A tripling of the rate of SOC sequestration on only half of the area of unmanaged pastureland would increase by one-half the amount of C sequestered on the total area, to 0.96 to 3.82 MMTC/yr (Table 16.1, Section C3a and b). Such dramatic changes within a single type of grazing land illustrate the potential for the U.S. to use soil more effectively as a reservoir to sequester C, while also providing an opportunity to increase the productivity of the resource.

Cold region systems (tundra and mountain meadow)

Little information exists on SOC sequestration in cold region environments such as tundra and mountain meadow. Vourlitis and Oechel (1997) reported that coastal tundra landscapes were net sources of 2.4 to 6.0 kg C/ha/day during the early season snowmelt period but were a net sink of 10.8 kg C/ha/day from mid-July to mid-August, and currently serve as a net sink for C. However, during the summer of 1996, Jones et al. (1998) observed that dry heath- and moist tussock-tundra were a net source amounting to 7.2 to 39 kg C/ha/day (900 to 4875 kg C/ha/yr, if during an assumed nominal active season of 125 days).

Anthropogenic changes may result in these regions' becoming a net source (Oechel et al., 1993). Chapter 8 (Povirk et al.) discusses the biogeochemical processes and the potential impacts of environmental changes on the C cycle. It provides an especially good discussion of the physiography, vegetation, and grazing effects encountered under cold region climatic conditions.

SOC sequestration in cold region systems has undergone large swings in net ecosystem flux during long time periods. The cold climate and nutrient limitations suggest that, with climate warming, these regions could become a net source of C to the atmosphere, because of increased rates of SOC decomposition with enhanced drainage and soil aeration, change in depth to water table, and increased soil temperature (Oechel et al., 1993).

The potential for increased C emissions to the atmosphere from soils in cold regions would be worsened if soils were shallow and fertility low. If sufficient nutrients and an adequate soil water supply were present, then increased plant growth and C inputs into the soil potentially could allow a net increase in SOC sequestration.

For the present, we assume that there would be a consistent and persisting cold region environment in which the SOC reserves were in a steady state, and there would be no net increase or decrease in SOC, but the annual fluctuations with time may perhaps be as much as 200 kg C/ha/yr (Table 16.1, Section C4).

Improved and Intensified Management

Because of the wide diversity of U.S. grazing land systems, their ecological status (Sobecki et al. Ch. 2), and their ecosystems, current management options vary greatly. The climate and the land resource itself also can limit severely the types of approaches possible to improve and intensify the land use. Widely different temperature and moisture regimes, soils, elevation, vegetation types, the physical land area occupied, resource availability, and effectiveness of management inputs are among the many factors to consider. Some grazing land ecosystems are more fragile or readily degraded, while others can be highly responsive.

In light of such considerations, we estimate that improved and intensified management on U.S. grazing land can sequester SOC in soil at 16.0 to 50.6 MMTC/yr (mean = 33.3 MMTC/yr) and that both rangeland and pastureland contribute to the total amounts of sequestered SOC (Fig. 16.5). Rangeland area is much larger, but pastureland can be more responsive, resulting in higher rates of SOC sequestration.

Improved rangeland management

Rangeland grazing management systems have been developed to sustain efficient use of forage by livestock, but there is generally little understanding of the effects of grazing on the redistribution of and cycling of C and N within the plant-soil system. In addition, estimating potential C gains for rangeland soils is complicated because of the limited number of long-term studies of soil C dynamics in rangeland.

Two long-term rangeland grazing studies did consider SOC (Frank et al.,1995; Schuman et al., 1999). Especially important in both was the effect of grazing intensity on the resulting dominant species and, in turn, on amounts of SOC and N in the surface soil where roots dominate. Grassland communities are diverse and have evolved with time in response to climate, soils, and management practices.

Chapter 11 (Schuman et al.) discusses these and other complexities. Chapter 5 (Bird et al.) explicitly recognizes that the capacity of rangeland ecosystems to sequester SOC is a function of interactions among the spatial distribution of plant production, availability of soil and water resources, and management inputs targeted to those parts of the landscape that already have a high potential for enhanced productivity and C storage because of higher resource availability. Based on 3 years (1995 to 1997) of CO_2 flux measurements representative of 51 Mha, Frank et al. (Ch. 7) estimated the annual SOC rates at Manden, ND, Woodward, OK, and Temple, TX, at 86, 87, and 459 kg C/ha/yr, respectively. These figures are similar to those of Schuman et al. (Ch. 11), who indicate that currently there are 107 Mha of private U.S. rangeland where improved management could sequester

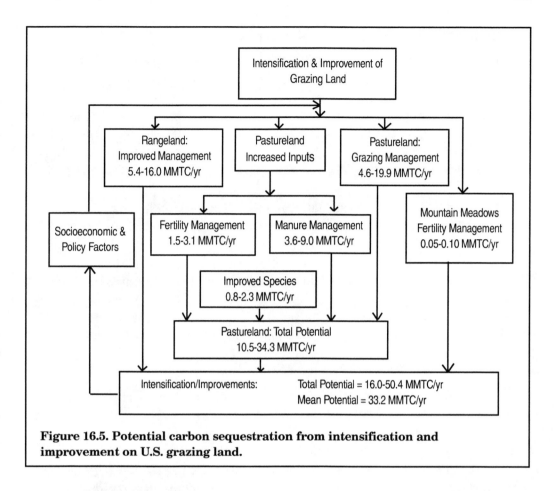

Figure 16.5. Potential carbon sequestration from intensification and improvement on U.S. grazing land.

50 to 150 kg C/ha/yr. Even though the rates are low, the large area involved yields total sequestration estimates of 5.4 to 16.0 MMTC/yr.

Improved pasture management

The majority of improved pastures in the U.S. are east of the Mississippi River. They occur on many soil types, with varying fertility, texture, and structure, and across a range of climatic conditions. Each of these soil properties and climatic factors have important, generally predictable effects (Ch. 12, Schnabel et al.). Overall, temperature and moisture are the most important climatic factors for determining soil C dynamics because of their effects on potential net primary productivity (NPP) and soil microbial activity.

Two types of management inputs to improve pasture production and thereby SOC sequestration are better nutrient availability in pastureland soils and the use of different plant species. We discuss grazing management, another way to improve pasture production, separately.

Managing fertility (liming and applying nitrogen)

Fertilization and improved soil fertility management practices usually improve forage production, increase feed, and improve the protein content of the forage for animal agriculture. Top yields of high quality forage requires adequate soil fertility and soil pH levels. Even though this discussion focuses on N fertility, the productivity potential of a pastureland soil, within its climatic regime, will not be reached if even one essential nutrient is limiting. Follett and Wilkinson (1995) provided a thorough discussion of the management of plant nutrients and liming.

For this discussion and to estimate the role of liming and fertilizer-N application (we made our estimates here in conference with one or more authors of Ch. 12), we estimated that 20% of the pasture area (10.2 Mha) in the U.S. currently receives lime and fertilizer N at rates of 200 to 300 kg N/ha/yr and that the rate of SOC sequestration that results is 100 to 200 kg/ha/yr. Consequently, the sequestration of SOC is 1.02 to 2.04 MMTC/yr. However, the potential exists to apply fertilizer to an additional 5.1 Mha, to result in an additional 0.51 to 1.02 MMTC/yr of SOC sequestration (Table 16.1, Sections A3 and D2a).

Applying manure from confined livestock

Nutrients from animal manure represent a valuable resource in the U.S. Here we focus on manure from confined livestock operations.

"Confined" livestock include livestock produced in confinement, or under any other type of operation where manure is typically recoverable, i.e., collected, stored, and available for use as a nutrient resource. The livestock categories that are included here are dairy or beef cattle (*Bos taurus*), swine (*Sus scrofa domesticus*), chickens (*Gallus gallus domesticus*), and turkeys (*Meleagris gallopavo*). These categories include dairy (milk cows and heifers); beef (steers, bulls, calves, and cows); hogs and pigs (growers and breeders); and poultry (broilers and layers, and turkeys) (USDA, 1995-1996).

The procedure for calculating annual weight of manure N produced is from Lander et al. (1998), as are the factors for estimating the area of pasture. Calculations here are only for applying manure in those regions of the U.S. where it can be applied to pastures (Lander et al., 1998). No estimates are made concerning applications to rangeland.

We estimate (in consultation with one or more authors of Ch. 12) that, if the manure were applied at a rate equivalent to 250 kg N/ha/yr, the manure could be used on 25% (12.75 Mha) of the area of U.S. pastures. Calculation of the amounts of manure available (Lander et al., 1998) indicate, in addition, adequate manure N to apply to an another 5.1 Mha. We estimate (in consultation with one or more authors of Ch. 12) that these applications of manure to pastures would sequester

C at 200 to 500 kg C/ha/yr, thus resulting in a potential total of 3.6 to 9 MMTC/yr (Table 16.1, Section d2b).

Planting improved species

Often pastures have been developed using existing grasses by allowing previously cropped land to revert to a grass sod (naturalized pastures) or by converting hayland to pasture. Improvement for grazing has resulted from applying fertilizer and manure and, possibly, from liming.

Existing pasture species can be quite tolerant to grazing but often low in productivity. Chapter 12 (Schnabel et al.) provides a number of examples that illustrate the improvement in pasture productivity that results from planting either mixtures or monocultures of improved plant species.

Current estimates are that improved plant species are planted on about 2.6 Mha annually, or 5% of U.S. pastureland. We estimate (in conference with one or more authors of Ch. 12) improved species to sequester SOC at 100 to 300 kg C/ha/yr and thus, using Equation 4, to sequester 0.26 to 0.78 MMTC/yr. We assumed it reasonable to increase the area of pastureland on which improved species are planted by another 5.1 Mha to increase the total potential SOC sequestration to 0.8 to 2.3 MMTC/yr (Table 16.1, Section D2c).

Managing grazing

As Chapter 12 (Schnabel et al.) discusses, SOC sequestration under cattle grazing results from the return of much of the plant-derived C and nutrients to the soil as feces, some of which rapidly enters the SOC pool. Pastures hayed or stocked continuously with few animals sequester less SOC than pastures with medium stocking rates. The heavier foot traffic associated with the generally higher stocking densities of this more intensive grazing probably enhances the breakdown of aboveground litter and its incorporation into the soil.

We estimated (in conference with one or more authors of Ch. 12) that 10.2 Mha of U.S. pastures use improved grazing management systems, with sequestration rates of 300 to 1300 kg C/ha/yr and total sequestration of 3.1 to 13.3 MMTC/yr. We assume it reasonable to increase the area of pastureland on which improved grazing practices are used by another 5.1 Mha, to increase the total potential SOC sequestration to 4.6 to 19.9 MMTC/yr (Table 16.1, Section D3).

Cold region systems (tundra and mountain meadow)

The only cold region system that we address in terms of its response to improved management is mountain meadow, which plays a vital role in livestock production in the West and occupies more than 1.6 Mha (Hart et al., 1980). Most mountain meadows are flooded during peak runoff in spring and early summer but are fairly dry in late summer and fall. Typically, the level of management of mountain meadows is low, and average yields are about 2.9 MT/ha (Follett et al., 1995). Vegetation is predominantly native sedges, rushes, and grasses, but some clovers, other forbes, and a few phreatrophic shrubs may be found. Vegetative composition can be manipulated by water control, fertilization, cutting, grazing management, and reseeding.

The only improved or intensified management that we consider here is N fertilization. Follett et al. (1995) reported that, in a 3-year study at two different locations, a rate of 150 kg N/ha resulted in yield increases of about 5 MT of hay/ha/yr above nonfertilized treatments. Assuming that hay contains 35% C, then the harvested hay contained 1750 kg C/ha/yr. Long-term studies with winter wheat showed that 5 to 10% of the biomass C (above- and belowground) produced was sequestered as SOC (Follett et al., 1997a).

We recognize that biomass C to SOC sequestration relationships in mountain meadows may be different from those in small grain production systems, but using this relationship could provide a conservative estimate of the rate of SOC sequestration with N fertilization in mountain meadows. This calculation suggests a sequestration rate of 88 to 175 kg C/ha/yr. Assuming as much as 20% of the 1.6 Mha of mountain meadows to be fertilized at the rate of 150 kg N/ha/yr, then SOC sequestered (Eq. 4) is 0.028 to 0.07 MMTC/yr. We estimate an additional 10% (0.16 Mha) of mountain meadow could be fertilized, for a total potential SOC sequestration by mountain meadow of 0.05 to 0.1 MMTC/yr (Table 16.1, Section D4).

Figure 16.6 illustrates the relative carbon sequestration potential for different components of U.S. grazing land systems.

Future Effects of Climate Change on C in U.S. Grazing Land

Most information currently available was developed under current or recently past climate conditions. Among the premises relating to "current climate conditions" is that, if consistent management and climate persist, then the amount of SOC in any ecosystem eventually approaches a steady state. A shift in management or environmental condition is necessary to change SOC. Much of the analysis in this book considers only the effects of management. However, because of the very large areas involved, the potential effects of environmental change need to be considered.

Parton et al. (Ch. 15) made such an analysis across the U.S. Great Plains, which includes a large area of rangeland soils. They provide projections of where and how much SOC would change during 70 years, with a doubling of atmospheric CO_2 concentration and an increase in the soil temperature of 2 to 4°C. The

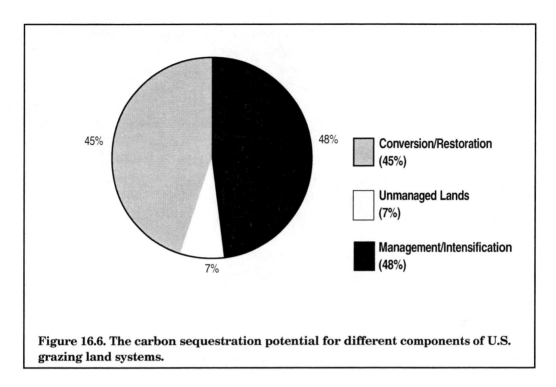

Figure 16.6. The carbon sequestration potential for different components of U.S. grazing land systems.

major factors that control soil C dynamics include the C inputs into the soil and the abiotic decomposition, which soil temperature and soil water level control.

The analyses show that SOC tends to increase with increasing plant production, but the increase is not direct, since decomposition rates are also greater in the high productivity zones. In addition, while mean annual temperature and annual precipitation have a substantial impact on observed SOC, soil texture exerts a dominant control on regional patterns in SOC content.

In summary, CENTURY model simulations of NPP, SOC, and DEFAC using the global circulation model, UKMO GCM, show soil C storage tended to increase with CO_2 and temperature increase, especially in the southern and central Great Plains. Decreases or neutral effects on SOC occurred for the northern Great Plains (Fig. 15.7 in Ch. 15, Parton et al.), primarily because of the increase in DEFAC that resulted from increased soil temperature.

We estimate (in consultation with one or more of the authors of Ch. 15, Parton et al.) the change across the southern, central, and northern sections of the area in Chapter 15's Figure 15.7 (Parton et al.). The areas of privately owned grazing land (grassland and pasture) were determined from USDA figures (1995-1996).

The "southern" section included grassland and pasture areas in Texas, Oklahoma, and New Mexico; the "central" section included grassland and pasture areas in Colorado, Kansas, Nebraska, and the southern half of Wyoming; and the "northern" section included grassland and pasture areas from North Dakota,

South Dakota, Montana, and the northern half of Wyoming. The areas of grazing land for the southern, central, and northern sections are 75.8, 37.7, and 44.8 Mha, respectively. The means for the additional amounts of SOC that would be sequestered in the southern, central, and northern sections were 56.25, 6.60, and 0.69 MMTC/70 yr, respectively, thus resulting in an increase of 63.5 MMTC/70 yr or 0.91 MMTC/yr on privately owned grazing lands across the entire region that Chapter 15 describes (Parton et al.).

Although relatively small annually, the change in SOC sequestration expected from climate change represents a significant addition to stocks of SOC within Great Plains grazing land systems in the U.S. These calculations provide an example of the impacts of climate change on SOC sequestration.

Conclusions

The net sequestration of C in U.S. grazing lands is 17.5 to 90.5 MMTC/yr and a mean of 54.0 MMTC/yr. Most of the potential for soil to sequester C is not being managed for, nor is it widely recognized that sequestration of C is occurring. As Table 16.1 makes evident, sequestration of C under grazing land results from numerous managed and unmanaged activities.

Both pastureland and rangeland contribute significantly to overall sequestration of C. Most pastureland is in the eastern U.S., receives more precipitation, has a higher potential to respond to management inputs, but occupies less total land area than rangeland.

The largest potential for decreasing the amount of CO_2 emitted from grazing land systems is from decreased soil erosion. Conversely, restoration of eroded and degraded soils provides a large potential to sequester SOC. The CRP is among the most important land use strategies for sequestering SOC and, in addition, contributes significantly to controlling soil erosion losses, restoring soil quality, providing wildlife habitat, and protecting air and water quality.

Even though the areas affected are relatively small, restoration of mined land soils is among the practices that provide the highest rates of SOC sequestration. Improved rangeland management has a high potential to sequester SOC in some soils, and pastureland soils have a very high potential to sequester SOC as a result of management inputs. Little information is available concerning the area and response to changes in management strategy of cold regions in the U.S.

Future evaluation of grazing lands' role in sequestering C needs to consider the potential effects of climate change. Estimates in this book generally indicate that higher atmospheric CO_2 results in a net increase in SOC sequestration within the Great Plains of the U.S. We also need predictions for pastureland, arid land, and cold regions that climate change may affect.

Table 16.2. The overall potential of U.S. grazing land for C-sequestration and fossil fuel offset.

Carbon Sequestration in Soil in Response to:		MMTC/yr Range
1. Land Conversion & Restoration		17.6 - 45.7
2. Unmanaged C-Sequestration		-4.1 - 13.9
3. Improved Management & Intensification		16.0 - 50.4
	Total	29.5 - 110.0
	Mean	69.8

Numerous gaps and uncertainties exist related to estimates of C emissions from agricultural activities on grazing lands and to estimates of these lands' potential to sequester C. Existing data were collected with a wide range of methods. Supporting data on soil bulk density and textural properties are often missing. Problems exist with aggregating the data from a point or plot scale to a regional or national scale.

In spite of these gaps, the information we have presented in this book supports the need for a coordinated program that uses scientifically proven technologies to improve sequestration of C in soils of grazing lands. Such a program could restore degraded land, enhance overall soil quality, increase biomass productivity, and improve water and air quality. Restorative practices create a win-win scenario that improves productivity and enhances the environment.

The overall potential of U.S. grazing lands to sequester C (not including emission losses) is between 29.5 and 110.0 MMTC/yr during a 25-year period (Table 16.2). This potential includes 45.3% from land conversion and restoration, 7.0% from unmanaged rates of C sequestration, and 47.7% due to adoption of improved practices. This total potential represents a sink that is 1.6 times the CO_2-C emission from all U.S. agriculture and 4.4 times the CO_2-C emission from all grazing lands agriculture, which therefore would help offset CO_2-C emissions from other sectors of the U.S. economy. The total potential for U.S. grazing land soils to sequester C, 29.5 to 110.0 MMTC/yr, represents about 5% of the total U.S. CO_2-C emissions of 1442 MMT/yr (Lal et al., 1998) and, coupled with the sequestration potential of U.S. cropland, would help offset about 14% of total U.S. CO_2-C emissions.

Acknowledgment

We are indebted to the authors of other chapters in this book for the groundwork that they have provided.

References

Batjes, N.H. 1996. Total C and N in soils of the World. *Eur. J. Soil Sci.* 47:151-163.

Brady, N.C. 1974. *The Nature and Property of Soils,* 8th ed. MacMillan. New York. 639p.

Eswaran, H., E. Van Den Berg, and P. Reich. 1993. Organic carbon in soils of the world. *Soil Sci. Soc. Am. J.* 57:192-194.

Follett, R.F. 1997. CRP and microbial biomass dynamics in temperate climates. *In* R. Lal, J. Kimble, R. Follett, and B.A. Stewart (eds), *Management of Carbon Sequestration in Soil,* Advan. Soil Sci., CRC Press, Boca Raton, FL, pp. 305-322.

Follett, R.F., E.A. Paul, S.W. Leavitt, A.D. Halvorson, D. Lyon, and G.A. Peterson. 1997a. The determination of the soil organic matter pool sizes and dynamics: $^{13}C/^{12}C$ ratios of Great Plains soils and in wheat-fallow cropping systems. *Soil Sci. Soc. Am. J.* 61:1068-1077.

Follett, R.F., E.G. Pruessner, S.E. Samson-Liebig, J.M. Kimble, and S.W. Leavitt. 1997b. Organic carbon storage in historic U.S.-grassland soils. *Agron. Abstr.* 89:208.

Follett, R.F., S.E. Samson-Liebig, J.M. Kimble, E.G. Pruessner, and S.W. Waltman. 2000. Carbon sequestration under CRP in the historic grassland soils of the USA. *In* R. Lal and K. McSweeney (eds), *Soil Management for Enhancing Carbon Sequestration,* SSSA Special Publication, Madison, WI, (in press).

Follett, R.F. and S.R. Wilkinson. 1995. Nutrient management of forages. *In* R.F. Barnes, D.A. Miller, and C.J. Nelson (eds), *Forages. Volume II: The Science of Grassland Agriculture,* Iowa State University Press. Ames, IA, pp. 55-82.

Follett, R.H., D.G. Westfall, J.F. Shanahan, and D.W. Lybecker. 1995. Nitrogen fertilization of mountain meadows. *J. Prod. Agric.* 8:239-243.

Frank, A.B., D.L. Tanaka, L. Hofmann, and R.F. Follett. 1995. Soil carbon and nitrogen on Northern great Plains grasslands as influenced by long-term grazing. *J. Range Manage.* 48:470-474.

Gebhart, D.L., H.B. Johnson, H.S. Mayeux, and H.W. Polley. 1994. The CRP increases soil organic carbon. *J. Soil Water Conserv.* 49:488-492.

Hart, R.H., H.R. Haise, D.D. Walker, and R.D. Lewis. 1980. Mountain meadow management: 12 years of variety, fertilization, irrigation, and renovation research. *Agricultural Research Results*, ARR-W-16.

Harte, P. 1999. Personal communication. Assistant to the Director, Farm Services Agency. Consolidated Farm Services Agency Computer Center, USDA-Computer Center, 8930 Ward Parkway, Kansas City, MO 64114.

Houghton, R.A. 1995. Changes in the storage of terrestrial carbon since 1850. *In* R. Lal, J. Kimble, E. Levine, and B.A. Stewart (eds), *Soils and Global Change,* CRC Press, Boca Raton, FL, pp. 45-65.

Houghton, R.A. 1999. The annual net flux of carbon to the atmosphere from changes in land use. *Tellus* 51:298-313.

Houghton, R.A., E.A. Davidson, and G.M. Woodwell. 1998. Missing sinks, feedbacks, and understanding the role of terrestrial ecosystems in the global carbon balance. *Global Biogeochem. Cycles* 12:25-34.

Jones, M.H., J.T. Fahnestock, D.A. Walker, M.D. Walker, and J.M. Welker. 1998. Carbon dioxide fluxes in moist and dry arctic tundra during the snow-free season: responses to increases in summer temperature and winter snow accumulation. *Arctic and Alpine Res.* 30:373-380.

Lal, R., J.M. Kimble, R.F. Follett, and C.V. Cole. 1998. *The Potential of U.S. Cropland to Sequester Carbon and Mitigate the Greenhouse Effect.* Ann Arbor Press. Chelsea, MI. 128 pp.

Lander, C.H., D. Moffitt, and K. Alt. 1998. Nutrients available from livestock manure relative to crop growth requirements. *Resource Assessment and Strategic Planning Working Paper 98.1.* USDA, Nat. Res. Cons. Svc. Web site: <*http://www.nhq.nrcs.usda.gov/land/pubs/nlweb.html*>

Machette, M.N. 1985. Calcic soils of the southwestern United States. *In* D.L. Weide (ed), *Soils and Quaternary Geology of the Southwestern United States*, Geological Soc. Am. Spec. Paper 203, Boulder, CO, pp. 1-22.

Monger, H.C., L.A. Daugherty, W.C. Lindemann, and C.M. Liddell. 1991. Microbial precipitation of pedogenic calcite. *Geol.* 19:997-1000.

NADP, National Acid Deposition Program. 1999. *http://nadp.sws.uiuc.edu/nadpdata/.* 1999. Website data report option for site numbers CA42, CO22, KS32, NE15, NM08, NV00, NVO3, SDO8, and TX02.

Oechel, W.C., S.J. Hastings, G. Vourlitis, M. Jenkins, G. Riechers, and N. Grulke. 1993. Recent change of arctic tundra ecosystems from a net carbon dioxide sink to a source. *Nature* 361:520-523.

Paul, E.A., and F.E. Clark. 1996. *Soil Microbiology and Biochemistry.* Academic Press. San Diego, CA.

Paul, E.A., R.F. Follett, S.W. Leavitt, A. Halvorson, G.A. Peterson, and D.J. Lyon. 1997. Radiocarbon dating for determination of soil organic matter pool sizes and dynamics. *Soil Sci. Soc. Am. J.* 61:1058-1067.

Paustian, K., C.V. Cole, E.T. Elliott, E.F. Kelly, C.M. Yonker, J. Cipra, and K. Killian. 1995. Assessment of the contributions of CRP lands to C sequestration. *Agron. Abstr.* 87:136.

Reeder, J.D., G.E. Schuman, and R.A. Bowman. 1998. Soil C and N changes on conservation reserve program lands in the Central Great Plains. *Soil and Tillage Res.* 47:339-349.

Sarmiento, J.L., and S.C. Wofsy (Co-Chairs). 1999. *A U.S. Carbon Cycle Plan. A report to the carbon and climate change working group of the U.S. Global Change Research Program.* 400 Virginia Ave. SW, Washington, D.C.

Schlesinger, W.H. 1995. An overview of the global carbon cycle. *In* R. Lal, J.M. Kimble, E. Levine and B.A. Stewart (eds), *Soils and Global Change*, CRC/Lewis Publishers. Boca Raton, FL, pp. 9-25. .

Schimel, D.S. 1995. Terrestrial ecosystems and the carbon cycle. *Global Change Biol.* 1:77-91.

Schuman, G.E., J.D. Reeder, J.T. Manley, R.H. Hart, and W.A. Manley. 1999. Impact of grazing management on the carbon and nitrogen balance of a mixed-grass rangeland. *Ecol. Appls.* 9:65-71.

Smith, J.L., and E.A. Paul. 1990. The significance of soil microbial biomass estimation. *In* J. Bollag and G. Stotzky (eds), *Soil Biochemistry* (Vol 6), Marcel Dekker, New York, pp. 357-396.

Stallard, R.F. 1998. Terrestrial sedimentation and the carbon cycle: Coupling weathering and erosion to carbon burial. *Global Biogeochem. Cycles* 12:231-257.

USDA. 1995-96. *Agricultural Statistics*. U.S. Government Printing Office. Washington, D.C.

USDA-ERS. 1994. *Agricultural Resources and Environmental Indicators, Agricultural Handbook. No. 705*. U.S. Govt Printing Office. Washington, D.C.

Vourlitis, G.L., and W.C. Oechel. 1997. Landscape-scale, CO_2, H_2O vapour energy flux of moist-wet coastal tundra ecosystems over two growing seasons. *J. Ecol.* 87:575-590.

CHAPTER **17**

Research and Development Priorities

R. Lal,[1] R.F. Follett,[2] and J.M. Kimble[3]

Introduction

The U.S. contains 212 Mha of privately owned and 124 Mha of publicly owned grazing lands. With the adoption of recommended land use, soil, and pasture management practices, the privately owned lands are estimated to potentially sequester 30 to 110 MMT C/yr (mean of 70 MMT/yr). This potential is about 50% of that for U.S. cropland, estimated at 75 to 208 MMTC/yr (mean of 142 MMT/yr) (Lal et al., 1998). Both estimates include CRP and mined lands, 6 to 13 MMTC/yr (mean of 10 MMTC/yr; Lal et al., 1998). Thus, the total potential to sequester C in agricultural soils, 99 to 305 MMTC/yr (mean of 202 MMTC/yr), offers an enormous opportunity for land managers and policy makers to enhance soil quality, decrease the risks of pollution of natural waters, and reduce the rate of emissions of greenhouse gases into the atmosphere.

The U.S. commitment under the Kyoto Protocol, if Congress ratifies the treaty, is to reduce all emissions by about 600 MMTCE/yr by the year 2010. Adopting recommended land use, soil, crop, and pasture management practices can let us meet much of this commitment. Managing cropland can fulfill the Kyoto commitment by 22%; grazing land, by 12%; for a total potential of 34%. We need to assess the feasibility of harnessing this enormous potential, and for that we need mechanisms and institutions that can explore this viable option fully

Grazing land systems can contribute to both the environmental and economic well being of U.S. agriculture while also sequestering C — provided that our national strategic goals include enhancing (1) grazing land condition, (2) environmental quality, and (3) the economic viability of producers. Numerous cultural,

[1] Prof. of Soil Science, School of Natural Resources, The Ohio State University, Columbus, OH, e-mail lal.1@osu.edu.

[2] Supervisory Soil Scientist, USDA-ARS, P.O. Box E, Fort Collins, CO 80522.

[3] Research Soil Scientist, USDA-NRCS-NSSC, Fed. Bldg. Rm 152 MS 34, 100 Centennial Mall North, Lincoln, NE 68506-3866, phone (402) 437-5376, fax (402) 437-5336, e-mail john.kimble@nssc.nrcs.usda.gov.

economic, policy, and social factors inhibit a more rapid and widespread adoption of sustainable grazing land practices.

Grazing land management systems often include mixed farming operations that incorporate livestock and grazing land with crop production. Low rates of economic return make it difficult to invest in grazing land improvement. Many of the diverse biological communities found across various grazing land systems are sensitive to such changes as those caused, for example, by invasive weeds, variable precipitation patterns, and grazing management practices. Changing climate patterns will lead to change in habitat extent and species mixtures for livestock and cropping activities. To enhance the condition of grazing land, we need research that will improve our overall understanding of ecosystem processes and thus provide opportunities to develop, manage, and use plant species and management systems to enhance grazing lands.

Environmental quality issues are increasingly important, and generally we can consider them in the context of on-site and off-site effects, both of which we will go into more detail about below. Economic considerations and the economic viability of agricultural enterprises require a thorough knowledge of the input-output relationships associated with viable grazing land agriculture. Optimizing the profitability of a grazing land system while reducing risks requires integrating grazing land, forage, and livestock production, which in some cases also requires integrating both grazing land and cropland management systems.

The strategy of sequestering C in agricultural soils is not only viable but also a uniquely win-win strategy. While decreasing the risks of accelerated greenhouse effects, sequestering C in agricultural soils also improves biomass productivity and water quality. The total U.S. emission of greenhouse gases (GHGs) is estimated at 1600 MMTCE/yr, including 116 MMTCE/yr from agricultural activities (Lal et al., 1998). Therefore, the potential C sequestration in grazing land amounts to 4.4% of the total U.S. emission of greenhouse gases; in cropland, 8.3%; for a total of 12.7%. Widespread adoption of recommended management practices can make U.S. agricultural soils a major sink for C sequestration. If the agricultural role includes both sequestering C in soil and producing biofuel crops on marginal soils to substitute for fossil fuels, this potential can increase (Follett, 1993).

Realizing this vast potential, however, depends on (1) obtaining research data on soil carbon dynamics, (2) assessing the cost of soil C by monitoring additional costs associated with adopting recommended management practices, (3) identifying policy issues that facilitate/encourage adopting the desired practices, and (4) putting in place the mechanisms needed.

Research Data on Soil Carbon Pools and Dynamics

We have an urgent need to strengthen the database required to provide credible estimates of the soil C pools and fluxes in grazing land for principal soils, predominant ecoregions, and major land uses and management practices. We must quantify those characteristics which affect the pools and fluxes of SOC (soil organic C) and SIC (soil inorganic C). Table 17.1 outlines the three parameters of specific research information needed.

Table 17.1. Research needed to strengthen the database.

Land Use/Processes	Research Needs
1. Land use, SOC, and SIC pools and fluxes	Assessment of the impact of land use and management on: a. SOC and SIC pools and fluxes b. Soil bulk density, porosity and pore size distribution, and compaction c. Soil structure, aggregation, stability, erodibility d. Infiltration capacity, available water capacity e. CEC, nutrient cycling f. Soil biological properties (e.g., microbial biomass carbon, earthworm activity)
2. Soil erosion	Assessment of the impact of erosion on soil C dynamics in relation to: a. aggregate slaking and disruption b. the magnitude of SOC and SIC displaced and the enrichment ratio of C under different management c. the delivery ratio of sediments and SOC in relation to land use and ecological factors d. the relative magnitude of C emitted vs. buried in depressional sites e. the net emission of CH_4 and N_2O for different erosional phases
3. Soil restoration	Assessment of the impact of soil restorative measures on: a. the rate of SOC and SIC sequestration b. improvement in soil quality and biomass productivity c. soil resilience characteristics, especially in relation to the impact of SOC sequestration on soil physical, chemical, and biological quality

The impact of land use and management on SOC pools and fluxes

We need to quantify the SOC pools and fluxes for principal soils in relation to land use and soil management. The SOC pool and its dynamics depend on several ancillary properties, including soil structure, water retention and movement, CEC and nutrient cycling (e.g., N, P, S), and soil biological properties and the relationships of the C, N, P, and S cycles.

Soil erosion and the fate of displaced C

In addition to quantifying the actual and potential risks of soil erosion under different management options, we urgently need to determine the fate of SOC displaced and redistributed over the landscape and of the SIC which erosional processes expose. Little experimental data exist on the fate of C displaced by water and wind erosion, and we need the information for erosional hot spots of grazing lands. In addition to assessing the adverse impact of soil erosion and the attendant breakdown of soil aggregates, we should assess the impact of other degradative processes on soil C pools and fluxes (e.g., salinization, compaction, nutrient depletion, etc.). It is important to identify soil, terrain, and climate factors that exacerbate soil degradation. We need to develop a "soil degradation index" based on biophysical factors.

Soil restoration and C dynamics

Restoring degraded soils and ecosystems is a high priority in terms of enhancing soil quality and sequestering carbon. Degraded soils have lost much of their original C pool, and the potential to sequester SOC is high in these soils and ecosystems. Adopting restorative measures (e.g., CRP) on degraded and marginal soils can restore the SOC pool and improve soil quality and biomass production (Follett, 1998).

Soil restoration depends on soil resilience and management. Therefore, assessing soil's resilience characteristics is important to determining strategies for soil restoration and SOC sequestration. Evaluating the impact of soil restorative processes on soil C dynamics requires quantifying soil degradative processes on soil C pools and fluxes. While the soil restorative process undergoes a marked hysteresis (Lal, 1994), understanding how degradation affects soil C dynamics is important to identifying restorative strategies.

Cost of Soil Carbon Sequestration

The value of soil C (Fig. 17.1), which depends on its both on-site and off-site benefits, will determine whether land managers adopt appropriate practices to sequester C. On-site benefits are those related to improvement in soil quality, e.g., enhancement of soil structure, increase in the available water capacity, improvement in soil fertility, and increase in actual and potential productivity. Off-site benefits are ancillary gains of soil C sequestration and are relevant to the society as a whole. The societal benefits of soil C sequestration include a decrease in risks of pollution of water, reduction in susceptibility to erosion, and decreased risk of damage by transport of sediment and sediment-borne pollutants to civil structures, waterways, and water bodies (e.g., lakes, reservoirs, oceans).

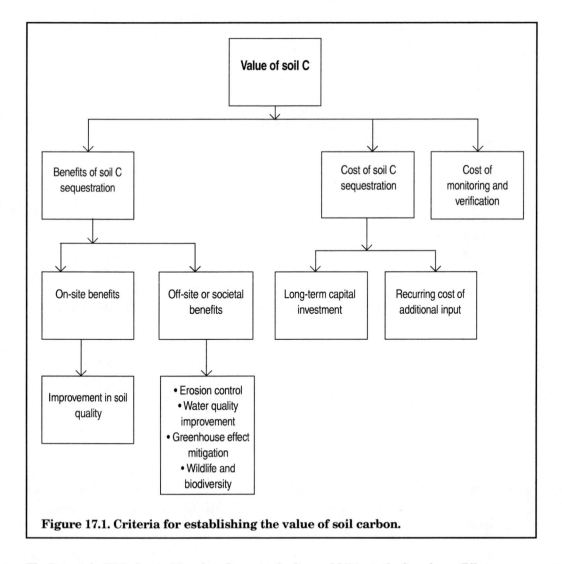

Figure 17.1. Criteria for establishing the value of soil carbon.

It is also important to determine the cost of sequestering C in terms of the management input needed to adopt recommended practices. Converting from conventional to recommended management practices may involve additional costs. Farmers and land managers may require compensation for that cost or for taking more risks by adopting these recommended practices. Grazing land scientists and agronomists need to work together with economists and social scientists to establish the guidelines for determining the value of soil C and the cost of C sequestration associated with adopting recommended practices (Fig. 17.1).

The cost of sequestering C also involves any expenses associated with monitoring and verifying temporal changes in the soil C pool. Such expenses may include those due to soil sampling, sample preparation, analyses of C and N, and the use of models and their inputs and use. In this regard, the need for standardized methods to assess C pools and fluxes has long been recognized, and attempts have been made to collate and synthesize the available information (Paul et al., 1997; Parton et al., 1987; Lal et al., 2000).

Policy Issues

Soil C sequestration is a biophysical process that socioeconomic and policy considerations affect. Realizing the potential of agricultural soils to sequester C depends on society's willingness to foster adoption of recommended management practices to restore degraded soils and judiciously manage lands already under production. Various governmental bodies may need to identify and implement appropriate policies to facilitate widespread adoption of recommended practices.

We need to develop interdisciplinary research teams (including biophysical and social scientists) to identify appropriate policies for soil C sequestration. Such policies may be based on rewarding land managers for adopting recommended practices and/or providing appropriate deterrents to those who do not. Because of differences in ecological factors (including soil properties and climatic parameters) the policies involved may be different for soil C sequestration in rangeland vs. pasturelands. We also need policies to reward good land management which has not lowered soil C levels but has resulted in good soils not suitable for a C sequestration program. Practices should not be encouraged that will rapidly reduce existing carbon levels so that more soils will qualify for a new program.

Soil Carbon as a Commodity

Soil C must be treated as a commodity, similar to other farm commodities (e.g., corn, soybean, beef, milk, poultry). All commodities, including soil C, can be traded. In addition to the cost of production and the societal value of soil C, its price also depends on the market forces governed by supply and demand. Commodifying soil C requires establishing institutions and mechanisms that facilitate C trading. Identifying criteria for trading C will involve considering (1) the rate of C sequestration under different management practices, (2) the cost of C sequestration, (3) the societal value of C, and (4) mechanisms for trading C credits.

Priority Issues

Soil C sequestration research has been ad hoc and piecemeal. We need to develop a systematic research program based on identified knowledge gaps and priorities of issues. The program has to be interdisciplinary and involve soil scientists, agronomists, economists, and policy makers (Fig. 17.2).

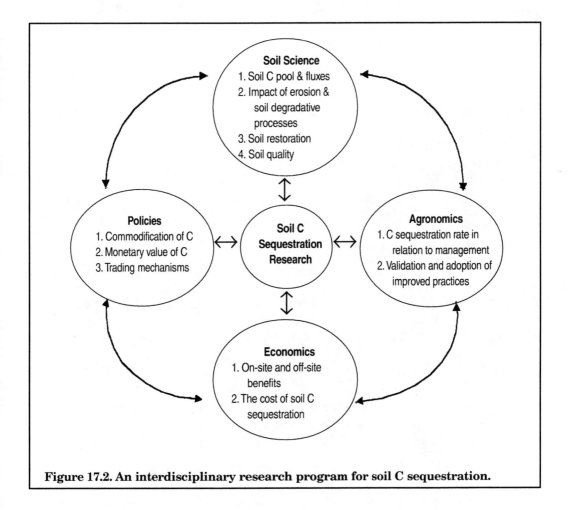

Figure 17.2. An interdisciplinary research program for soil C sequestration.

An important component of the interdisciplinary program is the land manager. Farmers and land managers need to be involved in program planning and policy development, from the beginning. A research program developed without involving farmers and land managers would be a counterproductive and futile effort.

Close cooperation between agricultural scientists (crop and range specialists) and foresters in their work also is crucial. Choosing appropriate land uses (whether agriculture or forestry) depends on identifying criteria that cover the entire spectrum from seasonal crops to perennials, and this broad spectrum is a continuum without disciplinary boundaries.

Future progress in soil C sequestration, in terms of both research and practical applications of proven technology, depends on (1) identifying researchable issues, (2) prioritizing research and development agenda, and (3) developing and implementing an interdisciplinary program at local, regional, and national scales.

References

Follett, R.F. 1993. Global climate change: U.S. agriculture and carbon dioxide. *J. Prod. Agric.* 6:181-190.

Follett, R. 1998. CRP and microbial biomass dynamics in temperate climates. *In* R. Lal, J. Kimble, R.F. Follett, and B.A. Stewart (eds), *Management of Carbon Sequestration*, CRC Press, Inc., Boca Raton, FL, pp. 305-322.

Lal, R. 1994. Sustainable land use systems and soil resilience. *In* D.J. Greenland and I. Szoblocs (eds), *Soil Resilience and Sustainable Land Use*, CAB International, Wallingford, U.K., pp. 41-67.

Lal, R, J.M. Kimble, R.F. Follett, and C.V. Cole. 1998. *The Potential of U.S. Cropland to Sequester Carbon and Mitigate the Greenhouse Effect.* Ann Arbor Press, Chelsea, MI.

Lal, R., J.M. Kimble, and R.F. Follett. 2000. Assessment Methods for Soil C Pools. CRC/Lewis Publishers, Boca Raton, FL (in press).

Parton, W.J., D.S. Schmil, C.V. Cole, and D.S. Ojima. 1987. Analysis of factors controlling soil organic matter levels in Great Plains Grasslands. *Soil Sci. Soc. Am. J.* 51:1173-1177.

Paul, E.A., R.F. Follett, S.W. Leavitt, A.D. Halvorson, G.A. Peterson, and D.J. Lyon. 1997. Radiocarbon dating for determination of soil organic matter pool sizes and dynamics. *Soil Sci. Soc. Am. J.* 61:1058-1067.

Appendix

Abbreviations and Units of Measure

Abbreviations

AWC = available water capacity

BMP = best management practices

CEC = cation exchange capacity

C:N = carbon:nitrogen ratio

CRP = Conservation Reserve Program

DOC = dissolved organic carbon

DOE = Department of Energy

EC = electrical conductivity

EIA = Energy Information Administration

EPA = Environmental Protection Agency

FAO = Food and Agriculture Organization of the United Nations

g C/m^2/yr = gram of carbon per square meter per year

GHGs = greenhouse gases

GT = gigaton = petagram = 1000 MMT

GWP = global warming potential

ha = hectare

IPCC = intergovernmental panel on climate change

kg = kilogram

m = meter

Mha = million hectares

mmhos/cm = millimhos per centimeter

MMT = million metric tons = Tg

MMTC = million metric tons of carbon

MMTCE = million metric tons of carbon equivalent

MT = metric ton = 1000 kg = 1 Mg

MT/ha = metric ton per hectare

MTC/ha = metric ton of carbon per hectare

MWD = mean weight diameter

The Potential of U.S. Grazind Lands to Sequester Carbon and Mitigate the Greenhouse Effect

NPP = net primary productivity
Pg = petagram = 10^{15} g = 1000 MMT = GT
ppbv = parts per billion by volume
ppmv = parts per million by volume
ppptv = parts per trillion by volume
RMP = recommended management practices
SIC = soil inorganic carbon
SOC = soil organic carbon = C
SOM = soil organic matter
Tg = teragram = 10^{12} g
USDA = United States Department of Agriculture
WRP = Wetland Reserve Program

Units of Measurement

Metric (SI) multipliers

Prefix	Abbreviation	Value
exa	E	10^{18}
peta	P	10^{15}
tera	T	10^{12}
giga	G	10^{9}
mega	M	10^{6}
kilo	k	10^{3}
hecto	h	10^{2}
deka	da	10^{1}
deci	d	10^{-1}
centi	c	10^{-2}
milli	m	10^{-3}
micro	μ	10^{-6}
nano	n	10^{-9}
pico	p	10^{-12}
femto	f	10^{-15}
atto	a	10^{-18}

Follett, Kimble, and Lal, editors